EX LIBRIS

Principles and Procedures
of Numerical Analysis

MATHEMATICAL CONCEPTS AND METHODS IN SCIENCE AND ENGINEERING

Series Editor: **Angelo Miele**
Mechanical Engineering and Mathematical Sciences
Rice University

Volume 1   **INTRODUCTION TO VECTORS AND TENSORS**
Volume 1: Linear and Multilinear Algebra
Ray M. Bowen and C.-C. Wang

Volume 2   **INTRODUCTION TO VECTORS AND TENSORS**
Volume 2: Vector and Tensor Analysis
Ray M. Bowen and C.-C. Wang

Volume 3   **MULTICRITERIA DECISION MAKING
AND DIFFERENTIAL GAMES**
Edited by George Leitmann

Volume 4   **ANALYTICAL DYNAMICS OF DISCRETE SYSTEMS**
Reinhardt M. Rosenberg

Volume 5   **TOPOLOGY AND MAPS**
Taqdir Husain

Volume 6   **REAL AND FUNCTIONAL ANALYSIS**
A. Mukherjea and K. Pothoven

Volume 7   **PRINCIPLES OF OPTIMAL CONTROL THEORY**
R. V. Gamkrelidze

Volume 8   **INTRODUCTION TO THE LAPLACE TRANSFORM**
Peter K. F. Kuhfittig

Volume 9   **MATHEMATICAL LOGIC**
An Introduction to Model Theory
A. H. Lightstone

Volume 10   **SINGULAR OPTIMAL CONTROLS**
R. Gabasov and F. M. Kirillova

Volume 11   **INTEGRAL TRANSFORMS IN SCIENCE
AND ENGINEERING**
Kurt Bernardo Wolf

Volume 12   **APPLIED MATHEMATICS: An Intellectual Orientation**
Francis J. Murray

Volume 13   **DIFFERENTIAL EQUATIONS WITH A SMALL
PARAMETER AND RELAXATION OSCILLATIONS**
E. F. Mishchenko and N. Kh. Rozov

Volume 14   **PRINCIPLES AND PROCEDURES OF NUMERICAL ANALYSIS**
Ferenc Szidarovszky and Sidney Yakowitz

A Continuation Order Plan is available for this series. A continuation order will bring delivery of each new volume immediately upon publication. Volumes are billed only upon actual shipment. For further information please contact the publisher.

Principles and Procedures of Numerical Analysis

Ferenc Szidarovszky

Eötvös Loránd University
Budapest, Hungary

and

Sidney Yakowitz

University of Arizona
Tucson, Arizona

PLENUM PRESS · NEW YORK AND LONDON

Library of Congress Cataloging in Publication Data

Szidarovszky, F
  Principles and procedures of numerical analysis.

  (Mathematical concepts and methods in science and engineering; v. 14)
  Bibliography: p.
  Includes index.
  1. Numerical analysis. I. Yakowitz, Sidney, J., 1937-   joint author. II.
Title.
QA297.S96   519.4   78-12070
ISBN 0-306-40087-1

© 1978 Plenum Press, New York
A Division of Plenum Publishing Corporation
227 West 17th Street, New York, N.Y. 10011

Printed in the United States of America

Preface

It is an incontestable fact that numerical analysis techniques are used routinely (although not always effectively) in virtually every quantitative field of scientific endeavor. In this book, which is directed toward upper-division and graduate level students in engineering and mathematics, we have selected for discussion subjects that are traditionally found in numerical analysis texts. But our choice of methodology rejects the traditional where analysis and experience clearly warrant such a departure, and one of our primary aspirations in this work is to equip the reader with the wherewithal to apply numerical analysis thinking to nontraditional subjects. For there is a plethora of computer-oriented sciences such as optimization, statistics, and system analysis and identification that are sorely in need of methods comparable to those related here for classical numerical analysis problems.

Toward uncovering for the reader the structure of numerical methods we have, for example, devoted a chapter to a metric space theory for iterative application of operators. In this chapter, we have collected those definitions and concepts of real and functional analysis that are requisite to a modern intermediate-level exposition of the principles of numerical analysis. Further, we derive the abstract theory (most notably, the contraction mapping theorem) for iteration processes. In later chapters we are able to appeal directly to this theory, as needed, to conclude without further derivation that standard iterative algorithms (such as the Gauss–Seidel method for linear equations, the Newton–Raphson method for nonlinear equations, and predictor–corrector methods for differential equations) are convergent. Furthermore, in the given numerical analysis setting, the general theory immediately implies the standard error bounds for these methods.

v

Through expository devices such as these, the reader will be able to see that seemingly diverse methods for different numerical analysis problems do have common structure.

In attempting to write a textbook covering traditional numerical analysis subjects, one must seek a balance between the desire, on the one hand, to provide a complete compilation of effective numerical procedures and the wish, on the other hand, to give a self-contained and thorough mathematical development of the major principles of numerical analysis. There are currently several weighty tomes that witness unbridled excursions in either of these directions and that are not wholly suitable for classroom use. Our compromise in this dichotomy of goals has been to include only those methods that are most important by virtue of their thoroughly proven reliability and efficiency and their attractive convergence properties. But we have taken care to reference comprehensive developments on specialized subjects so that readers can locate other techniques. Similarly, we have for the most part given complete and rigorous mathematical developments of the basic properties and error attributes of the methods discussed here, but, except in those cases where the mathematical development is particularly interesting or instructive, we have omitted derivations of those stated results that we believe are not vital, but that we think are nevertheless of value to the reader. We do include citations stating where the complete derivations may be found.

The present book has been used in mimeograph form with both engineering and mathematics students at the senior level. Essentially the entire book can be covered, without undue haste, in a two-semester course. Alternatively, by leaving out the chapters on eigenvalue methods and methods for partial differential equations and by omitting and summarizing the more sophisticated material (which generally lies toward the end of each chapter), we have used the book with success in one-semester courses.

The formal prerequisites for this study are calculus and some acquaintance with matrix theory. An advanced calculus or upper division engineering mathematics course would be helpful, but not vital. The informal prerequisites are willingness to deal with mathematical abstractions and a desire to look deeply at the underpinnings of numerical analysis.

The theoretical developments are liberally and systematically supplemented by examples. Where possible, in the examples we have applied different methods to the same problem so that the reader may have a basis for comparison of the accuracy and computational burden of the different algorithms.

It was through the auspices of a National Science Foundation U.S.–

Hungarian cooperative grant that the authors met and collaborated on research in the numerical aspects of hydrology. The first draft of this text was written while one of us (F.S.) was Visiting Professor of Systems and Industrial Engineering at the University of Arizona. We gratefully acknowledge this support as well as the assistance and cooperation afforded by our colleagues and especially the secretarial staff of the Systems and Industrial Engineering Department.

In preparation of this work, we have badgered out friends and acquaintances and tyrannized our graduate assistants. Among those graduate assistants, we owe our special thanks to Arik Kashper, Dan Murray, and Ed Zucker. Numerical analysts Warren Ferguson, Michael Golberg, Gene Golub, Dennis McCaughey, Olig Palusinski, and John Starner were particularly generous with their time. Additionally, Professor Golub let us riffle through his excellent files and Professor Ferguson gave us detailed and valued suggestions on initial and final drafts of the manuscript.

Budapest and Tucson   F. Szidarovszky
                                                         S. Yakowitz

Contents

1. **Preliminaries**   1
   1.1. Number Systems and Representations of Numbers   1
   1.2. Error Analysis   9
      1.2.1. Upper Bounds in Arithmetic   10
      1.2.2. Probabilistic Error Analysis   18
      1.2.3. Propagation of Errors   19
   1.3. Supplementary Notes and Discussion   23

2. **Approximation and Interpolation of Functions**   25
   2.1. Interpolating Polynomials   29
      2.1.1. Lagrange Interpolating Polynomials   29
      2.1.2. Error Bounds for Interpolating Polynomials   31
      2.1.3. Differences   35
      2.1.4. The Fraser Diagram   38
      2.1.5. Aitken's Method and Computational Requirements of
             Interpolation   46
      2.1.6. Hermite Interpolation   48
      2.1.7. Inverse Interpolation   49
   2.2. Uniform Approximations   49
   2.3. Least Squares Approximation   53
   2.4. Spline Functions   59
   2.5. Asymptotic Properties of Polynomial Approximations   63
   2.6. Supplementary Notes and Discussion   69

3. **Numerical Differentiation and Integration**   73
   3.1. Numerical Differentiation   75
   3.2. Numerical Quadrature   79

3.2.1. Interpolatory Quadrature Formulas   79
3.2.2. Error Analysis and Richardson Extrapolation   81
3.2.3. Gaussian Quadrature   88
3.2.4. The Euler–Maclaurin Formula   97
3.2.5. Romberg Integration 105
3.3. Supplementary Notes and Discussion 107

4. General Theory for Iteration Methods **111**

4.1. Metric Spaces 112
4.2. Examples of Metric Spaces 114
4.3. Operators on Metric Spaces 117
4.4. Examples of Bounded Operators 118
4.5. Iterations of Operators 121
4.6. Fixed-Point Theorems 124
4.7. Systems of Operator Equations 131
4.8. Norms of Vectors and Matrices 134
4.9. The Order of Convergence of an Iteration Process 137
4.10. Inner Products 138
4.11. Supplementary Notes and Discussion 139

5. Solution of Nonlinear Equations **141**

5.1. Equations in One Variable 141
     5.1.1. The Bisection Method 141
     5.1.2. The Method of False Position (Regula Falsi) 143
     5.1.3. The Secant Method 147
     5.1.4. Newton's Method 149
     5.1.5. Application of Fixed-Point Theory 154
     5.1.6. Acceleration of Convergence, and Aitken's δ^2-Method . . . 158
5.2. Solution of Polynomial Equations 160
     5.2.1. Sturm Sequences 161
     5.2.2. The Lehmer–Schur Method 163
     5.2.3. Bairstow's Method 165
     5.2.4. The Effect of Coefficient Errors on the Roots 167
5.3. Systems of Nonlinear Equations and Nonlinear Programming . . . 169
     5.3.1. Iterative Methods for Solution of Systems of Equations . . . 170
     5.3.2. The Gradient Method and Related Techniques 174
5.4. Supplementary Notes and Discussion 178

6. The Solution of Simultaneous Linear Equations **179**

6.1. Direct Methods 180
     6.1.1. Gaussian Elimination 180
     6.1.2. Variants of Gaussian Elimination 186
     6.1.3. Inversion by Partitioning 192
6.2. Iteration Methods 195
     6.2.1. Stationary Iteration Processes 196

6.2.2. Iteration Processes Based on the Minimization of Quadratic
Forms 201
6.2.3. Application of the Gradient Method 206
6.2.4. The Conjugate Gradient Method 207
6.3. Matrix Conditioning and Error Analysis 212
6.3.1. Bounds for Errors of Perturbed Linear Equations 212
6.3.2. Error Bounds for Rounding in Gaussian Elimination 216
6.4. Supplementary Notes and Discussion 219

7. The Solution of Matrix Eigenvalue Problems **221**

7.1. Preliminaries 222
7.1.1. Some Matrix Algebra Background 222
7.1.2. The Householder Transformation and Reduction to
Hessenberg Form 230
7.1.3. Matrix Deflation 234
7.2. Some Basic Eigenvalue Approximation Methods 234
7.2.1. The Power Method 236
7.2.2. The Inverse Power Method 239
7.2.3. The Rayleigh Quotient Iteration Method 240
7.2.4. Jacobi-Type Methods 245
7.3. The QR Algorithm 248
7.3.1. Principles and Convergence Rates 248
7.3.2. Implementation of the QR Algorithm 250
7.4. Eigenproblem Error Analysis 252
7.5. Supplementary Notes and Discussion 255

8. The Numerical Solution of Ordinary Differential Equations . . . **257**

8.1. The Solution of Initial-Value Problems 259
8.1.1. Picard's Method of Successive Approximation 259
8.1.2. The Power Series Method 260
8.1.3. Methods of the Runge–Kutta Type 263
8.1.4. Linear Multistep Methods 272
8.1.5. Step Size and Its Adaptive Selection 278
8.1.6. The Method of Quasilinearization 280
8.2. The Solution of Boundary-Value Problems 284
8.2.1. Reduction to Initial-Value Problems 284
8.2.2. The Method of Undetermined Coefficients 285
8.2.3. The Difference Method 287
8.2.4. The Method of Quasilinearization 289
8.3. The Solution of Eigenvalue Problems 290
8.4. Supplementary Notes and Discussion 292

9. The Numerical Solution of Partial Differential Equations **295**

9.1. The Difference Method 296
9.2. The Method of Quasilinearization 303

9.3. The Ritz–Galerkin Finite-Element Approach   304
    9.3.1. The Ritz Method   304
    9.3.2. The Galerkin Method   310
    9.3.3. The Finite-Element Method   311
9.4. Supplementary Notes and Discussion   318

REFERENCES   321
INDEX   327

1

Preliminaries

1.1. Number Systems and Representations of Numbers

In our conventional number system the symbol 649 designates the sum

$$6 \times 10^2 + 4 \times 10 + 9.$$

In other words a symbol denotes a value of a polynomial in the base 10, where the coefficients are from the set $\{0, 1, 2, \ldots, 9\}$. The reason for using a number system in the base 10 probably stems from people counting on their fingers. In other cultures, different number systems have been used. For instance, the Babylonian astronomers used a system having the base 60, and a consequence of this can be found in our dividing the circumference of the circle into 360 degrees.

Throughout this discussion, N will denote an integer greater than 1. The coefficients of a number system based on N are $0, 1, 2, \ldots, N - 1$, and so a representation of a positive integer in this number system must have the form

$$a_0 + a_1 N + a_2 N^2 + \cdots + a_k N^k,$$

where k is a nonnegative integer, and $a_0, a_1, \ldots, a_k$ are nonnegative integers not exceeding $N - 1$. The integer N is known as the *base* (or *radix*) of the number system.

Next we show how positive integers can be represented in any desired base. Let M_0 be any positive integer. The representation

$$M_0 = a_0 + a_1 N + a_2 N^2 + \cdots + a_k N^k \tag{1.1}$$

implies that a_0 is the remainder when we divide M_0 by N, because all the terms in the expansion of $M_0 - a_0$ are multiples of N. Let

$$M_1 = (M_0 - a_0)/N.$$

Then (1.1) implies

$$M_1 = a_1 + a_2N + \cdots + a_kN^{k-1},$$

and so a_1 is equal to the remainder of M_1 divided by N. By repeating this procedure, all the numbers $a_0, a_1, \cdots, a_{k-1}$ can be successively obtained. The algorithm terminates if $M_k < N$, and this condition implies $a_k = M_k$.

Letting $[x]$ denote the integer part of the real number x, we can summarize the above algorithm as

$$
\begin{aligned}
M_0 &= M, \\
M_1 &= [M_0/N], & a_0 &= M_0 - M_1N, \\
M_2 &= [M_1/N], & a_1 &= M_1 - M_2N, \\
&\;\;\vdots & &\;\;\vdots \\
M_k &= [M_{k-1}/N], & a_{k-1} &= M_{k-1} - M_kN, \\
a_k &= M_k,
\end{aligned}
\tag{1.2}
$$

and this discussion and uniqueness of quotient and remainder in division imply the following theorem:

Theorem 1.1. Let N and M be arbitrary integers, with $N > 1$, $M \geq 0$. Then the integer M can be represented uniquely in the number system having the base N.

Let us consider the representation of a number in a number system having base N, where N differs from 10. We shall write

$$(a_ka_{k-1} \cdots a_2a_1a_0)_N,$$

which is meant to indicate the number determined by the sum

$$M = a_kN^k + \cdots + a_2N^2 + a_1N + a_0.
\tag{1.3}$$

To compute the value of (1.3) we can use a recurrence relation. Construct the sequence

$$b_k = a_k,$$
$$b_{k-1} = b_k N + a_{k-1},$$
$$\vdots \qquad\qquad (1.4)$$
$$b_1 = b_2 N + a_1,$$
$$b_0 = b_1 N + a_0.$$

It is easy to verify that $M = b_0$. Sequence (1.4) is called *Horner's rule* and can be applied to evaluating polynomials by allowing N to be an arbitrary real number.

Example 1.1. (a) In the case $M = 649$ and $N = 3$, we have

$$a_0 = 1, \qquad M_1 = \frac{649 - 1}{3} = \frac{648}{3} = 216,$$

$$a_1 = 0, \qquad M_2 = \frac{216 - 0}{3} = \frac{216}{3} = 72,$$

$$a_2 = 0, \qquad M_3 = \frac{72 - 0}{3} = \frac{72}{3} = 24,$$

$$a_3 = 0, \qquad M_4 = \frac{24 - 0}{3} = \frac{24}{3} = 8,$$

$$a_4 = 2, \qquad M_5 = \frac{8 - 2}{3} = \frac{6}{3} = 2 < 3,$$

which implies

$$a_5 = 2,$$

and so

$$649 = (220001)_3.$$

(b) Now we calculate the value of the number $(220001)_3$. Using (1.4) we have

$$b_5 = 2,$$
$$b_4 = 2 \times 3 + 2 = 8,$$
$$b_3 = 8 \times 3 + 0 = 24,$$
$$b_2 = 24 \times 3 + 0 = 72,$$
$$b_1 = 72 \times 3 + 0 = 216,$$
$$b_0 = 216 \times 3 + 1 = 649.$$

We now consider representations of real numbers x, $0 \leq x < 1$. We wish to employ the representation

$$x = a_{-1}N^{-1} + a_{-2}N^{-2} + \cdots, \tag{1.5}$$

where the coefficients $a_{-1}, a_{-2}, \ldots$ are nonnegative integers not exceeding $N - 1$.

Let $x_0 = x$. By multiplying both sides of (1.5) by N, we have

$$Nx_0 = a_{-1} + a_{-2}N^{-1} + \cdots, \tag{1.6}$$

where $0 \leq Nx_0 < N$, which implies that a_{-1} is equal to the integer part of Nx_0; that is, $a_{-1} = [Nx_0]$. Let

$$x_1 = Nx_0 - a_{-1}.$$

Then using (1.5) we have

$$x_1 = a_{-2}N^{-1} + a_{-3}N^{-2} + \cdots,$$

and by repeatedly applying the above procedure, one may obtain the coefficients $a_{-2}, a_{-3}, \ldots$. The above algorithm can be written in the form

$$
\begin{aligned}
x_0 &= x, & a_{-1} &= [Nx_0], \\
x_1 &= Nx_0 - a_{-1}, & a_{-2} &= [Nx_1], \\
x_2 &= Nx_1 - a_{-2}, & a_{-3} &= [Nx_2].
\end{aligned}
\tag{1.7}
$$

Process (1.7) may continue indefinitely or may terminate through the occurrence of zero among the numbers $x_0, x_1, x_2, \ldots$. It is easy to prove that if $a_{-1}, a_{-2}, \ldots$ are the coefficients obtained by the above procedure, the value of the series in (1.5) is x. Consequently, the notation $(0.a_{-1}a_{-2}a_{-3}\cdots)_N$ is an unambiguous representation of x, and we have, through this discussion, verified the following theorem:

Theorem 1.2. Let N be an integer greater than 1. An arbitrary real number x, $0 \leq x < 1$, may be represented by the expression

$$x = a_{-1}N^{-1} + a_{-2}N^{-2} + a_{-3}N^{-3} + \cdots,$$

where the coefficients $a_{-1}, a_{-2}, a_{-3}, \ldots$ are nonnegative integers not exceeding $N - 1$.

Representation (1.5) need not be unique, as the following example (in the number system with base 10) shows:

$$0.1 = 0.09999\cdots,$$

because the value of the right-hand side is equal to

$$9[10^{-2} + 10^{-3} + \cdots] = 9\,\frac{10^{-2}}{1 - 10^{-1}} = 9\,\frac{10^{-1}}{9}$$

$$= 10^{-1} = 0.1.$$

For every set of coefficients satisfying the conditions of Theorem 1.2, the series in (1.5) is convergent since it can be majorized by the convergent sum

$$(N - 1)[N^{-1} + N^{-2} + N^{-3} + \cdots] = (N - 1)\,\frac{1}{N}\,\frac{1}{1 - 1/N} = 1.$$

In the case in which (1.5) is a finite sum, we can use Horner's rule to evaluate the number x from its coefficients $a_{-1}, \ldots, a_{-k}$, by replacing N in (1.4) by N^{-1} and a_j by a_{-j}, and setting $a_0 = 0$.

Theorems 1.1 and 1.2 imply that an arbitrary positive real number A can be represented in the number system with base N ($N > 1$, integer) by the expression

$$A = \sum_{l=-\infty}^{k} a_l N^l, \tag{1.8}$$

where the coefficients a_l are nonnegative integers not exceeding $N - 1$. We remark that the representation (1.8) is not necessarily unique. In order to obtain the coefficients of representation (1.8) we may use algorithms (1.2) and (1.7) for $M = [A]$ and $x = A - M$, respectively.

Example 1.2. (a) Let $x = 0.5$ and $N = 3$; then (1.7) implies

$$x_0 = 0.5, \qquad a_{-1} = [1.5] = 1,$$
$$x_1 = 0.5, \qquad a_{-2} = [1.5] = 1,$$
$$\vdots \qquad\qquad \vdots$$

and so $0.5 = (0.1111\ldots)_3$.

(b) The value of the representation $(0.111\ldots)_3$ can be obtained as the sum of the infinite series

$$3^{-1} + 3^{-2} + 3^{-3} + 3^{-4} + \cdots = \frac{1}{3}\,\frac{1}{1 - 1/3} = \frac{1}{2} = 0.5.$$

(c) By combining part (a) of Example 1.1 and this example we observe that $649.5 = (220001.1111\ldots)_3$.

Modern electronic computers use the number system with base 2 because in this case only two symbols are used, namely, 0 and 1, and they can be represented physically by the two stable states of a "flip–flop" circuit or the direction of flux in magnetic material. Thus the real numbers in the *binary system* (the system with base 2) can be represented as

$$A = \sum_{l=-\infty}^{k} a_l 2^l,$$

where the value of each coefficient a_l is either 0 or 1.

A negative real number can be denoted by a negative sign before its absolute value. Zero is denoted by 0 in every number system.

Only a finite number of symbols can be stored in the memory of electronic computers, and often the numbers represented in the computer only approximate the desired number. We will now discuss three types of arithmetic for approximating and operating on real numbers in computers.

The form of the *fixed-point arithmetic* discussed here requires that the number x must be less than one in magnitude; consequently the representation has the form

$$x_{\mathrm{fi}} = \pm \sum_{l=1}^{t} a_{-l} 2^{-l}, \tag{1.9}$$

where t is an integer depending on the structure of the computer, and x_{fi} is called the *fixed point machine number* (of word length t). Obviously the numbers representable by (1.9) lie in the interval

$$R = [-1 + 2^{-t-1}, 1 - 2^{-t-1}],$$

and an arbitrary real number x from the interval R can be approximated by the nearest number of the form (1.9) with an absolute error $|x - x_{\mathrm{fi}}|$ not exceeding 2^{-t-1}. The analysis to follow may easily be adapted for representations involving bases other than 2 or for different ranges of the exponent l. We have plotted x_{fi} against x for $t = 3$ in Figure 1.1.

If the result of an operation does not belong to the interval $[-1,1]$ then *overflow* occurs. The fact that overflow may occur after relatively few arithmetic operations is one of the serious disadvantages of fixed-point arithmetic. We note, however, that in the case of subtraction of two points in $[0,1]$ and multiplication of points in $[-1,1]$, no overflow can occur. If

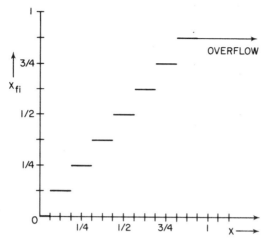

Fig. 1.1. Fixed-point approximation of x.

overflow does not occur, the result of addition and subtraction can be obtained exactly, without roundoff error. Multiplication and division, on the other hand, do entail roundoff error. This roundoff error is bounded by the accuracy of the fixed-point representation; that is, the absolute values of errors are not greater than 2^{-t-1}.

In *floating-point arithmetic* a nonzero number x is represented in the form

$$x_{fl} = \pm m \cdot 2^p,\tag{1.10}$$

where $p \in [-P, P]$ is an integer and

$$m = \sum_{l=1}^{t} a_{-l} 2^{-l}$$

is *normalized* by requiring that $a_{-1} = 1$. The integers t and P depend on the machine. Numbers representable in the form (1.10) lie in the interval

$$R = [-2^P(1 - 2^{-t-1}), 2^P(1 - 2^{-t-1})],$$

and numbers belonging to interval $(-2^p, 2^p)$ can be represented in the form (1.10) with error not exceeding $2^p 2^{-t-1}$. In the normalized representation, p is the *exponent* and m the *fraction part*.

We give a description of normalized floating-point number construction that minimizes the error of approximation. For $|x| < 2^{-P-2}$ we set $x_{fl} = 0$ and $p = 0$. For $2^{-P-2} \le |x| < 2^{-P-1}$, $p = -P$. For $|x| \ge 2^{-P-1}$, we have

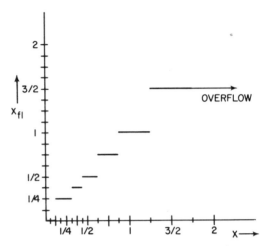

Fig. 1.2. Floating-point approximation of x.

$p = [\log_2| x |] + 1$, and since $| m | < 1$ the upper bound for the error cannot exceed $| x |2^{-t}$. Thus for $2^P \geq | x | \geq 2^{-P-1}$, the relative error (defined by $| x - x_{fl} |/| x |$) is bounded by 2^{-t}. Figure 1.2 shows x_{fl} as a function of x, with $t = 2$ and $P = 1$.

The reader may verify that the error induced by the operations of addition and subtraction cannot exceed 2^{-t+1} times the absolute value of the largest summand; in the case of multiplication and division the error does not exceed 2^{-t} times the exact value of the result of the operation provided that the absolute values of all the results of the operations lie in the interval $[2^{-P}, 2^P]$.

One may conclude that floating-point arithmetic affords more accuracy for a given dynamic range than fixed-point arithmetic, but the price for this is the extra computing burden of "updating" p at each stage of the computation. Still, most scientific calculations on computers and quality calculators are done in floating-point arithmetic.

The accuracy of the representation using either of the above arithmetics can be increased by using *multiple-precision arithmetic*, which is mathematically equivalent to single-precision t replaced by an integer multiple of t. Double precision is used most commonly and we remark that in FORTRAN programming, variables of double precision can be introduced by appropriate declaration-type statements. Most scientific computers offer the user all three types of arithmetic: fixed point, floating point, and double precision. Special-purpose machines and inexpensive hand calculators, for the purpose of simplicity, are often constrained to fixed-point arithmetic.

Problems

1.1. Suppose one evaluates (1.3) by successively evaluating each term $a_j N^j$ by multiplying a_j by N and that by N, and repeating this multiplication until $a_j N^j$ is evaluated. How many additions and multiplications are required in all to evaluate M? How many additions and multiplications are required by Horner's rule?

1.2. Convert $(178.8)_9$ to binary.

1.3. The integer word in the CDC 6400 is in octal and has a length of 20 symbols [i.e., k in (1.3) is 20, and $N = 8$]. What is the largest decimal number that can be thus represented?

1.4. Consider the "mixed-base" integer representation

$$M = \sum_{v=1}^{N} c_v v!,$$

where c_v is an integer such that $0 \le c_v \le v$. Show that every positive integer has a unique representation of the above form.

1.5. Give a method for evaluating x^N that, for $N > 3$, requires fewer than $N - 1$ multiplications. The rule is that you may multiply x with x or with any previously computed number. You may also take products of previously computed numbers.

1.6. Generally, computers perform exact arithmetic operations on floating-point numbers and then round off the exact answer to the nearest floating-point number. Assuming this is done, which of the following arithmetic laws remain valid for computer operations?

$$u - v = -(v - u),$$
$$(u + w) + v = u + (w + v),$$
$$u \cdot v = v \cdot u,$$
$$(u \cdot v) + (u \cdot w) = u \cdot (v + w).$$

1.2. Error Analysis

The data used in practical calculations are usually only approximate values. They are obtained from measurements, estimates, or previous calculations. As we have just seen, the representation of these numbers in the computer causes further error. Calculation with these numbers generally cannot be done exactly, and so even further errors arise. In this section we discuss the accumulation and propagation of different kinds of errors as the computation proceeds.

Error analysis can be characterized in several ways. If a certain guaranteed accuracy in the results is required, then the use of upper bounds is necessary. Let x denote an exact value and let a denote an approximation of x. The actual error is given by $\Delta a = x - a$, and any nonnegative number δa such that $|\Delta a| \leq \delta a$ can be considered to be an *upper bound* (or *absolute bound*) for the error of a. Of course, all values larger than δa also serve as upper bounds. Error estimates in terms of upper bounds are usually pessimistic in that they often are much larger than the actual errors realized, and for this reason the errors in the data and in the operations are sometimes profitably considered to be random variables with some assumed distribution. The distribution of the errors resulting from the calculations as well as functionals (such as expected squared error) of the induced distribution are then determined and used in place of absolute error bounds. Such bounds are termed *probabilistic bounds*.

In the first step of this discussion of error analysis we assume that the operations can be done exactly, and the errors of the results are caused exclusively by the error of input data approximation.

1.2.1. Upper Bounds in Arithmetic

First we give upper bounds for the errors in results of the four rules of arithmetic operating on inexact data. Let x and y denote two exact values, and let a and b denote their respective approximations. If δa and δb are error bounds, then

$$|x - a| = |\Delta a| \leq \delta a, \qquad |y - b| = |\Delta b| \leq \delta b.$$

The exact value of the sum of x and y is $x + y$ and its approximating value is $a + b$, and so we have

$$|(x + y) - (a + b)| = |(x - a) + (y - b)| = |\Delta a + \Delta b|$$
$$\leq |\Delta a| + |\Delta b| \leq \delta a + \delta b,$$

and therefore

$$|(x + y) - (a + b)| \leq \delta a + \delta b. \tag{1.11}$$

Similarly, the error of the result of subtraction can be bounded by

$$|(x - y) - (a - b)| = |(x - a) - (y - b)| = |\Delta a - \Delta b|$$
$$\leq |\Delta a| + |\Delta b| \leq \delta a + \delta b,$$

and consequently the error bound for subtraction is

$$|(x - y) - (a - b)| \leq \delta a + \delta b. \tag{1.12}$$

Like addition error, (1.12) is the best possible bound; that is to say, equality is actually achieved for certain a, b, x, and y, assuming δa and δb are as small as possible.

The exact and approximated values of the product of x and y are, respectively, xy and ab, and so

$$
\begin{aligned}
| xy - ab | &= |(a + \Delta a)(b + \Delta b) - ab | \\
&= | ab + a\,\Delta b + b\,\Delta a + \Delta a\,\Delta b - ab | \\
&\leq | a | \,\delta b + | b | \,\delta a + \delta a\,\delta b.
\end{aligned}
$$

If $| a | \gg \delta a$ and $| b | \gg \delta b$, the product error can be approximated by a more customary (but not strictly true) first-order approximation of the above inequality:

$$
| xy - ab | \leq | b | \,\delta a + | a | \,\delta b. \tag{1.13}
$$

Assume that y and $b \neq 0$. Then the error bound for division can be derived as follows:

$$
\begin{aligned}
\left| \frac{x}{y} - \frac{a}{b} \right| &= \left| \frac{a + \Delta a}{b + \Delta b} - \frac{a}{b} \right| = \left| \frac{ab + b\,\Delta a - ab - a\,\Delta b}{b(b + \Delta b)} \right| \\
&\leq \frac{| b | \,| \Delta a | + | a | \,| \Delta b |}{| b |^2 (1 - | \Delta b |/| b |)} \leq \frac{| b | \,\delta a + | a | \,\delta b}{| b |^2 (1 - \delta b/| b |)}.
\end{aligned}
$$

If $| b | \gg \delta b$, then the term $\delta b/| b |$ in the denominator can be neglected so as to obtain the approximating inequality

$$
\left| \frac{x}{y} - \frac{a}{b} \right| \leq \frac{| b | \,\delta a + | a | \,\delta b}{| b |^2}. \tag{1.14}
$$

We collect the above important formulas in Table 1.1.

Table 1.1. Absolute Error Bound Formulas for Arithmetic Operations

$$
\delta\,(a + b) = \delta a + \delta b
$$

$$
\delta\,(a - b) = \delta a + \delta b
$$

$$
\delta\,(ab) \simeq | a | \,\delta b + | b | \,\delta a
$$

$$
\delta\!\left(\frac{a}{b}\right) \simeq \frac{| a | \,\delta b + | b | \,\delta a}{| b |^2} \qquad (b \neq 0)
$$

As an illustration of the above relations we give two examples.

Example 1.3. We calculate the upper error bound of the quantity $a = 1.21 \times 3.65 + 9.81$ under the assumption that all of the numbers are accurate to two decimal places; that is, $\delta(1.21) = \delta(3.65) = \delta(9.81) = 0.005$. Using formulas of Table 1.1 we have

$$\delta(1.21 \times 3.65) \simeq 1.21 \times 0.005 + 3.65 \times 0.005$$
$$= 0.005 \times 4.86 = 0.0243,$$

and as a result

$$\delta(a) = \delta(1.21 \times 3.65) + \delta(9.81) \simeq 0.0243 + 0.005 = 0.0293.$$

Example 1.4. Assuming that all the numbers $a_i, b_i \ (1 \leq i \leq n)$ have the same error bound ε, we bound the error of the scalar product $\sum_{i=1}^{n} a_i b_i$.
For $i = 1, 2, \ldots, n$,

$$\delta(a_i b_i) \simeq |a_i|\,\delta b_i + |b_i|\,\delta a_i \leq \varepsilon(|a_i| + |b_i|),$$

which implies

$$\delta\left(\sum_{i=1}^{n} a_i b_i\right) = \sum_{i=1}^{n} \delta(a_i b_i) \leq \varepsilon \sum_{i=1}^{n} (|a_i| + |b_i|).$$

The *relative error bound* is defined by the quantity $\delta a/|x|$, where x is the exact value and δa is an error bound. Since x is unknown, instead of $\delta a/|x|$ we can compute only its approximation $\delta a/|a|$. It is easy to see that

$$\left|\frac{\delta a}{|x|} - \frac{\delta a}{|a|}\right| = \delta a \left|\frac{|a| - |x|}{|x||a|}\right| \leq \delta a \frac{|a - x|}{|x||a|} \leq \frac{(\delta a)^2}{|x||a|},$$

and consequently if $|x|$ is considerably larger than δa, the discrepancy in relative error caused by dividing by the approximate value instead of the exact value can be neglected.

Using formulas of Table 1.1 we can derive the bounds for the relative errors of resultants of arithmetic operations. First we prove that the relative error bound of a sum of two terms of the same sign is not greater than the greatest relative error bound in its terms. Assume that $\delta a/|a| \geq b/|b|$. This inequality is equivalent to the inequality

$$\delta a\,|b| \geq |a|\,\delta b.$$

Adding $|a|\,\delta a$ to both sides and dividing by $|a|(|a| + |b|)$, we have

$$\frac{\delta a}{|a|} \geq \frac{\delta b + \delta a}{|a| + |b|} = \frac{\delta a + \delta b}{|a + b|}$$

which implies our assertion.

Table 1.2. Relative Error Bounds for Arithmetic Operations

$$\frac{\delta(a+b)}{|a+b|} = \max\left\{\frac{\delta(a)}{|a|}, \frac{\delta(b)}{|b|}\right\}$$

$$\frac{\delta(a-b)}{|a-b|} = \frac{\delta(a)+\delta(b)}{|a-b|}$$

$$\frac{\delta(ab)}{|ab|} \simeq \frac{\delta(a)}{|a|} + \frac{\delta(b)}{|b|}$$

$$\frac{\delta(a/b)}{|a/b|} \simeq \frac{\delta(a)}{|a|} + \frac{\delta(b)}{|b|}$$

In subtraction of terms of the same sign, the relative errors can be arbitrarily large, because the denominator of the quotient

$$\frac{\delta a + \delta b}{|a-b|}$$

can be arbitrarily small, if a and b are sufficiently close to each other. In this circumstance, to avoid large relative error, subtraction should be replaced by other operations if at all possible. Example 1.5 will show a case in which subtraction can be replaced by addition.

In multiplication and division, the relative error bounds add, as the following calculations show:

$$\frac{\delta(ab)}{|ab|} \simeq \frac{|a|\,\delta b + |b|\,\delta a}{|a||b|} = \frac{\delta b}{|b|} + \frac{\delta a}{|a|}$$

and

$$\frac{\delta(a/b)}{|a/b|} \simeq \frac{(|a|\,\delta b + |b|\,\delta a)/|b|^2}{|a/b|} = \frac{|a|\,\delta b + |b|\,\delta a}{|a||b|} = \frac{\delta b}{|b|} + \frac{\delta a}{|a|}.$$

In Table 1.2, we have summarized the bounds for relative error induced by the arithmetic operations.

Example 1.5. Calculate the value of $x = (1972)^{1/2} - (1971)^{1/2}$, assuming that the values of the square roots are known to two decimal places, and estimate the absolute and relative error of subtraction.

We have $(1972)^{1/2} \simeq 44.41$ and $(1971)^{1/2} \simeq 44.40$, and so

$$x = (1972)^{1/2} - (1971)^{1/2} \simeq 0.01 = a.$$

The absolute error bound is given by the equation

$$\delta(a) = \delta(44.41) + \delta(44.40) = 0.01,$$

which implies that the relative error is 100%, because

$$\frac{\delta(a)}{|a|} = \frac{0.01}{0.01} = 1 = 100\%.$$

We can avoid the subtraction by using the identity

$$x = \frac{(1972)^{1/2} - (1971)^{1/2}}{1} = \frac{1972 - 1971}{(1972)^{1/2} + (1971)^{1/2}} = \frac{1}{(1972)^{1/2} + (1971)^{1/2}}$$
$$\simeq 0.01126.$$

The absolute error and approximate value of the denominator are, respectively, 0.01 and 88.79, and so the relative error of the denominator is $0.00011 = 0.011\%$. Since the numerator is exact with zero relative error and in division the relative errors are added, we conclude that the relative error of the quotient is also 0.011%, which is much less than the relative error resulting from direct subtraction.

Example 1.6. Now we solve for the relative error of the arithmetic calculations in Example 1.3. Since

$$\frac{\delta(1.21)}{1.21} = \frac{0.005}{1.21} \simeq 0.0041, \qquad \frac{\delta(3.65)}{3.65} = \frac{0.005}{3.65} \simeq 0.0014,$$

the relative error of the product 1.21×3.65 is equal to the sum of the relative errors of the factors: 0.0055. The relative error of the second term of the sum is equal to

$$\frac{\delta(9.81)}{9.81} = \frac{0.005}{9.81} \simeq 0.0005,$$

and so we obtain

$$\frac{\delta(a)}{|a|} = \max\{0.0055, 0.0005\} = 0.0055.$$

In the following portion of this section we bound errors of computed values of single and multivariate functions evaluated at an inexact data point. Let f denote a univariate function. The exact value x is unknown; only an approximate value a is known. We approximate $f(x)$ by $f(a)$. The absolute error $|f(a) - f(x)|$ can be estimated using Taylor's formula. Assuming that f is two times differentiable in an interval that contains the points a and x as interior points, we have

$$f(x) - f(a) = \frac{f'(a)}{1!}(x - a) + \frac{f''(\xi)}{2!}(x - a)^2,$$

where ξ is between x and a. Using absolute values, we have that

$$| f(x) - f(a) | \leq | f'(a) | \, \delta a + \frac{| f''(\xi) |}{2!} (\delta a)^2. \qquad (1.15)$$

Assuming that the second derivative is not too large in comparison to $| f'(a) |$, we may neglect the higher-order term, and consequently we may compute a bound $\delta f(x)$ for the difference $| f(x) - f(a) |$ by the formula

$$\delta(f(a)) \simeq | f'(a) | \, \delta a. \qquad (1.16)$$

If $f'(a)$ is zero or a small number, then the higher-order term should not be neglected. In this case, all the terms preceding the negligible terms must be considered. For instance if

$$f'(a) = f''(a) = \cdots = f^{(k-1)}(a) = 0,$$

but $f^{(k)}(a) \neq 0$ and the higher-order derivatives are not large values, then

$$\delta(f(a)) \simeq \frac{| f^{(k)}(a) |}{k!} (\delta a)^k. \qquad (1.17)$$

Example 1.7. One side and the two adjacent angles of a triangle shown in Figure 1.3 are approximately known:

$$a = 100, \qquad \beta = 45°, \qquad \gamma = 45°.$$

Calculate the other sides and angle of the triangle and give error bounds, assuming that $\delta(a) = 0.1$ and $\delta(\beta) = \delta(\gamma) = 0.1°$.

The value of α can be calculated very easily:

$$\alpha = 180° - (\beta + \gamma) \simeq 90°$$

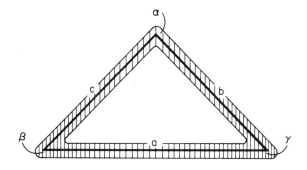

Fig. 1.3. A triangle with inexact specifications.

and

$$\delta(\alpha) = \delta(180°) + \delta(\beta) + \delta(\gamma) = 0.2°.$$

From the sine theorem of trigonometry, we have that

$$b = \frac{a \sin \beta}{\sin \alpha} \simeq \frac{100 \sin 45°}{\sin 90°} = \frac{100(2^{1/2}/2)}{1} \simeq 70.71.$$

Equation (1.17) implies, upon converting $\delta(\beta)$ to radians, that

$$\delta(\sin \beta) \simeq |\cos \beta| \, \delta(\beta) = |\cos 45°| \, 0.1 \times \frac{\pi}{180} = \frac{2^{1/2}}{3600} \pi.$$

The above formula should not be used for estimating the absolute error of $\sin \alpha$ because its derivative is $\cos \alpha = \cos 90° = 0$. Using formula (1.17) with $k = 2$ and converting $\delta(\alpha)$ to radians, we obtain

$$\delta(\sin \alpha) = \frac{|\sin \alpha|}{2!} (\delta\alpha)^2 = \frac{1}{2} \times 0.04 \times \left(\frac{\pi}{180}\right)^2$$

$$\simeq 6.1 \times 10^{-6}$$

By applying the equations in Table 1.1, $\delta(a \sin \beta)$ may be determined as follows:

$$\delta(a \sin \beta) = |a| \, \delta(\sin \beta) + |\sin \beta| \, \delta a$$

$$= 100 \times \frac{2^{1/2}}{3600} \pi + \frac{2^{1/2}}{2} \times 0.1$$

$$= \frac{2^{1/2}}{2} \left(\frac{\pi}{18} + 0.1\right) \simeq 0.1941,$$

which implies

$$\delta(b) = \frac{|a \sin \beta| \, \delta(\sin \alpha) + |\sin \alpha| \, \delta(a \sin \beta)}{|\sin \alpha|^2} \simeq 0.1945.$$

Since the calculations for c are identical, one has that

$$c = 70.71 \quad \text{and} \quad \delta(c) \simeq 0.1945.$$

The error bounds of multivariate functions can be estimated by a technique generalizing the univariate case. Let $\mathbf{x}$ be the unknown exact vector and $\mathbf{a}$ be a known approximation of $\mathbf{x}$. Assume that $\delta(a_i)$ is an upper bound for the modulus of ith component of the vector $\mathbf{x} - \mathbf{a}$. Let l denote the dimension of vectors $\mathbf{x}$ and $\mathbf{a}$. We calculate the value $f(\mathbf{a})$ instead of

$f(\mathbf{x})$, where $\mathbf{a}$ is thought of as an approximation of $\mathbf{x}$. Assuming that f is two times continuously differentiable and the second-order partial derivatives are not too large, by using the multivariate Taylor's formula we may conclude that

$$\delta\big(f(\mathbf{a})\big) \simeq \sum_{i=1}^{l} \left| \frac{\partial f(\mathbf{a})}{\partial x_i} \right| \delta(a_i). \tag{1.18}$$

If some of the first-order partial derivatives are small or zero, then the higher-order terms should not be neglected.

It is easy to prove that formula (1.18) implies the arithmetic error formulas in Table 1.1. For example, let $f(x, y) = xy$; then denoting by a and b the approximating values of x and y, we obtain

$$\delta\big(f(a, b)\big) \simeq |f_x(a, b)| \, \delta a + |f_y(a, b)| \, \delta b = |b| \, \delta a + |a| \, \delta b.$$

Example 1.8. Solve the problem in Example 1.3 using formula (1.18). By (1.18) we have

$$\delta(a) \simeq |1.21| \times \delta(3.65) + |3.65| \times \delta(1.21) + |1| \times \delta(9.81)$$
$$= (1.21 + 3.65 + 1)0.005 = 0.02930.$$

Problems

1.7. Bound the absolute error of the value

$$a = 2\pi(l/g)^{1/2},$$

where $l = 1$, $\delta l = 0.01$, $\pi \simeq 3.14$, $\delta\pi = 0.005$, $g = 9.81$, $\delta g = 0.005$. (This formula gives the time period of a swinging pendulum with g the acceleration of gravity and l the pendulum length.)

1.8. Give an upper error bound for the cosines of angles of two vectors given by the formula

$$\cos(\mathbf{a}, \mathbf{b}) = \sum_{i=1}^{n} a_i b_i \bigg/ \bigg(\sum_{i=1}^{n} a_i^2 \times \sum_{i=1}^{n} b_i^2 \bigg)^{1/2},$$

under the assumption that for $i = 1, 2, \ldots, n$, $\delta(a_i) = \delta(b_i) = \varepsilon$.

1.9. Bound the error in applying Horner's rule toward evaluating an nth-degree polynomial having inexact coefficients. Assume $\delta(a_i) = \varepsilon$, $1 \leq i \leq n$, and that $\delta(x) = \varepsilon'$. Compare this error with the error bound associated with evaluating each term $a_i x^i$ by performing the multiplication $a_i(x(x(\ldots)))$.

1.10. Using the Taylor's series technique of this section derive the error bounds for subtraction and division of inaccurate data given in Table 1.1.

1.11. Consider the following three expressions:

  (a) $F_1(x) = (1 - \cos x)/\sin x$,
  (b) $F_2(x) = \sin x/(1 + \cos x)$,
  (c) $F_3(x) = x/2 + x^3/24$.

Show that F_2 is identical to F_1 and that F_3 approximates F_1 in a neighborhood of zero. Evaluate the three expressions for small values of x using single and double precision, and compare. Do an error analysis of these three expressions by our Taylor's series technique. [This problem is discussed in Stegun and Abramowitz (1956) and mentioned in Dorn and McCracken (1972).]

1.2.2. Probabilistic Error Analysis

This section should be omitted by readers who have not covered the subject matter of a first course in probability theory.

The analysis given in the previous section provides bounds for the errors resulting from operations on inaccurate data. The actual errors are usually found to be much less than their upper bounds. Errors can often be approximated more accurately by using elementary probabilistic and statistical methods. In this approach, we must make suitable assumptions about the distributions of errors in the data. If the data are given by measurements, then it is often assumed that the errors follow the normal distribution with zero mean and a variance that depends on the accuracy of the apparatus. The distribution of rounding errors of operations executed by computer can often be considered to follow the uniform distribution in the interval $[-\varepsilon, \varepsilon]$, where 2ε is the discretization magnitude. The results of operations upon these randomized measurements are also considered to be random variables. The induced distribution of the randomized results can be determined by standard developments in probability theory. We illustrate such analysis by considering two specific cases.

Let $x = a + \xi_a$ denote the exact value, a being the approximate value, and let the probability density function (pdf) of the random variable ξ_a be denoted by g_a. Similarly, let $y = b + \xi_b$ and let g_b denote the pdf of ξ_b. Since $x + y = a + b + (\xi_a + \xi_b)$, the error of the addition is the sum of the random variables ξ_a and ξ_b. Assuming the independence of ξ_a and ξ_b, the pdf of the sum can be obtained by the convolution of g_a and g_b:

$$g_{a+b}(t) = \int_{-\infty}^{\infty} g_a(t - x) g_b(x) \, dx.$$

Analysis of other operations proceeds in a like fashion. For example, let f be a strictly monotone increasing function. We consider the effect of error ξ_a in the data when we evaluate $f(a)$ instead of the desired value $f(a + \xi_a)$. Then the distribution function of the error $\Delta f(x) = f(x) - f(a)$ is given by

$$\begin{aligned}
G_{\Delta f(a)}(t) &= P(f(a + \xi_a) - f(a) < t) \\
&= P(a + \xi_a < f^{-1}(t + f(a))) \\
&= P(\xi_a < f^{-1}(t + f(a)) - a) \\
&= G_a(f^{-1}(t + f(a)) - a),
\end{aligned}$$

where G_a denotes the cumulative distribution function of ξ_a.

The general case of real or multivariate functions is amenable to similar probabilistic analysis. The above convolution rule as well as other useful procedures for obtaining probabilistic bounds are to be found in Feller (1968, Section 11.2), for example.

Problems

1.12. Assuming the independence of ξ_a and ξ_b, express the pdfs of $\xi_a - \xi_b$, $\xi_a \cdot \xi_b$, and ξ_a/ξ_b in terms of the probability density functions of ξ_a and ξ_b. (Assume $\xi_a, \xi_b > 0$.)

1.13. Assume that the random variables ξ_i $(1 \le i \le r)$ are independent and uniformly distributed in the interval $[-\varepsilon, \varepsilon]$. Approximate the density function of $\sum_{i=1}^{r} \xi_i$, where r is assumed large. HINT: Use the central limit theorem, as stated in Feller (1968), Lindgren (1968), and other books on probability theory.

1.14. Derive the convolution representation of $g_{a+b}(t)$ from the elementary definitions of bivariate distributions and stochastic independence.

1.2.3. Propagation of Errors

In the previous two sections we assumed that the operations can be done exactly. Due to roundoff errors described in Section 1.1, as well as approximation, discretization, and truncation errors, generally each stage in a lengthy program augments and propagates errors introduced in earlier stages. In this section we shall see how, under certain assumptions, bounds for overall error of multistage calculations can be obtained.

Let $x_1, x_2, \ldots, x_r$ denote the initial exact data and $a_1, a_2, \ldots, a_r$ the corresponding known approximations to these data. The theoretical or ideal calculations determine the sequence $x_l = F_l(x_1, \ldots, x_{l-1})$, where $l = r + 1, \ldots, N$. In actuality, a sequence $a_l = G_l(a_1, \ldots, a_{l-1})$ is computed, where G_l approximates F_l but operates on the data that contain the error induced by the preceding computer operations.

Assume for all possible sequences $x_1, \ldots, x_{l-1}$ that

$$| F_l(x_1, \ldots, x_{l-1}) - G_l(x_1, \ldots, x_{l-1}) | \leq \sum_{k=1}^{l-1} A_{lk} | x_k |, \qquad (1.19)$$

where the numbers A_{lk} are suitable positive constants. It is also assumed for all possible sequences $x_1, \ldots, x_{l-1}$ and $a_1, \ldots, a_{l-1}$ that

$$| F_l(x_1, \ldots, x_{l-1}) - F_l(a_1, \ldots, a_{l-1}) | \leq \sum_{k=1}^{l-1} B_{lk} | x_k - a_k |, \qquad (1.20)$$

where the B_{lk} are given constants. Assumption (1.19) implies the error of the operation is linear in the data, and (1.20) is satisfied if, for example, F has continuous first partial derivatives and the values x_j and a_j are constrained to lie in closed, bounded sets.

Let $\delta(a_l)$ denote an upper bound for the error of a_l ($l = 1, 2, \ldots, k$, $k + 1, \ldots$). Then (1.19) and (1.20) imply

$$| x_l - a_l | \leq | F_l(x_1, \ldots, x_{l-1}) - F_l(a_1, \ldots, a_{l-1}) |$$
$$+ | F_l(a_1, \ldots, a_{l-1}) - G_l(a_1, \ldots, a_{l-1}) |$$
$$\leq \sum_{k=1}^{l-1} B_{lk} | x_k - a_k | + \sum_{k=1}^{l-1} A_{lk} | a_k |$$
$$\leq \sum_{k=1}^{l-1} A_{lk} | a_k | + \sum_{k=1}^{l-1} B_{lk} \delta a_k.$$

Thus we proved that the bound for the accumulated error of a_l can be obtained from the recursive [in $\delta(a_j)$] equation

$$\delta(a_l) = \sum_{k=1}^{l-1} A_{lk} | a_k | + \sum_{k=1}^{l-1} B_{lk} \delta(a_k). \qquad (1.21)$$

This formula depends on values a_j, $r \leq j \leq l$, which are not known before the calculation commences. Consequently, without modification, (1.21) can only be used to obtain *a posteriori bounds*, that is, bounds derivable only after calculations are completed and intermediate values made known. If the assumption that

$$| a_k | \leq U_k, \qquad r < k \leq l - 1,$$

is made, then (1.21) leads to *a priori bounds*. Specifically, (1.21) implies that

$$\delta(a_l) = \sum_{k=1}^{l-1} A_{lk} U_k + \sum_{k=1}^{l-1} B_{lk} \delta(a_k). \qquad (1.22)$$

As a final introductory study of error propagation, let us apply the method of probabilistic error analysis introduced in Section 1.2.2. Toward simplifying the notation and discussion, assume that function F_l is linear in all variables and for all possible values $a_1, a_2, \ldots$, the local errors

$$\varepsilon_l = F_l(a_1, \ldots, a_{l-1}) - G_l(a_1, \ldots, a_{l-1})$$

are independent random variables and the distribution of ε_l does not depend on the values $a_1, \ldots, a_{l-1}$. Because of our assumptions, for $l = r + 1$, $r + 2, \ldots$, we have

$$x_l - a_l = F_l(x_1, \ldots, x_{l-1}) - F_l(a_1, \ldots, a_{l-1})$$
$$+ F_l(a_1, \ldots, a_{l-1}) - G_l(a_1, \ldots, a_{l-1})$$
$$= \sum_{k=1}^{l-1} C_{lk}(x_k - a_k) + \varepsilon_l, \tag{1.23}$$

where the numbers C_{lk} are fixed constants. If the global error $x_l - a_l$ is denoted by ϕ_l, then using (1.23) we obtain

$$\phi_l = \sum_{k=1}^{l-1} C_{lk}\phi_k + \varepsilon_l. \tag{1.24}$$

If also the errors of the data $a_1, \ldots, a_r$ are denoted by $\varepsilon_1, \ldots, \varepsilon_r$ and they are considered to be random variables, then ϕ_l is a random variable and the distribution of ϕ_l can be estimated recursively using (1.24). For instance, the expectation of ϕ_l is given by

$$E(\phi_l) = \sum_{k=1}^{l-1} C_{ik}E(\phi_k) + E(\varepsilon_l).$$

In iterative calculations, it is often the case that F_j and G_j depend only on the most recently calculated intermediate number, a_{j-1}. In this case, ϕ_l satisfies the inhomogeneous time-varying linear difference equation

$$\phi_l = C_{l,l-1}\phi_{l-1} + \varepsilon_l,$$

whose explicit solution, as a function of the noise terms, is readily verified to be

$$\phi_l = \sum_{v=1}^{l-1} \left(\prod_{j=v}^{l-1} C_{j+1,j} \right) \varepsilon_v + \varepsilon_l \tag{1.25}$$

We let $W_{l,v}$ denote the product term $\prod_{j=v}^{l-1} C_{j+1,j}$ in the above formula and make the popular assumption that the ε_j have zero mean and are uncorrelated (that is, $E[\varepsilon_j] = 0$ and $E[\varepsilon_i\varepsilon_j] = 0$ if $i \neq j$). Then from (1.25) it is evident that $E[\phi_l] = 0$ and therefore the variance of ϕ_l [abbreviated $\text{Var}(\phi_l)$] is the expected squared error $E[(x_l - a_l)^2]$. Further, the reader will easily see that the zero-correlation assumption implies immediately that

$$\text{Var}(\phi_l) = \sum_{v=1}^{l-1} (W_{l,v})^2 \,\text{Var}(\varepsilon_v) + \text{Var}(\varepsilon_l).$$

Finally we remark that the propagation of errors in certain important iterative processes will be discussed in Section 4.6 and elsewhere.

Example 1.9. We illustrate error propagation analysis by means of a compound integration process. Let $f(t)$ denote some integrable function over the interval $[0, \pi]$, which we suppose to be divided up into N equal segments, each of length $h = \pi/N$. The grid points are denoted by $t_j = jh$, $0 \leq j \leq N$. Let

$$x_l = \int_0^{t_l} f(t)\, dt = F_l(x_0, \ldots, x_{l-1}) = x_{l-1} + \int_{t_{l-1}}^{t_l} f(t)\, dt.$$

For this problem, a_l is the approximation of the integral x_l obtained through use of the compound trapezoidal rule, which will be described in detail in Chapter 3. Thus

$$a_l = G_l(a_0, \ldots, a_{l-1}) = a_{l-1} + (h/2)[f(t_{l-1}) + f(t_l)] + \varepsilon_l,$$

where ε_l is the roundoff error due to finite word size of the computer. Then we have, corresponding to (1.20) and (1.19), respectively, the following error bounds:

(a) $|F_l(x_0, \ldots, x_{l-1}) - F_l(a_0, \ldots, a_{l-1})| = |x_{l-1} - a_{l-1}|$, $0 < l \leq N$,

and so $B_{lk} = 0$ for $k < l - 1$, and $B_{l,l-1} = 1$.

(b) $|F_l(x_0, \ldots, x_{l-1}) - G_l(x_0, \ldots, x_{l-1})| \leq Mh^3/12 + \varepsilon$, $0 < l \leq N$,

where M is a bound for $|(d^2/dt^2)f(t)|$; ε is a bound for the roundoff error. The bound (b) is a constant instead of linear, as in (1.19), but the propagation analysis still applies. In summary, we have for every number N of subdivisions, an error bound given by (1.22). Specifically, we have

$$\delta(a_N) = \text{error}(N) \leq N(Mh^3/12 + \varepsilon).$$

In Table 1.3, for various numbers of subdivisions N, we have tabulated the computed error bound Error(N) and the actual observed error for the case that $f(t) = \sin t + \cos t + 1$. In this case, $M = 1$. The calculation was done on an IBM 1130 having roundoff $\varepsilon = 2^{-16}$. In the tabulations, it is seen that we have overestimated the error, as is to be expected, but the minimum of the error bounds are in the vicinity of the minimum of the actual error.

Table 1.3. Example of Error Propagation and Bounding

N	Error (N)	Actual error
1	2.5839	2.0000
2	0.6460	0.4202
4	0.1615	0.1039
8	0.0405	0.0258
16	0.0103	0.0064
32	0.0030	0.0016
64	0.0016	0.0004
128	0.0021	0.0001
256	0.0039	2.7×10^{-5}
512	0.0078	-1×10^{-5}
1,024	0.0156	-4×10^{-5}
2,048	0.0312	-6×10^{-5}
4,096	0.0625	0.0001
8,192	0.1250	0.0010
16,384	0.2500	0.0033

Problems

1.15. Consider the sequence

$$a_k = a_{k-1} + Aa_{k-2} + Ba_{k-3},$$

where numbers A and B are constants given with error not exceeding δ. Assume that the right-hand side can be calculated with an error of at most ε. Bound the error of a_k for $k = 3, 4, 5, \ldots$, where a_0, a_1, a_2 are assumed exact.

1.16. Assume that the error of multiplication is less than or equal to ε and that a is exact. Bound the error of the number a^N, where the calculation is made by means of the series $a^2, a^3, a^4, \ldots$.

1.3. Supplementary Notes and Discussion

A primary reference for this chapter is Chapter 4 of *Seminumerical Algorithms* by Knuth (1969). This book provides elaboration and literature references to all subjects discussed here except propagation of error. In particular, Knuth (1969) outlines the history of number systems from the Babylonian sexagesimal system to present-day experimental representations such as the factorial base system and systems having negative and irrational bases. Horner's method is regarded as a useful algorithm for evaluating polynomials because of its combination of simplicity and efficiency, but as Knuth explains and documents, algorithms requiring fewer arithmetical operations exist. One also finds in this work supplementary discussion of Horner's method, with a description of modifications necessary for the sake of efficiency in the complex coefficient case, discussions of rationales for selecting noise distributions for error in data roundoff, and careful error analysis of floating-point and multiple-precision arithmetics.

There seems to be little unified theory for propagation of errors. The current situation, as we will see in later chapters, is that bounds for accumulated error are established on a case-by-case basis. Chapter 16 of Henrici (1964) is concerned with propagation of rounding error, and provides supplementary information to our discussion.

Among the interesting examples cited by Henrici is a difference equation whose upper bound for accumulated rounding grows proportionally to the number of iterations, but in the probabilistic analysis, the standard deviation for error grows as the square root of the number of iterations.

Another general approach (known as *backward error analysis*) to bounding error propagation is described in Fox and Mayers (1968). We shall use this approach in studying the effect of errors in Gaussian elimination (Chapter 6).

2

Approximation and Interpolation of Functions

Often functions arising in economics and engineering are not specified by explicit formulas, but rather by values at distinct points. Engineering and mathematical analysis cannot easily be applied to problems having partially specified functions, and so *interpolation* functions of explicit and uncomplicated form are sought that agree with the known function values. In other instances, functions are completely specified, but have a form that cannot be transmitted conveniently to a computer. For in order to be describable to a computer, a function must be *finitely representable*. That is, it must be determined, with respect to whatever computer language is being used, by some finite set of numbers or instructions. The only real functions that are finitely representable in any of the computer languages are piece wise rational functions (ratios of polynomials). Ultimately then, if computer analysis involving other than a rational function is to be done, one must face the task of finding a rational function that provides a good *approximation* to the desired function.

It turns out that polynomials are sufficiently accurate for many approximation and interpolation tasks. For that reason, the bulk of the literature of numerical approximation, as well as this chapter, focuses on polynomial approximation and interpolation. We remark, however, that once an accurate approximation scheme for a class of transcendental functions has been devised, these functions in turn can be used to obtain finitely representable approximations. We mention Fourier series approximations as a prominent example of this notion.

Our study begins with a constructive demonstration that any continuous function on a bounded interval can be approximated as accurately as desired by a polynomial. We may, without loss of generality, assume the interval to be the unit interval [0,1].

Theorem 2.1. (Weierstrass) Let f be any continuous function on the closed bounded interval [0,1] and let ε be an arbitrary positive number; then there exists a polynomial p such that for all $x \in [0,1]$ the inequality

$$| f(x) - p(x) | \leq \varepsilon$$

holds.

Proof. We promise the reader that this is our last development that appeals to probability theory. A proof that does not explicitly entail probabilistic notions is to be found in Section 5.1 of Isaacson and Keller (1966).

Let X denote the Bernoulli random variable with parameter x, $0 \leq x \leq 1$. That is,

$$P[X = 1] = x = 1 - P[X = 0].$$

The reader may easily confirm that the mean and variance of X are, respectively, x and $x(1 - x)$. If S_n is the sum of n independent observations of X, then, since means and variances add, S_n has mean and variance, respectively, nx and $nx(1 - x)$. Also, being the sum of n independent Bernoulli variables, S_n is a binomial random variable with parameter (n, x). As a consequence of these observations, we obtain the identities

$$\sum_{k=0}^{n} \binom{n}{k} x^k (1 - x)^{n-k} \quad = 1 = P[0 \leq S_n \leq n],$$

$$\sum_{k=0}^{n} \frac{k}{n} \binom{n}{k} x^k (1 - x)^{n-k} = x = \frac{1}{n} E[S_n],$$

$$\sum_{k=0}^{n} \frac{k^2}{n^2} \binom{n}{k} x^k (1 - x)^{n-k} = \left(1 - \frac{1}{n}\right) x^2 + \frac{1}{n} x = \left(\frac{1}{n}\right)^2 E[S_n^2]$$

$$= \left(\frac{1}{n}\right)^2 [\text{Var}(S_n) + E[S_n]^2].$$

[See Feller (1968), pp. 223–230, for supplementary discussion, if necessary.] From now on, no probabilistic ideas are employed.

For each n, let the approximating polynomial be

$$B_n(x) = \sum_{k=0}^{n} \binom{n}{k} x^k (1 - x)^{n-k} f\left(\frac{k}{n}\right).$$

It is obvious that B_n, which is called the *Bernstein polynomial*, is a polynomial of degree n. Consider the difference between f and B_n:

$$f(x) - B_n(x) = \sum_{k=0}^{n} \left[f(x) - f\left(\frac{k}{n}\right) \right] \binom{n}{k} x^k (1 - x)^{n-k}.$$

Since f is continuous on $[0,1]$, it is uniformly continuous and bounded. Consequently, for each positive ε there exist numbers δ and M such that if $|x_1 - x_2| < \delta$, then

$$|f(x_1) - f(x_2)| \leq \varepsilon/2,$$

and

$$|f(x)| \leq M \qquad (x \in [0,1]).$$

For any x, the points k/n ($k = 0,1, \ldots, n$) can be divided into two sets A and B such that

$$\frac{k}{n} \text{ is in } \begin{cases} A & \text{if } \left|\dfrac{k}{n} - x\right| < \delta, \\ B & \text{otherwise.} \end{cases}$$

Then using the above identities we have

$$|f(x) - B_n(x)| \leq \sum_{k=0}^{n} \left| f(x) - f\left(\frac{k}{n}\right) \right| \binom{n}{k} x^k (1 - x)^{n-k}$$

$$\leq \sum_{A} \frac{\varepsilon}{2} \binom{n}{k} x^k (1 - x)^{n-k} + \sum_{B} 2M \binom{n}{k} x^k (1 - x)^{n-k}$$

$$\leq \frac{\varepsilon}{2} + 2M \sum_{B} \frac{(k/n - x)^2}{(k/n - x)^2} \binom{n}{k} x^k (1 - x)^{n-k}$$

$$\leq \frac{\varepsilon}{2} + \frac{2M}{\delta^2} \sum_{B} \left[\frac{k^2}{n^2} - 2x \frac{k}{n} + x^2 \right] \binom{n}{k} x^k (1 - x)^{n-k}$$

$$\leq \frac{\varepsilon}{2} + \frac{2M}{\delta^2} \sum_{k=0}^{n} \left[\frac{k^2}{n^2} - 2x \frac{k}{n} + x^2 \right] \binom{n}{k} x^k (1 - x)^{n-k},$$

whence our earlier identities imply

$$|f(x) - B_n(x)| \leq \frac{\varepsilon}{2} + \frac{2M}{\delta^2} \left[\left(1 - \frac{1}{n}\right) x^2 + \frac{1}{n} x - 2x^2 + x^2 \right].$$

When the x^2 terms are cancelled, the above equation reduces to

$$|f(x) - B_n(x)| \leq \frac{\varepsilon}{2} + \frac{2M}{\delta^2} \frac{x(1 - x)}{n} \leq \frac{\varepsilon}{2} + \frac{M}{\delta^2 2n}. \qquad (2.1)$$

where we have used that for $x \in [0,1]$, $x(1-x) \leq 1/4$. Choosing n such that the final term of (2.1) is less than $\varepsilon/2$, we have that for every x in the unit interval,

$$| f(x) - B_n(x) | \leq \frac{\varepsilon}{2} + \frac{\varepsilon}{2} = \varepsilon,$$

which completes the proof of the theorem.   ■

From this theorem we know that any continuous function on a closed interval can be approximated with arbitrary accuracy by polynomials. It turns out that the Bernstein approximating polynomial is not popular because increasing the degree requires repeating the entire computation, and because the function values at the fractions k/n, $0 \leq k \leq n$, must be known. Furthermore, as we shall experimentally observe later, convergence is typically slow. Finally, error analysis is more difficult than with alternative constructs. We now examine more efficient approximating and interpolating polynomials.

Let $x_1, x_2, \ldots, x_n$ be different real numbers and let $f_1, f_2, \ldots, f_n$ be the corresponding values of a function f. The approximating polynomial is denoted by p. The bulk of the methods for polynomial approximations and interpolations can be divided into the following four classifications:

1. The *method of Lagrange interpolation* determines the unique polynomial p of least degree, which has the values $f_1, \ldots, f_n$ at the points $x_1, \ldots, x_n$, i.e., for which $p(x_i) = f_i$, $1 \leq i \leq n$. We shall see later that, in some troublesome cases, Lagrange interpolation polynomials p based on n interpolation points will not converge to a continuous target function f as n increases, despite the Weierstrass theorem.

In many practical applications we wish to approximate the function f by a polynomial p_m of degree m much less than n, the number of data points. In this case the problem is the following: find the polynomial or polynomials of degree m that are "closest" to the function values at the given distinct points. The construction of the required approximating polynomial depends on the chosen measure of distance between given functions and their approximating polynomials. Items 2 and 3 below give the most popular distance measures in polynomial approximation applications.

2. Let us define the distance between the given function and an mth-degree polynomial p_m by the discrete uniform distance formula

$$\max_{1 \leq i \leq n} | f_i - p_m(x_i) |. \qquad (2.2)$$

In this case, we choose the polynomial of given degree m for which the quantity (2.2) has a minimum. Such a polynomial is called a *best-approximating polynomial* (BAP) of degree m based on the n points $x_1, \ldots, x_n$.

3. In the least squares method we measure the distance between the given function and polynomial p_m by the value of

$$\left\{\sum_{i=1}^{n} [f_i - p_m(x_i)]^2\right\}^{1/2}, \tag{2.3}$$

and the *least squares approximation* of the function is defined by the polynomial of degree m for which the value (2.3) has a minimum. This chapter presents the details of approximating polynomial construction using both of the above criteria.

4. The main ideas and properties of *piecewise polynomial approximating functions* are also presented in this chapter. Among this class of approximating functions, splines are the most widely studied and used. The theory of spline functions is much more recent than the other approximating techniques, and it is of great and increasing importance in the numerical solution of differential equations and the analysis of random functions.

2.1. Interpolating Polynomials

2.1.1. Lagrange Interpolating Polynomials

Let $x_1, \ldots, x_n$ denote distinct real numbers and let $f_1, \ldots, f_n$ be arbitrary real numbers. The existence and uniqueness of interpolating polynomials are stated in the following theorem.

Theorem 2.2. There exists a unique polynomial of degree less than n for which $p(x_1) = f_1, \ldots, p(x_n) = f_n$.

Proof. Let us consider the functions

$$l_k(x) = \frac{(x - x_1)(x - x_2) \cdots (x - x_{k-1})(x - x_{k+1}) \cdots (x - x_n)}{(x_k - x_1)(x_k - x_2) \cdots (x_k - x_{k-1})(x_k - x_{k+1}) \cdots (x_k - x_n)}, \tag{2.4}$$

for $k = 1, 2, \ldots, n$. It is evident that the functions l_k are polynomials of degree $n - 1$ and

$$l_k(x_k) = 1 \quad \text{and for } k \neq i, \quad l_k(x_i) = 0.$$

Using the polynomials l_k, construct the function

$$p(x) = \sum_{i=1}^{n} f_i l_i(x), \tag{2.5}$$

which is a polynomial of degree less than or equal to $n - 1$. For $i = 1, 2,$ $\ldots, n$, we have that it interpolates. That is,

$$p(x_i) = \sum_{\substack{j=1 \\ j \neq i}}^{n} f_j l_j(x_i) + f_i l_i(x_i) = \sum_{\substack{j=1 \\ j \neq i}}^{n} f_j \times 0 + f_i \times 1 = f_i.$$

Thus the existence of the interpolating polynomial is proved.

To prove the uniqueness, assume that both p and q are interpolating polynomials of degree less than n, i.e., for $i = 1, 2, \ldots, n$, $p(x_i) = q(x_i) = f_i$. Consider the function $r(x) = p(x) - q(x)$. Then r is a polynomial of degree less than or equal to $n - 1$, and for $i = 1, 2, \ldots, n$,

$$r(x_i) = p(x_i) - q(x_i) = f_i - f_i = 0.$$

Thus the $(n - 1)$th-degree (or less) polynomial r has at least n different zeros, and so by the fundamental properties of polynomials, r must be the zero polynomial, which implies $p \equiv q$. ∎

The polynomial (2.5) is called the *Lagrange interpolating polynomial*. As a consequence of equations (2.4) and (2.5), the polynomial p can be written in the form

$$p(x) = \sum_{i=1}^{n} \left(f_i \prod_{\substack{j=1 \\ j \neq i}}^{n} \frac{x - x_j}{x_i - x_j} \right). \tag{2.6}$$

Formula (2.6) shows the explicit structure of p.

Example 2.1. Using the values $(100)^{1/2} = 10$, $(121)^{1/2} = 11$, $(144)^{1/2} = 12$, calculate and tabulate the Lagrange interpolation polynomial.

Let $x_1 = 100$, $x_2 = 121$, $x_3 = 144$, and $f_1 = 10$, $f_2 = 11$, $f_3 = 12$. The Lagrange interpolating polynomial has the form

$$p(x) = f_1 l_1(x) + f_2 l_2(x) + f_3 l_3(x),$$

where

$$l_1(x) = \frac{(x - 121)(x - 144)}{(100 - 121)(100 - 144)},$$

$$l_2(x) = \frac{(x - 100)(x - 144)}{(121 - 100)(121 - 144)},$$

$$l_3(x) = \frac{(x - 100)(x - 121)}{(144 - 100)(144 - 121)}.$$

From this approximation we have tabulated the values in Table 2.1.

Table 2.1. Lagrange Interpolation of Square Root Function

x	Lagrange interpolation	$x^{1/2}$
100	10.0000	10.0000
104	10.1969	10.1980
108	10.3907	10.3923
112	10.5816	10.5830
116	10.7694	10.7703
120	10.9543	10.9545
124	11.1361	11.1355
128	11.3149	11.3137
132	11.4907	11.4891
136	11.6635	11.6619
140	11.8332	11.8322
144	12.0000	12.0000

2.1.2. Error Bounds for Interpolating Polynomials

Let f be a real function and let x_k, $1 \le k \le n$, be different real numbers. Let $f_k = f(x_k)$ for $k = 1, 2, \ldots, n$. By Theorem 2.2 there exists a unique interpolating polynomial p of degree less than or equal to $n - 1$ for which $f_k = p(x_k)$ ($k = 1, 2, \ldots, n$). Since we shall approximate functions using interpolating polynomials, we provide below an estimate for the difference of the function and interpolating polynomial at points outside the set x_k, $1 \le k \le n$.

Theorem 2.3. Let f be n times differentiable in an interval $I = [a, b]$ that contains the points $x_1, \ldots, x_n$ and x, and let p be as above. Then there exists an interior point ξ in I such that

$$f(x) - p(x) = \frac{1}{n!} f^{(n)}(\xi)w_n(x), \tag{2.7}$$

where $f^{(n)}$ denotes the nth derivative of f and

$$w_n(x) = (x - x_1)(x - x_2) \cdots (x - x_n).$$

Proof. If x belongs to the set of interpolating points, then both sides of (2.7) are equal to zero, and so we can assume that x is not an interpolation point. Consider the function of z defined by

$$g(z) = f(z) - p(z) - [f(x) - p(x)] \frac{w_n(z)}{w_n(x)},$$

where the values $x_1, \ldots, x_n, x$ are assumed to be fixed numbers.
   For $z = x$ we have

$$g(x) = f(x) - p(x) - [f(x) - p(x)] \frac{w_n(x)}{w_n(x)} = 0,$$

and for $z = x_k$ $(1 \leq k \leq n)$,

$$g(x_k) = f(x_k) - p(x_k) - [f(x) - p(x)] \frac{w_n(x_k)}{w_n(x)} = 0,$$

which implies that the function g has at least $n + 1$ different zeros in I.
Using Rolle's theorem [to be found in any elementary calculus book or
in Kaplan (1959, p. 21)], we conclude that g' has at least n different zeros
in I, $g^{(2)}$ has at least $n - 1$ zeros in I, $\ldots$, and $g^{(n)}$ has at least one zero in I.
Let this zero of $g^{(n)}$ be denoted by ξ. Since

$$g^{(n)}(z) = f^{(n)}(z) - p^{(n)}(z) - [f(x) - p(x)] \frac{w_n^{(n)}(z)}{w_n(x)}$$

$$= f^{(n)}(z) - [f(x) - p(x)] \frac{n!}{w_n(x)}$$

we have

$$0 = g^{(n)}(\xi) = f^{(n)}(\xi) - [f(x) - p(x)] \frac{n!}{w_n(x)},$$

which is equivalent to the desired equation (2.7).   ■
   As ξ is not explicitly known, formula (2.7) cannot be used for the calculation of an exact value of $f(x) - p(x)$ but, as we shall now see, it can
be used for obtaining upper bounds for the absolute value of $f(x) - p(x)$.

   Corollary 2.1. If

$$M_n = \sup_{x \in I} |f^{(n)}(x)|, \qquad \Omega_n = \max_{x \in I} |w_n(x)|,$$

then

$$|f(x) - p(x)| \leq \frac{M_n}{n!} \Omega_n. \tag{2.8}$$

   For the case $n = 2$, p is called a *linear interpolation polynomial*. Let
x_1, x_2 denote the interpolating points and f_1, f_2 the corresponding values of

a function. Let $h = x_2 - x_1$ and, by defining the new variable

$$t = \frac{x - x_1}{h},$$

we have

$$x - x_1 = x_1 + th - x_1 = th,$$
$$x - x_2 = x_1 + th - x_1 - h = h(t - 1),$$

which implies

$$w_2(x) = (x - x_1)(x - x_2) = h^2 t(t - 1),$$
$$l_1(x) = \frac{x - x_2}{x_1 - x_2} = \frac{h(t - 1)}{-h} = 1 - t,$$
$$l_2(x) = \frac{x - x_1}{x_2 - x_1} = \frac{th}{h} = t.$$

Thus the linear interpolation polynomial has the form

$$p(x) = f_1 l_1(x) + f_2 l_2(x) = f_1(1 - t) + f_2 t.$$

If $x \in [x_1, x_2]$ then from inequality (2.8) we have

$$|f(x) - p(x)| \leq \frac{M_2}{2} \max_{x_1 \leq x \leq x_2} |w_2(x)|$$
$$\leq \frac{M_2}{2} \max_{0 \leq t \leq 1} |h^2 t(t - 1)|$$
$$\leq \frac{M_2 h^2}{2} \max_{0 \leq t \leq 1} [t(1 - t)].$$

The function $t(1 - t)$ has a maximum of $\frac{1}{4}$ at $t = 0.5$. Consequently,

$$|f(x) - p(x)| \leq \frac{M_2 h^2}{8}. \tag{2.9}$$

A table of a function of one variable is said to be *well suited for linear interpolation* if the error due to interpolation does not exceed the rounding error of the entries.

Example 2.2. We determine the greatest permissible step of a well-suited interpolation table for $\sin x$ if the entries are given to four decimal places.

In the case of the function $\sin x$,

$$M_2 = \max |(d^2/dx^2) \sin x| = \max |-\sin x| = 1,$$

and consequently the table is well suited for linear interpolation if the right-hand side of inequality (2.9) does not exceed $\frac{1}{2} \times 10^{-4}$, i.e.,

$$\frac{1 \times h^2}{8} \leq \frac{1}{2} \times 10^{-4}$$

which implies $h \leq 0.02$.

In case the functional values $f_1, \ldots, f_n$ are not available, but only approximate values $\tilde{f}_1, \ldots, \tilde{f}_n$ are known, a further error of interpolation arises. Assume for $k = 1, 2, \ldots, n$ that

$$| f_k - \tilde{f}_k | \leq \delta,$$

where δ is a fixed positive number. Let p denote the interpolating polynomial that uses the functional values $f_1, \ldots, f_n$ and let $\tilde{p}$ denote the interpolating polynomial with functional values $\tilde{f}_1, \ldots, \tilde{f}_n$. By use of equation (2.5) the difference of p and $\tilde{p}$ can be bounded by

$$| p(x) - \tilde{p}(x) | = \left| \sum_{k=1}^{n} f_k l_k(x) - \sum_{k=1}^{n} \tilde{f}_k l_k(x) \right|$$

$$= \left| \sum_{k=1}^{n} (f_k - \tilde{f}_k) l_k(x) \right|$$

$$\leq \sum_{k=1}^{n} | f_k - \tilde{f}_k | \, | l_k(x) |$$

$$\leq \delta \sum_{k=1}^{n} | l_k(x) |,$$

where the second term of the last formula depends on x and the interpolating points $x_1, \ldots, x_n$, but not on δ. If $\delta \to 0$, then this error bound also converges to zero, but the error factor $\sum_{k=1}^{n} | l_k(x) |$ typically grows quite rapidly with increasing n. The implication is that if data or roundoff error is not negligible, high-degree interpolation polynomials can be dangerous.

Problems

2.1. Prove that $\sum_{j=1}^{n} l_j(x) = 1$.

2.2. Prove that $l_k = w_n(x)/(x - x_k)w_n'(x_k)$, where $w_n(x) = \prod_{k=1}^{n}(x - x_k)$.

2.3. Let $f(x) = e^x$. Then $f(0.2) = 1.2214$. One may evaluate f at multiples of 0.2 by taking powers of $f(0.2)$. In this fashion use the points $x_j = 0.2j, 0 \leq j \leq 5$, to obtain a fifth-degree Lagrange interpolation polynomial for $f(x)$ and use it to approximate the value $f(0.56)$. Bound your error under the assumption that your values of $f(0.0), \ldots, f(1.0)$ are exact. It turns out that $f(0.56) = 1.7507$. What was your actual error?

2.4. (Continuation of Problem 2.3.) Use the methods of Chapter 1 to bound the roundoff error due to multiplication in evaluating $f(0.0), \ldots, f(1.0)$. With this bound, use the developments in the preceding section to bound the error of your approximation of $f(0.56)$, accounting for error in function values.

2.5. What discretization size should be used to tabulate $\exp(\sin x)$ in the interval $[0,2]$ so that linear interpolation will be accurate to four decimal places?

2.6. Assume $n > 2$. Let the evenly spaced interpolation points x_j be ordered so that $x_1 < x_2 < \cdots < x_n$. Let $w_n(x) = \prod_{i=1}^{n}(x - x_i)$. Show that

$$\max_{x_2 < x < x_3} |w_n(x)| < \max_{x_1 < x < x_2} |w_n(x)|$$

and conclude that one may anticipate the maximum interpolation error to occur toward the ends of the interval containing the interpolation points.

2.7. Prove that for arbitrary integers $n > 1$, $0 \le k \le n - 2$, the following identity holds:

$$\sum_{i=1}^{n} \frac{i^k}{\displaystyle\prod_{\substack{j=1 \\ j \ne i}} (i - j)} = 0.$$

2.8. Prove that the interpolating polynomials can be represented in the form

$$p(x) = w_n(x) \sum_{k=1}^{n} \frac{f_k}{(x - x_k)w_n{}'(x_k)}.$$

2.9. Construct a function f and points in I for which inequality (2.8) becomes an equality.

2.1.3. Differences

A disadvantage of the Lagrange representation (2.6) of the interpolating polynomial is that if one wishes to use additional interpolating points, the entire calculation must be repeated. In the following section, more computationally efficient representations of the interpolation polynomial will be derived in terms of differences. Differences also have some intrinsic interest as well as relevance to other aspects of numerical analysis, especially numerical solution of differential equations. We shall note later that it affords more economical computation of interpolation polynomials than direct application of (2.5).

Let $\ldots, x_{-3}, x_{-2}, x_{-1}, x_0, x_1, x_2, x_3, \ldots$ denote the interpolating points and let $\ldots, f_{-3}, f_{-2}, f_{-1}, f_0, f_1, f_2, f_3, \ldots$ denote the corresponding values of function f. Assume for all k that $x_{k+1} - x_k = h$, where h is a fixed positive number. We define *first-order differences* by the equation

$$\Delta f_i = f_{i+1} - f_i,$$

and the *higher-order differences* are defined recursively by the relation

$$\Delta^j f_i = \Delta^{j-1} f_{i+1} - \Delta^{j-1} f_i.$$

First we prove the following theorem:

Theorem 2.4. For $j \geq 1$ and all i, the equation

$$\Delta^j f_i = \sum_{k=0}^{j} (-1)^{j-k} \binom{j}{k} f_{i+k} \tag{2.10}$$

holds.

Proof. The proof will be by induction. For $j = 1$, obviously

$$\Delta f_i = f_{i+1} - f_i = (-1)^1 \binom{1}{0} f_i + (-1)^0 \binom{1}{1} f_{i+1}$$

$$= \sum_{k=0}^{1} (-1)^{1-k} \binom{1}{k} f_{i+k}.$$

Assume that equation (2.10) is true for some number j, and consider $\Delta^{j+1} f_i$. Using the recursive definition of higher-order differences and identity (2.10) for j, we have

$$\Delta^{j+1} f_i = \Delta^j f_{i+1} - \Delta^j f_i$$

$$= \sum_{k=0}^{j} (-1)^{j-k} \binom{j}{k} f_{i+1+k} - \sum_{k=0}^{j} (-1)^{j-k} \binom{j}{k} f_{i+k}$$

$$= -(-1)^j \binom{j}{0} f_i + \sum_{k=1}^{j} \left[(-1)^{j-k+1} \binom{j}{k-1} - (-1)^{j-k} \binom{j}{k} \right] f_{i+k}$$

$$+ (-1)^{j-j} \binom{j}{j} f_{i+j+1}.$$

Since the coefficient of f_{i+k} in the sum has the form

$$-(-1)^{j-k} \left[\binom{j}{k-1} + \binom{j}{k} \right] = (-1)^{j-k+1} \binom{j+1}{k},$$

we have

$$\Delta^{j+1}f_i = (-1)^{j+1-0}\binom{j+1}{0}f_i + \sum_{k=1}^{j}(-1)^{j+1-k}\binom{j+1}{k}f_{i+k}$$

$$+(-1)^{j+1-(j+1)}\binom{j+1}{j+1}f_{i+j+1} = \sum_{k=0}^{j+1}(-1)^{j+1-k}\binom{j+1}{k}f_{i+k},$$

which is equation (2.10) for $j + 1$. This completes the proof. ∎

By using the values of the f_k and higher-order differences and constructing Table 2.2, we obtain the *difference table* of function f.

Before constructing interpolating polynomials using differences, some numerical properties of differences and difference tables are discussed.

From subsequent developments, one will be able to verify that

$$\Delta^n f_i = h^n f^{(n)}(\xi_{in}),$$

where ξ_{in} is a point in the interval $[x_i, x_{i+n}]$, assuming that function f is n times continuously differentiable on a suitable interval (see Problem 2.17). As a consequence of the preceding equation, we have that if f is a polynomial of degree m, then $\Delta^{m+1}f_i = 0$.

Table 2.2. Difference Table

f	Δ	Δ^2	Δ^3	Δ^4	Δ^5	Δ^6
⋮						
f_{-3}		⋮				
	Δf_{-3}		⋮			
f_{-2}		$\Delta^2 f_{-3}$		⋮		
	Δf_{-2}		$\Delta^3 f_{-3}$		⋮	
f_{-1}		$\Delta^2 f_{-2}$		$\Delta^4 f_{-3}$		⋮
	Δf_{-1}		$\Delta^3 f_{-2}$		$\Delta^5 f_{-3}$	
f_0		$\Delta^2 f_{-1}$		$\Delta^4 f_{-2}$		$\Delta^6 f_{-3}$
	Δf_0		$\Delta^3 f_{-1}$		$\Delta^5 f_{-2}$	
f_1		$\Delta^2 f_0$		$\Delta^4 f_{-1}$		⋮
	Δf_1		$\Delta^3 f_0$		⋮	
f_2		$\Delta^2 f_1$		⋮		⋮
	Δf_2		⋮			
f_3		⋮				
⋮	⋮					

Problems

2.10. Demonstrate that for every k, $\Delta^k(f_i - g_i) = \Delta^k f_i - \Delta^k g_i$, all i.

2.11. Suppose

$$g_i = \begin{cases} \varepsilon, & i = 0 \\ 0, & i \neq 0. \end{cases}$$

Develop a formula for $\Delta^k g_i$. Make a difference table with four columns. Suppose the kth differences of a sequence $\{f_i\}$ are known to be fairly small. Explain, in view of Problem 2.10, how one can locate an error in *one* of the tabulated values of $\{f_i\}$.

2.12. Values of third-degree polynomials are tabulated as follows:

x:	-3	-2	-1	0	1	2	3
$p(x)$:	-28	-9	-2	-1	1	7	26

This tabulation has one error in the second row. Correct the table, using the idea in Problem 2.11.

2.13. Suppose $g_j = \varepsilon(-1)^j$, for every integer j. Calculate $\Delta^6 g_j$. Make a difference table for this function.

2.1.4. The Fraser Diagram

In this section a general method for computing interpolating polynomials is studied. If to Table 2.2 we add connecting lines and binomial coefficients as in Table 2.3, where $t = (x - x_0)/h$, then we obtain a modified difference tableau called the *Fraser diagram*. In this table,

$$\binom{t}{k} = \frac{t(t-1)(t-2) \cdots (t-k+1)}{k!}.$$

To generate interpolating formulas, the procedure is the following:

1. Start at an entry in the first column (it is a function value) and proceed along any path in the diagram (i.e., if a segment terminates on a difference, the path may be continued along any of the four paths leading from the difference, and if a segment terminates on a function value, the path may be continued along either of the two paths leading from the functional value), and end the path at any difference.

2. Construct the formula by (a) writing down the functional value at which the path started, and (b) for every additional path segment adding an additional term to the formula in the following way:

(i) for every left-to-right segment with positive slope, the additional term is the product of the difference on which the segment terminates and

Table 2.3. The Fraser Diagram

f	Δf	$\binom{\cdot}{1}$	$\Delta^2 f$	$\binom{\cdot}{2}$	$\Delta^3 f$	$\binom{\cdot}{3}$	$\Delta^4 f$	$\binom{\cdot}{4}$
f_{-3}			$\Delta^2 f_{-4}$				$\Delta^4 f_{-5}$	
	Δf_{-3}			$\binom{t+3}{2}$	$\Delta^3 f_{-4}$			$\binom{t+4}{4}$
f_{-2}		$\binom{t+2}{1}$	$\Delta^2 f_{-3}$			$\binom{t+3}{3}$	$\Delta^4 f_{-4}$	
	Δf_{-2}			$\binom{t+2}{2}$	$\Delta^3 f_{-3}$			$\binom{t+3}{4}$
f_{-1}		$\binom{t+1}{1}$	$\Delta^2 f_{-2}$			$\binom{t+2}{3}$	$\Delta^4 f_{-3}$	
	Δf_{-1}			$\binom{t+1}{2}$	$\Delta^3 f_{-2}$			$\binom{t+2}{4}$
f_0		$\binom{t}{1}$	$\Delta^2 f_{-1}$			$\binom{t+1}{3}$	$\Delta^4 f_{-2}$	
	Δf_0			$\binom{t}{2}$	$\Delta^3 f_{-1}$			$\binom{t+1}{4}$
f_1		$\binom{t-1}{1}$	$\Delta^2 f_0$			$\binom{t}{3}$	$\Delta^4 f_{-1}$	
	Δf_1			$\binom{t-1}{2}$	$\Delta^3 f_0$			$\binom{t}{4}$
f_2		$\binom{t-2}{1}$	$\Delta^2 f_1$			$\binom{t-1}{3}$	$\Delta^4 f_0$	
	Δf_2			$\binom{t-2}{2}$	$\Delta^3 f_1$			$\binom{t-1}{4}$
f_3			$\Delta^2 f_2$				$\Delta^4 f_1$	

the binomial coefficients directly below this difference, and directly above if the slope of the segment is negative;

(ii) for every right-to-left segment, the additional term is the product of -1 and the additional term associated with the oppositely directed segment (which was defined above).

The most popular difference formulas for the interpolation polynomial are derived below from our Fraser diagram method. In these formulas, since $t = (x - x_0)/h$, the polynomial $p_n(t)$ is related to the Lagrange formula (2.5) for p by $p_n(t) = p(x_0 + th)$.

1. Newton's Forward Formula. By starting at f_0, proceeding along lines sloping downward to the right and terminating at the nth difference (see Figure 2.1), we obtain the formula

$$p_n(t) = f_0 + \binom{t}{1} \Delta f_0 + \binom{t}{2} \Delta^2 f_0 + \cdots + \binom{t}{n} \Delta^n f_0. \qquad (2.11)$$

2. Newton's Backward Formula. Starting at f_0 and proceeding along lines sloping upward and to the right, we have

$$p_n(t) = f_0 + \binom{t}{1} \Delta f_{-1} + \binom{t+1}{2} \Delta^2 f_{-2} + \cdots + \binom{t+n-1}{n} \Delta^n f_{-n}. \qquad (2.12)$$

3. Gauss' Forward Formula. Here we proceed in a zigzag, downward and to the right, then upward and to the right, and again downward and to the right, etc. The corresponding formula is

$$p_n(t) = f_0 + \binom{t}{1} \Delta f_0 + \binom{t}{2} \Delta^2 f_{-1} + \binom{t+1}{3} \Delta^3 f_{-1} + \binom{t+1}{4} \Delta^4 f_{-2}$$

$$+ \cdots + \binom{t+m-1}{2m-1} \Delta^{2m-1} f_{-m+1} + \binom{t+m-1}{2m} \Delta^{2m} f_{-m}. \qquad (2.13)$$

4. Gauss' Backward Formula. Here we proceed similarly to the Gauss' forward formula but the first step is upward and to the right, as shown in Figure 2.1. Letting $[a]$ denote the integer part of a, the result is

$$p_n(t) = f_0 + \binom{t}{1} \Delta f_{-1} + \binom{t+1}{2} \Delta^2 f_{-1} + \binom{t+1}{3} \Delta^3 f_{-2} + \binom{t+2}{4} \Delta^4 f_{-2}$$

$$+ \cdots + \binom{t+[n/2]}{n} \Delta^n f_{-[(n+1)/2]}. \qquad (2.14)$$

5. Stirling's Formula. If we take the mean of Gauss' formulas (2.13) and (2.14), we have

$$p_n(t) = f_0 + \frac{1}{2} \binom{t}{1} (\Delta f_0 + \Delta f_{-1}) + \frac{1}{2} \left[\binom{t}{2} + \binom{t+1}{2} \right] \Delta^2 f_{-1} + \cdots . \qquad (2.15)$$

6. Bessel's Formula. Taking the mean of formulas (2.13) and (2.14) launched from f_1 rather than f_0, we obtain

$$p_n(t) = \frac{1}{2} (f_0 + f_1) + \left[\binom{t}{1} - \frac{1}{2} \right] \Delta f_0 + \frac{1}{2} \binom{t}{2} [\Delta^2 f_0 + \Delta^2 f_1] + \cdots . \qquad (2.16)$$

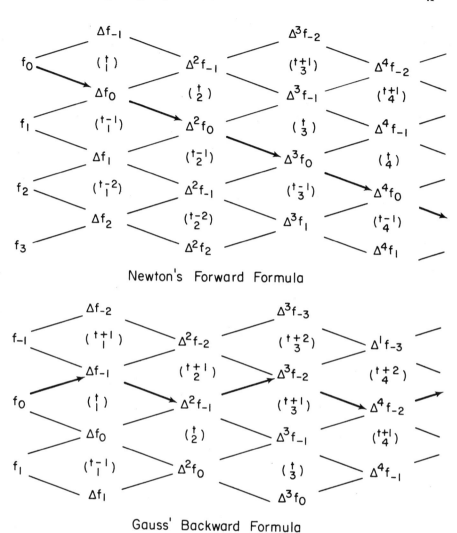

Newton's Forward Formula

Gauss' Backward Formula

Fig. 2.1. Application of the Fraser diagram.

Remark 2.1. Bessel's and Stirling's formulas are interpolating polynomials only if the endpoints of the corresponding Gauss formulas are the same.

In Figure 2.1, we show the paths from which Newton's forward and Gauss' backward formulas were constructed.

To show that the formulas directly obtained from the Fraser diagram are really interpolating polynomials, we prove the following theorem:

Theorem 2.5. All the Fraser diagram formulas that terminate on the same difference (independently of the path by which they reach that difference) are algebraically equivalent to the same equal-interval Lagrangian polynomial in the variable $t = (x - x_0)/h$. This formula interpolates the values involved in the terminating difference. (In general, the difference $\Delta^j f_i$ depends on the values $f_i, f_{i+1}, \ldots, f_{i+j}$.)

Proof. First we prove that all the formulas with the same terminating difference are equal. The proof consists of several steps.

(a) The sum of additional terms of segments from $\Delta^j f_i$ to $\Delta^{j+1} f_i$ and back, or from $\Delta^j f_i$ to $\Delta^{j+1} f_{i-1}$ and back is equal to zero. This follows directly from the rule for incorporating additional terms in the Fraser diagram.

(b) The formula corresponding to the path from f_i to Δf_i to f_{i+1} is equivalent to f_{i+1} because

$$f_i + \binom{t-i}{1} \Delta f_{i+1} - \binom{t-i-1}{1} \Delta f_{i+1} = f_i + \Delta f_{i+1}$$
$$= f_i + f_{i+1} - f_i = f_{i+1}.$$

Consequently any additional zigzag consisting of function values and first-order differences has no effect on the formula.

(c) Any closed path of the form

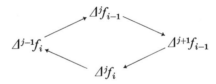

results in a contribution of zero. To show this, we calculate the sum of the corresponding additional terms:

$$\Delta^j f_{i-1} \binom{t-i}{j} + \Delta^{j+1} f_{i-1} \left[\binom{t-i+1}{j+1} - \binom{t-i}{j+1} \right] - \Delta^j f_i \binom{t-i}{j}.$$

After using the well-known additive property of binomial coefficients, we see that the above formula is equal to

$$\Delta^j f_{i-1} \binom{t-i}{j} + \Delta^{j+1} f_{i-1} \binom{t-i}{j} - \Delta^j f_i \binom{t-i}{j}$$
$$= \binom{t-i}{j} [\Delta^{j+1} f_{i-1} - (\Delta^j f_i - \Delta^j f_{i-1})]$$
$$= \binom{t-i}{j} (\Delta^{j+1} f_{i-1} - \Delta^{j+1} f_{i-1}) = 0$$

(d) The result in (c) implies that closed paths do not change the value of the formulas, because any closed path can be divided into paths discussed in the previous point.

(e) Let I_1 and I_2 be two formulas that terminate on a given difference. If the formula I_2 starts at a different functional value than formula I_1, we can connect the starting points with a zigzag consisting of functional values and first-order differences and assume that this zigzag belongs to I_2 (as we proved in part b); this has no effect on formula I_2. Then if the path corresponding to I_1 is continued in the direction opposite to that of path I_2, we get a closed path with formula equivalent to zero, and so $I_1 - I_2 = 0$. consequently, $I_1 = I_2$.

Now we prove that the formulas obtained from the diagram are equivalent to interpolating polynomials with interpolating points involved in the terminating difference. For convenience, we show this for Newton's forward formula. Applying Theorem 2.4, we have

$$\sum_{j=0}^{n} \binom{t}{j} \Delta^j f_0 = \sum_{j=0}^{n} \binom{t}{j} \sum_{k=0}^{j} (-1)^{j-k} \binom{j}{k} f_k$$

$$= \sum_{k=0}^{n} \sum_{j=k}^{n} (-1)^{j-k} \binom{j}{k} \binom{t}{j} f_k$$

$$= \sum_{k=0}^{n} f_k \left[\sum_{j=k}^{n} (-1)^{j-k} \binom{j}{k} \binom{t}{j} \right]. \qquad (2.17)$$

Since Newton's forward formula terminates on the difference $\Delta^n f_0$ involving the interpolating points $x_0, x_1, x_2, \ldots, x_n$, and for $x = x_k$

$$t = \frac{x - x_0}{h} = \frac{x_k - x_0}{h} = \frac{x_0 + kh - x_0}{h} = k,$$

it is sufficient to show that in (2.17) the coefficient of f_k is equal to 1 for $t = k$ and is equal to 0 for integer $t \neq k$.

(a) If $k > t$, then the binomial coefficients $\binom{t}{j}$ are equal to zero, and so the coefficient of f_k is also zero.

(b) If $k = t$, then only the one term

$$(-1)^0 \binom{k}{k} \binom{t}{k} = (-1)^0 \binom{k}{k} \binom{k}{k} = 1$$

differs from zero.

(c) Assume that $k < t$. Since the binomial coefficients $\binom{t}{j}$ are zeros for $j > t$, and

$$\binom{j}{k}\binom{t}{j} = \frac{j!}{k!(j-k)!} \times \frac{t!}{j!(t-j)!}$$

$$= \frac{t!}{k!(t-k)!} \times \frac{(t-k)!}{(j-k)!(t-j)!}$$

$$= \binom{t}{k}\binom{t-k}{j-k},$$

then the coefficient of f_k can be written in the form

$$\sum_{j=k}^{t} (-1)^{j-k}\binom{t}{k}\binom{t-k}{j-k} = \binom{t}{k}\sum_{J=0}^{M} (-1)^M \binom{M}{J}$$

$$- \binom{t}{k}(1 \quad 1)^M$$

$$= 0,$$

where

$$M = t - k, \quad J = j - k.$$

This completes the proof of the theorem. ∎

For approximating the value of function f at point $x \neq x_k$ by interpolating polynomials, we have to choose the formula that involves those interpolating points closest to x. Only in this way can it be guaranteed that the second term of the right-hand side of inequality (2.8) is minimal. This problem arises if the number of interpolating points is large, and in the polynomial approximation not all of the points are to be used. For instance, Newton's forward formula is recommended if x is close to the first interpolating point; Newton's backward formula should be used at the end of the sequence of interpolating points and the other formulas can be used at the middle of the interpolating table.

In order to increase the accuracy of interpolation, it is often necessary to use additional interpolation points. We do not need to repeat the calculations from the beginning because the introduction of a new interpolating point is equivalent to the addition of a new term into the interpolating polynomial.

Example 2.3. Calculate the value of 0.5^3 using the interpolating points 0, 1, 2, 3, 4. The corresponding difference table of function x^3 is given in Table 2.4. In this case $x = 0.5$, $t = (0.5 - 0)/1 = 0.5$. Using Newton's forward formula

Table 2.4. Difference Table for x^3

f	Δ	Δ^2	Δ^3	Δ^4
0				
	1			
1		6		
	7		6	
8		12		0
	19		6	
27		18		
	37			
64				

we have

$$p(0.5) = 0 + \binom{0.5}{1} \cdot 1 + \binom{0.5}{2} \cdot 6 + \binom{0.5}{3} \cdot 6 + \binom{0.5}{4} \cdot 0$$

$$= 0.5 + \frac{0.5(-0.5)}{2} \cdot 6 + \frac{0.5(-0.5)(-1.5)}{6} \cdot 6$$

$$+ \frac{0.5(-0.5)(-1.5)(-2.5)}{24} \cdot 0$$

$$= 0.125.$$

Problems

2.14. The *divided difference of order* k, $k > 1$, of function f is defined recursively by

$$f[x_1, \ldots, x_k] = \frac{f[x_2, \ldots, x_k] - f[x_1, \ldots, x_{k-1}]}{x_k - x_1},$$

with $f[x_j] = f(x_j)$, $1 \le j \le k$. Prove that

$$f[x_1, \ldots, x_k] = \sum_{i=1}^{k} f(x_i) \times \left[\prod_{\substack{j=1 \\ j \ne 1}}^{k} (x_i - x_j) \right]^{-1}.$$

2.15. Show that

$$f[x_1, \ldots, x_{k-1}, x] = f[x_1, \ldots, x_k] + (x - x_k)f[x_1, \ldots, x_k, x],$$

and apply this identity to prove that

$$f(x) = p(x) + E(x),$$

where

$$p(x) = f[x_1] + (x - x_1)f[x_1, x_2]$$
$$+ \cdots + (x - x_1)(x - x_2) \cdots (x - x_{n-1})f[x_1, x_2, \ldots, x_n]$$

and

$$E(x) = w_n(x)f[x_1, x_2, \ldots, x_n, x].$$

The polynomial p is called *Newton's divided-difference interpolating polynomial*.

2.16. Show that polynomial p in Problem 2.15 is equal to the Lagrange interpolating polynomial that uses the points $x_1, \ldots, x_n$.

2.17. Prove that

$$f[x_1, x_2, \ldots, x_n, x] = \frac{f^{(n)}(\xi)}{n!},$$

where ξ is in the interval spanned by $x_1, \ldots, x_n$ and x. HINT: Use Problems 2.15 and 2.16, and equate interpolation error terms.

2.18. Verify that for order 2, Newton's forward formula (2.11) and Stirling's formula (2.15) are indeed interpolation polynomials by directly evaluating the differences and then comparing the coefficients of the f_j terms with the Lagrange interpolation formula (2.6).

2.1.5. Aitken's Method and Computational Requirements of Interpolation

In applying Lagrange interpolating polynomials according to the preceding technique, a new interpolating point requires the repetition of the calculations. If we calculate a sequence of Lagrange interpolating polynomials by the method to be discussed, this disadvantage of Lagrange interpolation can be eliminated.

Let $p_{n_1, n_2, \ldots, n_k}$ denote the Lagrange interpolating polynomial that uses the interpolating points $x_{n_1}, x_{n_2}, \ldots, x_{n_k}$. Letting the vertical lines denote the determinant of the enclosed matrix, we prove that

$$p_{1,2,\ldots,k,n}(x) = \frac{1}{x_n - x_k} \begin{vmatrix} p_{1,2,\ldots,k}(x) & x_k - x \\ p_{1,2,\ldots,k-1,n}(x) & x_n - x \end{vmatrix}. \qquad (2.18)$$

The right-hand side of equation (2.18) is a polynomial of degree k and for $x = x_i$, $1 \leq i \leq k - 1$, we have

$$p_{1,2,\ldots,k,n}(x_i) = \frac{1}{x_n - x_k} [f_i \times (x_n - x_i) - f_i \times (x_k - x_i)]$$

$$= f_i;$$

for $x = x_k$ we get

$$p_{1,2,\ldots,k,n}(x_k) = \frac{1}{x_n - x_k}\,[f_k \times (x_n - x_k) - 0] = f_k;$$

and if $x = x_n$, then

$$p_{1,2,\ldots,k,n}(x_n) = \frac{1}{x_n - x_k}\,[0 - f_n \times (x_k - x_n)] = f_n.$$

By using equation (2.18), which we have just established, Table 2.5 can be constructed, row by row. The last polynomial of the nth line is the required interpolating polynomial. This procedure is known as *Aitken's method.*

Knuth asserts (1969, pp. 428–430) that efficient evaluation of the Lagrange representation (2.6) of the nth-degree interpolation polynomial requires n additions, $2n^2 + 2$ subtractions, $2n^2 + n - 1$ additions, and $n + 1$ divisions. Newton's forward formula (2.11) or its generalization, Newton's divided difference formula in Problem 2.15, requires using Horner's method, n multiplications, and $2n$ additions, once the coefficients of the factorial polynomials have been found. This, in turn, requires $\frac{1}{2}(n^2 + n)$ divisions and $n^2 + n$ subtractions, implying a savings of about three-fourths of the work over the Lagrange formula. For an nth-degree interpolation polynomial, the table for Aitken's method will have $n + 1$ rows. Each term in the table must be evaluated by using formula (2.18). There are $n(n + 1)/2$ such terms, and each will require two multiplications, one subtraction, and one division. As we see from this discussion, regardless of interpolation method, we can anticipate that the number of arithmetic operations will be (asymptotically) proportional to n^2, but that the constant of proportionality can be reduced through difference formulas or Aitken's method.

Table 2.5. Aitken's Method

Given data		Constructed polynomials		
x_1	$p_1(x)$			
x_2	$p_2(x)$	$p_{1,2}(x)$		
x_3	$p_3(x)$	$p_{1,3}(x)$	$p_{1,2,3}(x)$	
$\vdots$	$\vdots$	$\vdots$	$\vdots$	
x_n	$p_n(x)$	$p_{1,n}(x)$	$p_{1,2,n}(x) \cdots p_{1,2,3,\ldots,n}(x)$	

Problem

2.19. Determine the Lagrangian interpolating polynomial associated with the following tabulation, by using the Aitken method:

$$
\begin{array}{c|cccc}
x: & -3 & -2 & -1 & 0 \\
f(x): & 10 & 5 & 0 & 1
\end{array}
$$

2.1.6. Hermite Interpolation

Hermite interpolation is a generalization of Lagrangian interpolation. Let $x_1, \ldots, x_n$ be different interpolating points and let $f_k^{(0)}, f_k^{(1)}, \ldots, f_k^{(\alpha_k-1)}, 1 \le k \le n$, be function values and the derivatives up to the $(\alpha_k - 1)$th of the $\alpha_k - 1$ times differentiable function at the point x_k. The problem is to find a polynomial p of least degree such that

$$ p^{(l)}(x_k) = f_k^{(l)}, \qquad 1 \le k \le n, \quad 0 \le l \le \alpha_k \quad 1. $$

It can be proved in a way similar to the proof of Theorem 2.2 that there exists exactly one polynomial, the Hermite polynomial, of degree $\alpha = (\sum_{k=1}^n \alpha_k) - 1$, that satisfies the above condition, and it can be written in the form

$$ p(x) = \sum_{k=1}^{n} \sum_{l=0}^{\alpha_k-1} f_k^{(l)} H_{kl}(x), $$

where the functions H_{kl} are polynomials of degree α satisfying the equation

$$ H_{kl}^{(i)}(x_j) = \begin{cases} 1, & \text{if } i = l \text{ and } k = j, \\ 0, & \text{otherwise.} \end{cases} $$

Problems

2.20. Prove that $\sum_{k=1}^n H_{k0}(x) = 1$.

2.21. (This problem relates to Problem 2.3.) Let $f(x) = e^x$. Observe that $f'(x) = f(x)$. Construct the Hermite polynomial of the fifth degree using values $f(0.4)$, $f(0.5)$, $f(0.6)$, and the derivatives at these points, and use the Hermite interpolating polynomial to estimate $f(0.56)$. Compare your error with that arising in the problem referred to.

2.22. Extend the proof of Theorem 2.2 to show that if $p(x)$ is the $(2n - 1)$th-degree Hermite polynomial such that $p(x_i) = f(x_i)$, $p'(x_i) = f'(x_i)$, $1 \le i \le n$, then provided f has a continuous $2n$th derivative

$$ f(x) - p(x) = \frac{[w_n(x)]^2 f^{(2n)}(\xi)}{(2n)!}. $$

2.23. Prove that

$$\sum_{k=1}^{n} [x_k H_{k0}(x) + H_{k1}(x)] = x.$$

2.1.7. Inverse Interpolation

If the values of a function f at points $x_1, \ldots, x_n$ are $f_1, \ldots, f_n$ and f has an inverse, then the values of f^{-1} at points $f_1, \ldots, f_n$ are $x_1, \ldots, x_n$. Using the interpolating points $f_1, \ldots, f_n$ and functional values $x_1, \ldots, x_n$ we estimate the value of $f^{-1}(c)$ for a real number c by interpolation, obtaining a value x that is an approximate solution of equation $f(x) = c$. If $c = 0$ then the above procedure can be used for computing zeros of one variable functions.

Problem

2.24. Suppose the following values of f are known:

x:	-1.3	-1.4	-1.5	-1.6	-1.7
$f(x)$:	3.033	1.776	0.375	-1.176	-2.883

Approximate the root of f that lies in the interval $(-1.7, -1.3)$. In constructing this table, we actually took $f(x)$ to be the polynomial $f(x) = x^3 - 3x^2 - x + 9$, which serves as a test problem in the examples of Chapter 5. In Chapter 5, it is seen that to seven places, the desired root is -1.525102.

2.2. Uniform Approximations

Let $f_1, \ldots, f_n$ be values of a function f on the set of interpolation points $S_n = \{x_1, \ldots, x_n\}$, and let $m \leq n - 1$. If $m = n - 1$, then Lagrange interpolation affords us an mth-degree polynomial p for which $p(x_k) = f_k$, $1 \leq k \leq n$. In the case that $m < n - 1$, usually the polynomial of degree m is to be determined whose values at points x_k are "closest," in some given sense, to the functional values f_k. The *best approximating polynomial* with respect to S_n is defined to be the mth degree polynomial p for which the value

$$E_d(f, p) = \max_{x_k \in S_n} | f_k - p(x_k) | \tag{2.19}$$

is minimal. It is easy to see that equation (2.19) is equivalent to the system of inequalities

$$| f_k - a_0 - a_1 x_k - a_2 x_k^2 - \cdots - a_m x_k^m | \leq E_\mathrm{d}(f, p) \qquad (k = 1, 2, \ldots, n),$$

(2.20)

where $p(x) = a_0 + a_1 x + \ldots + a_m x^m$ and, for at least one k, (2.20) holds with equality. We now discuss the computation of best approximating polynomials through the use of linear programming.

Consider the linear programming problem [as developed in Luenberger (1973), for example] with unknowns $E, a_0, \ldots, a_m$, which is specified by

minimize E,
subject to

$$E + a_0 + a_1 x_k + a_2 x_k^2 + \cdots + a_m x_k^m \geq f_k,$$
$$-E + a_0 + a_1 x_k + a_2 x_k^2 + \cdots + a_m x_k^m \leq f_k \qquad (k = 1, 2, \ldots, n).$$

(2.21)

Since the constraints are equivalent to the system of inequalities (2.20) and the value $E_\mathrm{d}(f, p)$ is to be minimal, the linear programming problem (2.21) is equivalent to the determination of the minimization problem associated with (2.19). If $E, a_0, a_1, \ldots, a_m$ are the solution values to the linear programming problem then $p(x) = \sum_{i=0}^{m} a_i x^i$ is the BAP with respect to S_n, and if $q(x)$ is any other polynomial of degree m (or less), then

$$E_\mathrm{d}(f, q) \geq E_\mathrm{d}(f, p) = E.$$

In Blum (1972, Section 7.5), for example, some interesting and useful properties of best approximating polynomials are developed. For instance, if $m = n - 2$, and p is the BAP with respect to S_n, then for $r(x)$ the residual [i.e., $r(x) = f(x) - p(x)$]

$$| r(x_i) | = E_\mathrm{d}(f, p), \qquad x_i \in S_n.$$

If the x_i are ordered to increase with i, an assumption that will be in force throughout this discussion, then additionally the residual $r(x_i)$ alternates in sign with increasing i. As Blum further demonstrates, the property "constant magnitude–alternating sign" on S_n characterizes the BAP for $n > m + 2$ in the sense that if E is any subset of S_n and E contains $m + 2$ elements, and if $r(x_i)$ alternates in sign for increasing i, $x_i \in E$, then, if $| r(x_i) |$ is not constant on I, p is not the BAP. One implication of this development is that if p is the BAP, then, since $r(x)$ alternates in sign on E, r must be 0 at $m + 1$ points, which implies that p is a Lagrange interpolation polynomial. Thus the set of mth-degree interpolation polynomials for f

contains the BAP polynomial for any continuous function f. Unfortunately, the optimal interpolation points depend on f.

Assume now that f is a continuous function on a closed interval $I = [a, b]$ and let m be a positive integer. For an arbitrary polynomial p of degree not exceeding m, the continuous version of formula (2.19) can be defined by

$$E(f, p) = \max_{x \in [a,b]} |f(x) - p(x)|.$$

It can be proven (see Isaacson and Keller, 1966, Section 5.4) that there exists exactly one polynomial p_m that minimizes the value $E(f, p)$. This minimizing polynomial p_m is called the *Chebyshev approximation* of function f on the interval I. If $E_m(f) \equiv E(f, p_m)$, then the Weierstrass theorem implies that for $m \to \infty$, $E_m(f) \to 0$.

There is no effective procedure (that is, a method that eventually terminates successfully) for calculating the Chebyshev approximation for every given continuous function f. However, there are techniques for calculating approximations of the Chebyshev approximation. For example, if $p_{n,m}$ is the BAP mth-degree polynomial for f with respect to S_n, and if for increasing n, S_n becomes dense in $[a, b]$, that is, if for every $x \in [a, b]$,

$$\min_{y \in S_n} |y - x| \to 0, \qquad \text{as } n \to \infty,$$

then $p_{n,m}$ converges uniformly in n to the mth-degree Chebyshev approximation for f. (This convergence, incidentally, implies that the Chebyshev approximation is also an interpolation polynomial, since the roots of polynomials are continuous functions of the coefficients and since the BAPs are interpolating polynomials.) This convergence result as well as more efficient constructions (especially the Remes variable-exchange method) of the Chebyshev approximation are described in Blum (1972, Section 7.5).

In certain specific cases, there are effective procedures for calculating the Chebyshev approximation. We now give the most well known and useful of these constructs.

Theorem 2.6. Let $I = [-1, 1]$. Then the $(n - 1)th$-degree Chebyshev approximation of x^n on I is the interpolation polynomial for x^n whose interpolation points are the roots of the nth-degree polynomial

$$T_n(x) = \cos(n \cos^{-1} x).$$

$T_n(x)$ is called the nth *Chebyshev polynomial*.

Proof. First we prove that the function T_n is really a polynomial of degree n.

By using the DeMoivre formula [see, e.g., Kaplan (1959, page 489)] for complex numbers having an arbitrary real value of α, we have for $i = \sqrt{-1}$ that

$$(\cos \alpha + i \sin \alpha)^n = \cos n\alpha + i \sin n\alpha.$$

By applying the binomial theorem to the left-hand side, calculating the real part, and comparing it with the real part of the right-hand side, we find

$$\binom{n}{0} \cos^n \alpha - \binom{n}{2} \cos^{n-2} \alpha \sin^2 \alpha + \binom{n}{4} \cos^{n-4} \alpha \sin^4 \alpha - \cdots = \cos n\alpha,$$

which implies

$$\cos n\alpha = \cos^n \alpha - \binom{n}{2} \cos^{n-2} \alpha \, (1 - \cos^2 \alpha)$$

$$+ \binom{n}{4} \cos^{n-4} \alpha \, (1 - \cos^2 \alpha)^2 - \cdots .$$

If we take $\alpha = \arccos x$, we have

$$T_n(x) = \cos(n \arccos x)$$

$$= x^n - \binom{n}{2} x^{n-2}(1 - x^2) + \binom{n}{4} x^{n-4}(1 - x^2)^2 - \cdots .$$

The coefficient of x^n in $T_n(x)$ is $\binom{n}{0} + \binom{n}{2} + \binom{n}{4} + \cdots = 2^{n-1}$; consequently, T_n is a polynomial of degree n.

If we approximate x^n by the $(n-1)$th-degree interpolation polynomial p_{n-1} having as interpolation points the zeros of $T_n(x)$, then in the notation of Section 2.1.2, $w_n(x) = (1/2^{n-1})T_n(x)$ and consequently the right-hand side of the error equation (2.7) has the form

$$x^n - p_{n-1}(x) = \frac{1}{n!} \, n! w_n(x) = w_n(x) = \frac{1}{2^{n-1}} \, T_n(x).$$

The function $| w_n(x) |$ has an extreme at $n \arccos x = k\pi$, i.e., at $x_k = \cos(k\pi/n)$, $0 \le k \le n$, and the maximal value of $| w_n(x) |$ is equal to $1/2^{n-1}$, but $w_n(x_k)$ oscillates in sign with increasing k.

Suppose that there exists another polynomial q of degree not exceeding $n - 1$ and

$$E(x^n, q) < E(x^n, p_{n-1}) = \frac{1}{2^{n-1}}.$$

Then the polynomial $h(x) = w_n(x) - [x^n - q(x)]$ of degree not exceeding $n - 1$ has the same sign as $w_n(x)$ at $n + 1$ points x_k; consequently, polynomial h has n different zeros without vanishing identically. This is a contradiction. ∎

One application of the construction given in the theorem is to provide a heuristic basis for a procedure for finding a polynomial of low dimension that gives an adequate fit for any given continuous function f.

Let $g_n(x)$ be an nth-degree polynomial approximation of $f(x)$, this approximation giving acceptable accuracy. One may verify that $g_n(x)$ admits a unique representation

$$g_n(x) = \sum_{j=0}^{n} C_j T_j(x)$$

in terms of Chebyshev polynomials T_v of increasing degree v. By Theorem 2.6,

$$p_{n-1}^*(x) = \sum_{j=0}^{n-1} C_j T_j(x)$$

is the $(n - 1)$th-degree Chebyshev approximation of $g_n(x)$. In the procedure called *Chebyshev economization*, $g_n(x)$ is approximated by the kth-degree polynomial

$$p_k^*(x) = \sum_{j=0}^{k} C_j T_j(x),$$

where k is the least integer such that $p_k^*(x)$ affords an acceptable approximation of $f(x)$. Details and an example of the Chebyshev economization method are to be found in Conte and de Boor (1972, Section 4.12).

Problem

2.25. Construct in detail the parameters of the linear programming problem associated with finding the linear BAP associated with the following function pairs:

x:	-3	-2	-1	0
$f(x)$:	10	5	0	1

2.3. Least Squares Approximation

As in the previous section, let $f_1, \ldots, f_n$ denote the values of function f on a set S_n of points $\{x_1, \ldots, x_n\}$, and let $m < n - 1$. The discrete least squares approximation of f with respect to S_n is a polynomial of degree not exceeding m for which the sum of the squares of the errors

$$Q(f, p) = \sum_{k=1}^{n} [f_k - p(x_k)]^2 \qquad (2.22)$$

is minimal over the set of all polynomials of degree not exceeding m. Let $p(x) = c_0 + c_1 x + c_2 x^2 + \cdots + c_m x^m$. Let us introduce the following notation:

$$\mathbf{f} = (f_1, \ldots, f_n)^{\mathrm{T}}, \qquad \mathbf{c} = (c_0, c_1, \ldots, c_m)^{\mathrm{T}},$$

$$\mathbf{X} = \begin{pmatrix} 1 & x_1 & x_1^2 & \cdots & x_1^m \\ 1 & x_2 & x_2^2 & \cdots & x_2^m \\ \cdot & \cdot & \cdot & & \cdot \\ \cdot & \cdot & \cdot & & \cdot \\ \cdot & \cdot & \cdot & & \cdot \\ 1 & x_n & x_n^2 & \cdots & x_n^m \end{pmatrix},$$

$$\mathbf{A} = \mathbf{X}^{\mathrm{T}}\mathbf{X}, \qquad \mathbf{g} = \mathbf{X}^{\mathrm{T}}\mathbf{f}, \qquad (2.23)$$

where T denotes transpose. Matrix $\mathbf{X}$ has linearly independent columns (Problem 2.31). First we prove that for all vectors $\mathbf{a}$ the inequality

$$\mathbf{a}^{\mathrm{T}}\mathbf{A}\mathbf{a} \geq 0 \qquad (2.24)$$

holds and it becomes an equality if and only if $\mathbf{a} = \mathbf{0}$. Since

$$\mathbf{a}^{\mathrm{T}}\mathbf{A}\mathbf{a} = \mathbf{a}^{\mathrm{T}}\mathbf{X}^{\mathrm{T}}\mathbf{X}\mathbf{a} = (\mathbf{X}\mathbf{a})^{\mathrm{T}}(\mathbf{X}\mathbf{a}),$$

the value of $\mathbf{a}^{\mathrm{T}}\mathbf{A}\mathbf{a}$ is equal to the sum of the squares of coefficients of vector $\mathbf{X}\mathbf{a}$; consequently, it must be nonnegative. It is zero if and only if all components of $\mathbf{X}\mathbf{a}$ are zero, i.e., the vector $\mathbf{X}\mathbf{a}$ is zero. Because of the linear independence of the columns of matrix $\mathbf{X}$, $\mathbf{X}\mathbf{a} = \mathbf{0}$ if and only if $\mathbf{a} = \mathbf{0}$.

Elementary calculations show that

$$Q(f, p) = (\mathbf{f} - \mathbf{X}\mathbf{c})^{\mathrm{T}}(\mathbf{f} - \mathbf{X}\mathbf{c}) = \mathbf{f}^{\mathrm{T}}\mathbf{f} - \mathbf{c}^{\mathrm{T}}\mathbf{X}^{\mathrm{T}}\mathbf{f} - \mathbf{f}^{\mathrm{T}}\mathbf{X}\mathbf{c} + \mathbf{c}^{\mathrm{T}}\mathbf{X}^{\mathrm{T}}\mathbf{X}\mathbf{c}.$$

Upon using the definitions in (2.23) and adding and subtracting $\mathbf{g}^{\mathrm{T}}\mathbf{A}^{-1}\mathbf{A}\mathbf{A}^{-1}\mathbf{g} = \mathbf{g}^{\mathrm{T}}\mathbf{A}^{-1}\mathbf{g}$, we have

$$Q(f, p) = \mathbf{f}^{\mathrm{T}}\mathbf{f} - \mathbf{c}^{\mathrm{T}}\mathbf{A}\mathbf{A}^{-1}\mathbf{g} - \mathbf{g}^{\mathrm{T}}\mathbf{A}\mathbf{A}^{-1}\mathbf{c} + \mathbf{c}^{\mathrm{T}}\mathbf{A}\mathbf{c}$$
$$+ \mathbf{g}^{\mathrm{T}}\mathbf{A}^{-1}\mathbf{A}\mathbf{A}^{-1}\mathbf{g} - \mathbf{g}^{\mathrm{T}}\mathbf{A}^{-1}\mathbf{g},$$

which, when rearranged, yields

$$Q(f, p) = (\mathbf{c} - \mathbf{A}^{-1}\mathbf{g})^{\mathrm{T}}\mathbf{A}(\mathbf{c} - \mathbf{A}^{-1}\mathbf{g}) + \mathbf{f}^{\mathrm{T}}\mathbf{f} - \mathbf{g}^{\mathrm{T}}\mathbf{A}^{-1}\mathbf{g}.$$

Since only the first term of the right-hand side depends on the unknown vector **c**, inequality (2.24) implies that the only minimum of $Q(f, p)$ is at the value **c** for which

$$\mathbf{c} - \mathbf{A}^{-1}\mathbf{g} = \mathbf{0},$$

that is,

$$\mathbf{c} = \mathbf{A}^{-1}\mathbf{g} = (\mathbf{X}^{\mathrm{T}}\mathbf{X})^{-1}\mathbf{X}^{\mathrm{T}}\mathbf{f}. \tag{2.25}$$

We note in passing that the nonsingularity of **X** implies the nonsingularity of **A**, which in turn implies the uniqueness of **c** and consequently the existence and uniqueness of the least squares approximation.

In practice, instead of computing the inverse matrix in (2.25), one solves for **c** by solving the linear system

$$(\mathbf{X}^{\mathrm{T}}\mathbf{X})\mathbf{c} = \mathbf{X}^{\mathrm{T}}\mathbf{f}. \tag{2.26}$$

This system of linear equations is popularly referred to as the *normal equations*. It is well known that for large m, the normal equations are very often numerically intractible because the matrix $\mathbf{A} = \mathbf{X}^{\mathrm{T}}\mathbf{X}$ is ill conditioned, a subject to be discussed in Chapter 6. The gist of the problem is that roundoff errors and slight errors in the data lead to large and sometimes astronomical errors in the solution vector **c**. This numerical stability problem is customarily overcome through an alternative analysis employing orthogonal polynomials. For brevity of notation, we shall employ inner-product notation. Let g and h be any functions. We define

$$(g, h) \equiv \sum_{x \in S_n} g(x)h(x).$$

With this notation, the least squares approximation is that polynomial P of degree m or less that minimizes $(f - P, f - P)$. Let P_i, $0 \le i \le m$, be *orthogonal polynomials* of degree less than or equal to m. That is,

$$(P_i, P_j) = 0 \qquad \text{if } i \ne j.$$

We shall speak of their construction presently. Orthogonality implies linear independence, and since we have $m + 1$ polynomials they must span the vector space of mth-degree polynomials. Consequently, the least squares approximation P can be written as

$$P(x) = \sum_{j=0}^{m} b_j P_j(x).$$

We can minimize the square error of approximation $(f - \sum_{j=0}^{m} b_j P_j, f - \sum_{j=0}^{m} b_j P_j)$ by taking the partial derivatives with respect to the b_j and setting these partial derivatives equal to zero. Then we find

$$b_j = (P_j, f)/(P_j, P_j). \tag{2.27}$$

Note that the square error expression is quadratic in the b_j, and thus (2.27) must provide a minimum. This is a more accurate computational procedure than solving system (2.26). We now outline the construction of orthogonal sets of polynomials. Let $p_0, p_1, \ldots, p_m$ be any polynomials of degree less than or equal to m such that the vectors $\mathbf{p}_j = (p_j(x_1), \ldots, p_j(x_n))$ are linearly independent. (We may always take $p_j = x^j$.) Construct P_j recursively by setting $P_0 = (p_0, p_0)^{-1} p_0$ and

$$G_j(x) = p_j(x) - (P_{j-1}, p_j)P_{j-1} - (P_{j-2}, p_j)P_{j-2} - \cdots - (P_0, p_j)P_0,$$

$$P_j(x) = (G_j, G_j)^{-1} G_j(x).$$

One may readily verify that the P_j so constructed are orthogonal. This procedure is an instance of the *Gram–Schmidt* method, a well-known technique for orthogonalization in linear algebra (e.g., Nering 1963). Further discussion of this approach to least squares approximation computation is to be found in Young and Gregory (1972) and Isaacson and Keller (1966).

As in the case of uniform approximation discussed in the preceding section, discrete least squares approximation has a continuous counterpart, namely, finding the polynomial p in the class of polynomials of degree not exceeding m that minimizes $(f - p, f - p)_2 = \int_b^a [f(x) - p(x)]^2 \, dx$. Here, the inner product $(g, h)_2$ is defined by

$$(g, h)_2 = \int_a^b g(x)h(x) \, dx.$$

By expanding the quadratic form (with respect to the c_j) and taking first partial derivatives, we derive the linear system of equations for the least squares polynomials $p(x) = \sum_{i=0}^{m} c_i x^i$:

$$(f - \sum c_i x^i, f - \sum c_i x^i)_2 = (f, f)_2 - 2 \sum c_i (x^i, f)_2 + \sum c_i c_j (x^i, x^j)_2,$$

and so

$$\frac{\partial}{\partial c_j} (f - \sum c_i x^i, f - \sum c_i x^i)_2 = -2(x^j, f)_2 + 2 \sum c_i (x^i, x^j)_2.$$

Thus for $p = \sum c_i x^i$ to be the least squares polynomial, it is necessary and sufficient that $\mathbf{c} = (c_0, c_1, \ldots, c_m)^{\mathrm{T}}$ satisfy

$$\mathbf{Ac} = \mathbf{w},$$

where $(\mathbf{A})_{ij} = (x^i, x^j)_2$ and $(x^i, f)_2$ is the ith coordinate of $\mathbf{w}$. One may see that this matrix equation is closely related to (2.26). An interesting observation is that if $a = 0$ and $b = 1$ then, as the reader may confirm, $\mathbf{A}$ is the matrix whose i, j coordinate is $1/(i + j + 1)$, $0 \leq i, j \leq m$. This matrix is the *Hilbert segment matrix* and is famous as an example of a highly ill-conditioned matrix. For example, Young and Gregory (1972, p. 324) assert that, "with single precision on most machines, it is impractical to solve accurately a system involving a Hilbert matrix of order more than 8." This observation provides evidence for our earlier claim that the linear system associated with large or even moderate least squares problems is poorly behaved.

One may confirm that the orthogonal polynomial least squares construction is directly applicable to the continuous least squares approximation problem. While it is relatively easy to find orthogonal families of polynomials, computation of the integral $(f, P_i)_2$ usually must be done numerically and bothersome error ensues. Consequently, the discrete least squares criterion is the one most often employed in practice.

Example 2.4. Let

$$x_1 = -2, \quad x_2 = -1, \quad x_3 = 0, \quad x_4 = 1, \quad x_5 = 2,$$
$$f_1 = 0, \quad f_2 = 1, \quad f_3 = 2, \quad f_4 = 1, \quad f_5 = 0.$$

We determine the best second-degree least squares approximation for f, using these points.

In our case, $m = 2$, $n = 5$,

$$\mathbf{X} = \begin{pmatrix} 1 & -2 & 4 \\ 1 & -1 & 1 \\ 1 & 0 & 0 \\ 1 & 1 & 1 \\ 1 & 2 & 4 \end{pmatrix}, \quad \mathbf{X}^{\mathrm{T}}\mathbf{X} = \begin{pmatrix} 5 & 0 & 10 \\ 0 & 10 & 0 \\ 10 & 0 & 34 \end{pmatrix}, \quad \mathbf{g} = \begin{pmatrix} 4 \\ 0 \\ 2 \end{pmatrix},$$

and so system (2.26) has the form

$$5c_0 + 0c_1 + 10c_2 = 4,$$
$$0c_0 + 10c_1 + 0c_2 = 0,$$
$$10c_0 + 0c_1 + 34c_2 = 2,$$

Table 2.6. RMS Error in Approximation of $f(x) = e^x$, $x \in [0, 1]$

Polynomial	N	$I(p)$
Tenth-degree Bernstein	11	0.1567884×10^{-1}
Tenth-degree Chebyshev	41	$0.1174307 \times 10^{-12}$
Tenth-degree Lagrange	11	$0.1935452 \times 10^{-12}$
Fifth-degree least squares	21	0.7891922×10^{-6}

which implies $c_0 = 58/35$, $c_1 = 0$, $c_2 = -3/7$, and so

$$p(x) = -\frac{3}{7} x^2 + \frac{58}{35}.$$

Example 2.5. For $f(x) = e^x$, $x \in [0,1]$, we compute the approximation error associated with tenth-degree Bernstein, Lagrange, and BAP approximating polynomials as well as fifth-degree least squares approximation, the tenth-degree approximation leading to a linear system too ill conditioned for our linear equation algorithm (Gauss elimination). The error was the root mean square (RMS) error calculated at grid points having spacing 0.01. Specifically the error of an approximating polynomial p is given by

$$I(p) = \left\{ \frac{1}{101} \sum_{i=0}^{100} \left[e^{i/100} - p\left(\frac{i}{100}\right) \right]^2 \right\}^{1/2}$$

In Table 2.6, N is the number of equally spaced data points x_i used in computing the indicated approximation polynomial. The data point spacing $h = 1/(N - 1)$.

In Example 2.6, more numerical experimentation will be reported.

Problems

2.26. Let

$$x_k = -1 + \frac{k}{n} \quad (k = 0, 1, 2, \ldots, 2n), \qquad f(x) = \tfrac{1}{2}(x + |x|).$$

Find the best BAP and discrete least squares approximation of f with constant functions.

2.27. Let $p_0, p_1, \ldots, p_m$ be multivariate functions and $x_1, \ldots, x_n$ different vectors. Assume that $m \leq n$, and the columns of the matrix

$$\mathbf{X} = \begin{pmatrix} p_0(\mathbf{x}_1) & p_1(\mathbf{x}_1) & \cdots & p_m(\mathbf{x}_1) \\ \vdots & \vdots & & \vdots \\ p_0(\mathbf{x}_n) & p_1(\mathbf{x}_n) & \cdots & p_m(\mathbf{x}_n) \end{pmatrix}$$

are linearly independent. Let $\mathbf{f} = (f_1, f_2, \ldots, f_n)^{\mathrm{T}}$ be a real vector. Prove that the vector $\mathbf{c} = (c_0, c_1, \ldots, c_m)^{\mathrm{T}}$ that minimizes the value

$$\sum_{k=1}^{n} \left[f_k - \sum_{i=0}^{m} c_i p_i(\mathbf{x}_k) \right]^2$$

can be computed by solving the linear system $(\mathbf{X}^{\mathrm{T}}\mathbf{X})\mathbf{c} = \mathbf{X}^{\mathrm{T}}\mathbf{f}$.

2.28. (For readers who have studied statistics.) Assuming that $p_0(\mathbf{x}) = 1$ and for $i = 1, 2, \ldots, m$, $p_i(\mathbf{x})$ is the ith component of $\mathbf{x}$, show that the procedure of the previous problem is equal to the method of multivariate regression.

2.29. Assume that in Problem 2.27 the columns of matrix $\mathbf{X}$ are orthogonal to each other, and let the kth column of $\mathbf{X}$ be denoted by $\mathbf{p}_k$. Prove that, for $k = 0, 1, 2, \ldots, m$,

$$c_k = \mathbf{p}_k^{\mathrm{T}}\mathbf{f} / \mathbf{p}_k^{\mathrm{T}}\mathbf{p}_k.$$

2.30. Generalize the method of constructing BAPs for multivariate functions in an analogous fashion to the way we generalized the least squares algorithm in Problem 2.27.

2.31. Prove that if $m \leq n$, the columns of $\mathbf{X}$ in (2.23) are linearly independent. [HINT: There are many ways of showing the principal minor consisting of the first m rows and columns is nonzero.] The matrix composed of the first m rows of $\mathbf{X}$ is called the *Vandermonde matrix* and it shows up in many contexts (e.g., Bellman, 1970, p. 359). (It is known from matrix theory that a square matrix is non-singular if and only if its columns are linearly independent.)

2.32. Construct the fifth-degree least squares polynomial approximation for $f(x) = e^x$. Use the values f_i at $x_i = (0.05)i$, $0 \leq i \leq 20$. Compute an approximation to $f(0.56)$. This problem is related to Problem 2.3. What is the error in your answer?

2.4. Spline Functions

In the last few years, piecewise polynomial approximations have become very important in theoretical studies as well as in engineering applications. The most popular class of such approximating functions are spline functions. Basically, instead of trying to approximate a function over the entire interval

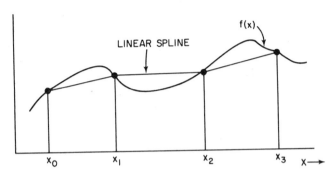

Fig. 2.2. A linear spline.

by one polynomial of high degree, in the method of splines, one partitions the interval I into smaller intervals, and approximates the function f on each of these smaller intervals by a different low-degree polynomial. We shall see in Chapter 3 that high-order Newton–Cotes rules for numerical integration are integrals of interpolation polynomials, whereas composite quadrature formulas are closely related to splines. Figure 2.2 illustrates a linear spline approximation.

In Section 2.5, it will be pointed out that Lagrange interpolation procedures in general do not provide arbitrarily good fit, even over the set of continuous functions, whereas splines do not suffer this deficiency. Splines are therefore an effective method for approximating functions f that are not smooth or well behaved, or for approximating derivatives of f, when they do exist. Finally, we shall note that splines have been playing a prominent role in certain recent theoretical studies in numerical analysis and statistics.

Let $[a, b]$ be a finite interval and let $a = x_0 < x_1 < \cdots < x_n = b$ be partition points for this interval. A spline function of degree m with nodes at the points x_i, $0 \le i \le n$, is a function $s(x)$ with the following properties:

(a) s is $m - 1$ times continuously differentiable on $[a, b]$;

(b) on each subinterval $[x_i, x_{i+1}]$, $0 \le i \le n - 1$, s is a polynomial of degree not exceeding m.

In the special case that $m = 1$, the spline function is a piecewise linear function. For $1 \le i \le n - 1$, let

$$\psi_i(x) = \begin{cases} 0, & x \notin [x_{i-1}, x_{i+1}], \\ (x - x_{i-1})/(x_i - x_{i-1}), & x \in [x_{i-1}, x_i], \\ (x_{i+1} - x)/(x_{i+1} - x_i), & x \in [x_i, x_{i+1}], \end{cases} \qquad (2.28)$$

and let $y_0, y_1, \ldots, y_n$ be given real numbers. Then the spline function of degree one having the property $s(x_i) = y_i$, $0 \leq i \leq n$, is unique and can be written in the form

$$s(x) = \sum_{i=0}^{n} y_i \psi_i(x). \qquad (2.29)$$

The identity (2.29) follows immediately from the fact that

$$\psi_i(x_j) = \begin{cases} 0, & j \neq i, \\ 1, & j = i, \end{cases} \quad 1 \leq i \leq n-1.$$

Observe that equation (2.29) bears some resemblance to Lagrangian interpolation. In particular, it interpolates and is linear at the function values.

The cubic spline functions are the most widely used splines in applications and are therefore discussed in more detail.

Let $a = x_0 < x_1 < x_2 < \cdots < x_{n-1} < x_n = b$ be a grid of points for the interval $[a, b]$. Let the values $y_0, y_1, \ldots, y_n$ be given real numbers. We show how a spline function of degree three having the properties $s(x_i) = y_i$, $0 \leq i \leq n$, is computed.

For $i = 1, 2, \ldots, n$, let

$$h_i = x_i - x_{i-1}, \qquad d_i = \frac{y_i - y_{i-1}}{h_i},$$

and

$$t = \frac{x - x_{i-1}}{x_i - x_{i-1}} \qquad \text{for } x \in [x_{i-1}, x_i].$$

Then we shall show that for $x \in [x_{i-1}, x_i]$, $s(x) = s_i(x)$ satisfies the above properties (a) and (b) of a spline, provided

$$s_i(x) = ty_i + (1 - t)y_{i-1} + h_i t(1 - t)[(k_{i-1} - d_i)(1 - t) - (k_i - d_i)t],$$

and $k_0, k_1, \ldots, k_n$ satisfy the linear system of equations,

$$h_{i+1}k_{i-1} + 2(h_i + h_{i+1})k_i + h_i k_{i+1}$$
$$= 3(h_i d_{i+1} + h_{i+1} d_i), \qquad 1 \leq i \leq n-1. \qquad (2.30)$$

Observe first that for $x = x_{i-1}$, $t = 0$, and for $x = x_i$, $t = 1$. Then simple calculations show that

$$s_i(x_{i-1}) = y_{i-1}, \qquad s_i(x_i) = y_i,$$
$$s_i'(x_{i-1}) = k_{i-1}, \qquad s_i'(x_i) = k_i,$$

Thus the parameters k_i, $0 \le i \le n$, are the derivatives of s at the distinct points. Since $s_i'(x_i) = k_i = s_{i+1}'(x_i)$ for $i = 0, 1, 2, \ldots, n - 1$, s is continuously differentiable on $[a, b]$. It can also be easily shown that (2.30) is equivalent to the condition $s_{i+1}''(x_i) = s_i''(x_i)$.

Thus the construction of a cubic spline function requires the solution of a linear system of equations. Observe that the number of equations is equal to $n - 1$ and the number of unknowns is $n + 1$, which implies that for the uniqueness of the solution, two additional conditions have to be given. Usually the additional conditions are given by $s_1''(a) = s_n''(b) = 0$, which are equivalent to the equations

$$2k_0 + k_1 = 3d_1, \qquad k_{n-1} + 2k_n = 3d_n. \qquad (2.31)$$

It is obvious that by putting equations (2.30) between the two equations of (2.31), the following complete linear tridiagonal system of equations can be obtained.

$$
\begin{bmatrix}
2 & 1 & 0 & 0 & \cdots & & 0 \\
h_2 & 2(h_1 + h_2) & h_1 & 0 & \cdots & & 0 \\
0 & h_3 & 2(h_2 + h_3) & h_2 & \cdots & & 0 \\
0 & 0 & h_4 & 2(h_3 + h_4) & h_3 & & \\
\vdots & \vdots & & \ddots & \ddots & \ddots & \\
& & & h_n & 2(h_{n-1} + h_n) & h_{n-1} \\
0 & 0 & \cdots & 0 & 1 & 2
\end{bmatrix}
\begin{bmatrix}
k_0 \\ k_1 \\ \\ \vdots \\ \\ \\ k_n
\end{bmatrix}
$$

$$
= 3
\begin{bmatrix}
d_1 \\
h_1 d_2 + h_2 d_1 \\
h_2 d_3 + h_3 d_2 \\
h_3 d_4 + h_4 d_3 \\
\vdots \\
d_n
\end{bmatrix}
\qquad (2.32)
$$

Observe that the absolute values of the diagonal elements of the matrix of coefficients are greater than the sum of the off-diagonal elements of the same row, and so it can be proved very easily (see Section 6.1.2) that the matrix of coefficients is nonsingular, and consequently the linear system of equations has a unique solution. This proves the existence and uniqueness of the cubic spline functions under the additional constraints (2.31). Gauss elimination (Chapter 6) gives efficient solutions to such tridiagonal systems.

Table 2.7. RMS Error in Runge Function Approximation

Method	Degree	N	$I(p)$
Cubic splines	—	21	0.98×10^{-3}
	—	41	0.05×10^{-3}
Least squares	5	21	0.1324
	10	41	0.0408
BAP	5	21	0.154
	10	41	0.047
Bernstein	5	6	0.220
	10	11	0.158
Lagrange	5	6	0.149
	10	11	0.611

Example 2.6. Various methods for approximation were applied to the function $f(x) = 1/(1 + x^2)$, which, as will be seen in the next section, is troublesome. The interval for approximation was $[-5,5]$, and N interpolation points were evenly spaced on this interval. Our measure of error was the root mean square error criterion with error computed at points $x_j = -5 + 0.1j$, $1 \leq j \leq 100$. This was the root mean square criterion in $I(p)$ used in Example 2.5. The results are given in Table 2.7.

Problems

2.33. Show that the function

$$s(x) = \begin{cases} x^3, & \text{if } x \geq 0, \\ 0, & \text{if } x < 0. \end{cases}$$

is a cubic spline.

2.34. Assume that for $i = 1, 2, \ldots, n$, $h_i = 1$. Show that the matrix of coefficients of the linear tridiagonal system (2.32) is positive definite.

2.5. Asymptotic Properties of Polynomial Approximations

In the beginning of this chapter, we established, by means of the Weierstrass theorem, that if f is continuous on the unit interval, then there exists a sequence of polynomials P_n that converges uniformly to f. That is,

$$\max_{0 \leq x \leq 1} | P_n(x) - f(x) | \to 0 \quad \text{as } n \to \infty.$$

Our analysis did not allow us to give a lower bound to the rate of this convergence. The difficulty stems from the fact that the modulus of continuity $\delta = \delta(\varepsilon)$ in the proof on the Weierstrass theorem can decrease arbitrarily slowly with ε. On the other hand, if f has a continuous derivative, then we can conclude that there is some constant L such that

$$| f(x) - f(y) | < L\,| x - y |, \qquad \text{for all } x, y \in [0,1], \qquad (2.33)$$

and consequently the modulus of continuity $\delta(\varepsilon)$ satisfies $\delta(\varepsilon) = \varepsilon/L$. We shall let ε and $\delta = \delta(\varepsilon)$ in (2.1) depend on n.

Let K be an arbitrary positive constant and, for each n, set $\varepsilon_n = Kn^{-1/4}$. Then we may take $\delta_n = (K/L)n^{-1/4}$ and thus from (2.1) we have that for every x in the unit interval,

$$2\,| f(x) - B_n(x) | \leq Kn^{-1/4} + M/[(K/L)^2 n^{-1/2} n].$$

This implies the following statement:

Theorem 2.7. Let f be a function with a continuous first derivative on the unit interval and B_n the nth-degree Bernstein approximating polynomial of f. Then for some constant C and for all n,

$$\sup_{0 \leq x \leq 1} |f(x) - B_n(x) | \leq C/n^{1/4}.$$

It turns out that for a fixed system of interpolating points, the Lagrange interpolation polynomial does not share with the Bernstein approximation the property of convergence to any continuous function. A famous instance (due to Runge) is the function $1/(1 + x^2)$ over the interval $I = [-5, 5]$. For each n, the Lagrange interpolation polynomial is determined by the set S_n of evenly spaced interpolation points $\{x_j = -5 + 10j/n,\ 0 \leq j \leq n\}$. A demonstration of divergence is given in Isaacson and Keller (1966, Section 6.3.4).

In Figure 2.3, we have plotted fifth- and tenth-degree approximations of the Runge function over the interval $[-5, 5]$, the approximations being obtained by the Lagrange interpolation polynomial, and the Bernstein, BAP, and least squares polynomial approximations. These are, in fact, the polynomials described in Example 2.6, which may be consulted for construction details. Notice that the error in the tenth-degree Lagrange approximation is appreciably greater than that of the fifth degree.

Another prominent example of divergence for evenly spaced interpolation points is the function $| x |$ on the interval $[-1, 1]$. One might hope

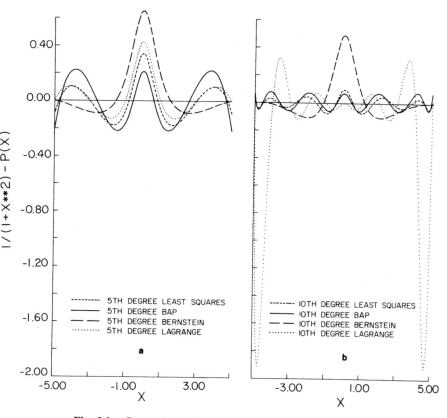

Fig. 2.3. Comparison of approximators for the Runge function.

that a more informed choice of interpolation point sets might somehow lead to convergence of Lagrange interpolation polynomials over the set of continuous functions. However, this hope is dashed by the negative results (Davis, 1963, p. 78) that for any fixed system S_n, $n = 1, 2, 3, \ldots$, of interpolation points contained in any fixed interval, there is a continuous function f such that the interpolation polynomials P_n based on S_n do not converge uniformly to f.

On the other hand, there are useful sufficient conditions for convergence of interpolation polynomials using equally spaced interpolation points. One such condition is now derived.

Theorem 2.8. For $n = 1, 2, \ldots$, let the interpolation points $S_n = \{x_0, x_1, \ldots, x_n\}$ be such that $x_j = j/n$, and assume that all derivatives $f^{(r)}$

of f exist and the terms $f^{(r)}(x)r^r$ are uniformly bounded on the unit interval and for all r by a constant M. Then the interpolating polynomials P_n based on S_n converge uniformly to f on the unit interval.

Proof. From the interpolation polynomial error bound (2.7) we have

$$f(x) - P_n(x) = \frac{1}{(n+1)!} f^{(n+1)}(\xi)w_n(x),$$

where $w_n(x) = \prod_{i=0}^{n}(x - i/n)$. Letting $t = nx$ (and so t ranges over $[0, n]$), we have $w_n(x) = n^{-(n+1)} \prod_{i=0}^{n}(t - i)$. It is not difficult to show that a maximum of the factorial polynomial $| \psi_n(t) | = \prod_{i=0}^{n} | t - i |$ occurs for $t \in [0, 1]$, and consequently

$$| \psi_n(t) | \leq n!.$$

For, observe that for $t + 1 \leq n/2$, t not integer,

$$| \psi_n(t + 1) | - | \psi_n(t) | = \prod_{i=-1}^{n-1} | t - i | \, [| t + 1 | - | t - n |].$$

The bracketed term is nonpositive since $t + 1 \leq n/2$. By symmetry of $| \psi_n(t) |$ about $n/2$, we have established the assertion that $| \psi_n |$ is maximized on $[0, 1]$. Having bounded $| w_n(x) |$ by $n!/n^{n+1}$, we have left the rest of the demonstration, which is easy, to the reader. ■

Note that we have proved that the maximum of $| w_n |$ occurs within a distance of $1/n$ of the endpoints of $[0, 1]$. A well-known rule of thumb is that the maximum interpolation error can be expected to occur near the extremes of the interpolation interval, as exemplified by the Lagrange curves in Figure 2.3.

We have established at this point that the Bernstein approximation converges uniformly to the target function f, provided f is continuous. Furthermore if f has a continuous first derivative, the rate of convergence is at least as fast as $C/n^{1/4}$. It will be apparent that any result concerning uniform convergence we can prove about any polynomial approximations must also hold for Chebyshev approximations. A natural question to ask is whether the $n^{-1/4}$ rate can be improved upon by some other scheme. It turns out that the answer is affirmative, and the approximation is afforded by (trigonometric) Fourier series analysis. We collect in the following lemma some diverse facts we shall need.

Lemma 2.1. (i) For every positive integer n and number $r \geq 2$,

$$\sum_{k>n} \frac{1}{k^r} < \frac{1}{n^{r-1}}.$$

(ii) If $g(\theta)$ is integrable and $g(\theta) = g(2\pi - \theta)$ for $0 \leq \theta \leq 2\pi$, then

$$\int_0^{2\pi} g(\theta) \sin n\theta \, d\theta = 0, \qquad \text{for all } n \geq 0.$$

(iii) For all $n \geq 0$, $\cos n\theta$ is expressible as a linear combination of terms $(\cos \theta)^k$, $0 \leq k \leq n$.

Proof. (i) $\sum_{k \geq n+1} 1/k^2 < \int_n^{\infty} 1/x^2 \, dx = 1/n$. Then for any $r > 2$

$$\sum_{k \geq n+1} \frac{1}{k^r} < \left[\frac{1}{n^{r-2}} \right] \times \sum_{k \geq n+1} \frac{1}{k^2} < \frac{1}{n^{r-1}}.$$

(ii) Note that $\sin n\theta = -\sin(2\pi - n\theta)$, and so

$$I \triangleq \int_0^{2\pi} g(\theta) \sin n\theta \, d\theta = - \int_0^{2\pi} g(2\pi - \theta) \sin(2\pi - n\theta) \, d\theta.$$

Substitute the variable $t = 2\pi - \theta$ into this last equation. Then, noting that for any integer k, $\sin(2k\pi + \alpha) = \sin \alpha$,

$$I = - \int_0^{2\pi} g(t) \sin(2\pi - 2n\pi + nt) \, dt = -I.$$

To satisfy this equation, I must be 0.

(iii) Observe that if $\operatorname{Re} \zeta$ denotes the real part of ζ, then

$$\cos n\theta = \operatorname{Re}(e^{in\theta}) = \operatorname{Re}[(e^{i\theta})^n]$$
$$= \operatorname{Re}[(\cos \theta + i \sin \theta)^n].$$

Any real terms involving $\sin \theta$ must be even powers of $\sin \theta$; and thus are expressible as powers (including, perhaps, the zeroth power) of $\cos \theta$. ∎

Definition 2.1. Let g be an integrable function on $[0, 2\pi]$. The nth-order *Fourier approximation* of g is defined by

$$\pi_n(\theta) = \sum_{k=0}^{n} (A_k \cos k\theta + B_k \sin k\theta),$$

where

$$A_k = \frac{1}{\pi} \int_0^{2\pi} g(\theta) \cos k\theta \, d\theta,$$

$$B_k = \frac{1}{\pi} \int_0^{2\pi} g(\theta) \sin k\theta \, d\theta.$$

We refer the reader to Apostol (1957, Chapter 15) for supplementary discussion of Fourier series representation.

Lemma 2.2. Let r be a positive integer and $g(\theta)$ a periodic function of period 2π having a continuous $(r + 1)$th derivative on $[0, 2\pi]$. The function $\pi_n(\theta)$ denotes the nth-order Fourier approximation of $g(\theta)$. Then for some constant C, and every n,

$$\max_{0 \leq \theta \leq 2\pi} \mid g(\theta) - \pi_n(\theta) \mid \leq C/n^r.$$

Proof. First we show that if $g^{(r)}$ exists, then

$$\mid A_n \mid + \mid B_n \mid \leq K/n^{r+1}, \tag{2.34}$$

for some K and all n. Our demonstration is for $\mid A_n \mid$ and $r = 1$, but the method is general. Using integration by parts, we obtain

$$A_n = \frac{1}{\pi} \int_0^{2\pi} g(\theta) \cos n\theta \, d\theta$$

$$= -\frac{1}{\pi n} \int_0^{2\pi} g'(\theta) \sin n\theta \, d\theta.$$

Also, changing the variable of integration to $\theta_1 = \theta - \pi/n$ and observing that $\sin n\theta = -\sin n\theta_1$, we have

$$A_n = \frac{1}{\pi n} \int_0^{2\pi} g'(\theta_1 + \pi/n) \sin n\theta_1 \, d\theta_1.$$

Averaging the above representations, we have

$$A_n = \frac{1}{2\pi n} \int_0^{2\pi} [g'(\theta + \pi/n) - g'(\theta)] \sin n\theta \, d\theta. \tag{2.35}$$

If L is an upper bound for $\mid g^{(2)} \mid$ we have that $\mid g'(\theta + \pi/n) - g'(\theta) \mid < L\pi/n$. Thus from (2.35)

$$\mid A_n \mid \leq L\pi/n^2, \quad \text{for all } n.$$

Under the conditions of the lemma, g has a Fourier series representation and so, for all θ,

$$| g(\theta) - \pi_n(\theta) | \leq \sum_{k \geq n+1} (| A_k | + | B_k |)$$

$$< 2L \sum_{k \geq n+1} 1/k^2.$$

The lemma follows for $r = 2$, upon recalling the inequality of Lemma 2.1. For larger r, iterate the integration by parts step. ∎

Theorem 2.9. Let P_n be the Chebyshev approximation of degree n for f on $I = [-1, 1]$, and for some $r \geq 1$ assume f has a continuous $(r + 1)$th derivative on I. Then for some C and all n,

$$\max_{-1 \leq x \leq 1} |f(x) - P_n(x) | \leq C/n^r.$$

Proof. We perform the change of variables $\theta = \cos^{-1} x$, and restrict θ to be in $[0, 2\pi]$. Define $g(\theta) = f(\cos \theta)$, and observe that the hypotheses of the theorem imply that g is periodic and has a continuous $(r + 1)$th derivative. Notice also that $g(\theta) = g(2\pi - \theta)$, and so by part (ii) of Lemma 2.1 the B_k terms in the Fourier expansion of $g(\theta)$ are 0, and by part (iii) of that lemma, then $\pi_n(\theta)$, the nth-order Fourier approximation of g, is a polynomial P_n in $\cos \theta$. By the previous lemma then, we have that for some C and all $n > 0$,

$$\max_{-1 \leq x \leq 1} |f(x) - P_n(x) | = \max_{0 \leq \theta \leq 2\pi} | g(\theta) - P_n(\cos \theta) |$$

$$\leq C/n^r.$$ ∎

2.6. Supplementary Notes and Discussion

At the present time, interpolation polynomials are given more attention than alternative approximation methods in introductory numerical analysis texts. There are some good reasons for this: interpolation polynomials are reasonably easy to conceptualize, their computational formulas are relatively easy to describe and program, and error analysis is relatively straightforward.

There are serious drawbacks to Lagrange interpolation, on the other hand. We have seen that an interpolation scheme need not converge to the target function. It is a well-known phenomenon among practitioners to find that high-degree interpolation formulas can give large errors, either because the target function is not regular enough or because there are some random errors in the function values. Thus, in practice, only low-order interpolation is used, and one seeks alternative methods if the relative error is required to be less than around 10^{-6}, say.

On the surface, least squares approximation would seem to provide a promising alternative. But as we have seen this approach requires care and sophistication because, for high order, the normal equation is ill conditioned. As we have noted, orthogonal polynomials provide a means for achieving greater computational stability, but care must be exercised here too because the Gram–Schmidt orthogonalization is recursive and error propagates. Techniques for improving the accuracy of the Gram–Schmidt procedure, such as reorthogonalization [described in Blum (1972, p. 300)] can help. Young and Gregory (1972) show that the linear system that is associated with orthogonal polynomials is still well behaved and useful even in the presence of error in the orthogonalization procedure. Golub (1965), Golub and Reinsch (1971), and Golub and Wilkinson (1966) give another technique known as *singular-value decomposition* for overcoming the ill-conditioning problem. One may intuitively anticipate that for a polynomial of given degree, BAP methods ought to be most effective. Indeed, the uniform approximation ideas are powerful and widely used when a high-degree of accuracy is needed. The drawback is that one must employ a simplex routine to solve the associated linear programming problem, and simplex codes tend to get involved. The library routines for the common functions (such as trigonometric, exponential, and their inverses) for the 6000 series of the CDC computers are most frequently Chebyshev approximations, or economizations. They guarantee a relative error of about 10^{-14}, which is commensurate with the word-length, by means of polynomials of degree about 10 (depending on the function and range). For specifics, see the Control Data Corporation Manual (1974). One will see in this manual that some use is made of rational function approximations. Methods for rational function approximation are more difficult yet. [See Blum (1972) and Young and Gregory (1972) for some development of this subject.]

In some cases the library routines given in the CDC manual use different approximations for different ranges of the input value. This procedure is in the spirit of spline approximation. It is fair to say that of the various areas of approximation, the theory and application of splines is presently the most active area of research. The book by Ahlberg *et al.* (1967) gives a comprehensive development of the theory. A monograph by Varga (1971) is a representative study exploiting properties of splines to develop new and interesting numerical procedures for solving partial differential equations. Papers by Bellman *et al.* (1972, 1973) present a dynamic programming scheme for iteratively evaluating splines without having to solve a large simultaneous equation as required by the conventional spline procedure such as in Section 2.4. Wahba (1973) and Wold (1974) use splines to approximate a function f on the basis of noisy pairs $\{(x_i, f(x_i) + n_i)\}_{i=1}^{n}$ (instead of $\{(x_i, f(x_i))\}_{i=1}^{n}$), where the n_i are random errors. The subject (known as *nonparametric regression*) of approximating functions on the basis of noisy function pairs, is cogent to this chapter and is analytically interesting and of great practical consequence. Work in this area includes Watson (1964), Rosenblatt (1969), Fisher and Yakowitz (1976), and Schuster and Yakowitz (1979).

In addition to the subject of approximation by rational functions, another important topic not discussed in this chapter is approximation of multivariate functions. See Isaacson and Keller (1966) and Blum (1972) for information and references. To get a flavor for some of the modern directions of research in approximation theory, the reader may wish to consult the book edited by Law and Schney (1976).

3

Numerical Differentiation and Integration

By using the methods of Chapter 2, we have means at our disposal for constructing good polynomial approximations of given functions. In this chapter, we seek to approximate the derivatives and integrals of given functions by taking the derivatives and integrals of approximating functions. Interpolation polynomials are comparatively easily computed approximations, and consequently, formulas for numerical differentiation and integration of polynomials are the basis of the methods to follow.

Let f be a real function; let $x_1, \ldots, x_n$ be different numbers, and let $f_k = f(x_k)$, $1 \leq k \leq n$. Theorem 2.3 asserts that if f is n times differentiable and if p denotes the interpolation polynomial constructed from the n function pairs (x_i, f_i), $1 \leq i \leq n$, then

$$f(x) - p(x) = \frac{1}{n!} f^{(n)}(\xi) w_n(x), \tag{3.1}$$

where ξ is a point in the smallest interval containing $x_1, \ldots, x_n$, and x, and

$$w_n(x) = (x - x_1)(x - x_2) \cdots (x - x_n).$$

In equation (3.1) ξ depends on x, and we make this dependency explicit through the notation $\xi = \xi(x)$. Differentiating both sides of equation (3.1), we have formally

$$f'(x) = p'(x) + \frac{1}{n!} \frac{d}{dx} [f^{(n)}(\xi(x)) w_n(x)]$$

$$= p'(x) + \frac{1}{n!} [f^{(n+1)}(\xi(x)) \cdot \xi'(x) w_n(x) + f^{(n)}(\xi(x)) \cdot w_n'(x)].$$

But the function $\xi(x)$ is unknown and there is no assurance of its differentiability. Thus we seek another way for differentiating the right-hand side of equation (3.1). A similar difficulty is encountered when we take the integral of both sides of equation (3.1).

The following theorem can be used for computing the errors of numerical differentiation and integration formulas.

Theorem 3.1. If a function f is $n + j$ times continuously differentiable in an interval $I = [a, b]$ containing the points $x_1, x_2, \ldots, x_n, x$, then for $j \geq 1$.

$$\frac{1}{n!} \frac{d^j}{dx^j} f^{(n)}(\xi(x)) = \frac{1}{(n+j)!} f^{(n+j)}(\eta_j),$$

where η_j is an interior point of the interval $[a, b]$.

Proof. Assume that x is not an interpolation point. For simplification, assume that $j = 1$; the general case follows by induction.

If we represent p by $\sum_{i=1}^{n} f_i l_i(x)$ as in (2.5), a simple calculation shows that (see Problem 2.2)

$$l_k(x) = \frac{w_n(x)}{(x - x_k)w_n'(x_k)}. \tag{3.2}$$

Choosing a new interpolation point x_{n+1} in the interval $[a, b]$, we have

$$w_{n+1}(x) = (x - x_{n+1})w_n(x),$$

$$w_{n+1}'(x) = w_n(x) + (x - x_{n+1})w_n'(x),$$

which implies

$$w_{n+1}'(x_j) = \begin{cases} w_n(x_{n+1}), & \text{if } j = n + 1, \\ (x_j - x_{n+1})w_n'(x_j), & \text{if } j < n + 1. \end{cases} \tag{3.3}$$

Using the Lagrangian interpolation formula and equations (3.1)–(3.3) for n and $n + 1$, we have

$$f(x) = \sum_{k=1}^{n} \frac{w_n(x)f(x_k)}{(x - x_k)w_n'(x_k)} + \frac{f^{(n)}(\xi)}{n!} w_n(x), \tag{3.4}$$

and

$$f(x) = \sum_{k=1}^{n} \frac{f(x_k)w_n(x)(x - x_{n+1})}{(x - x_k)(x_k - x_{n+1})w_n'(x_k)} + \frac{f(x_{n+1})w_n(x)(x - x_{n+1})}{(x - x_{n+1})w_n(x_{n+1})}$$

$$+ \frac{f^{(n+1)}(\eta)}{(n + 1)!} w_{n+1}(x). \tag{3.5}$$

Dividing equation (3.4) by $w_n(x)$ and differentiating, we see that

$$\frac{d}{dx}\left[\frac{f(x)}{w_n(x)}\right] = \sum_{i=1}^{n} \frac{f(x_k)}{-(x-x_k)^2 w_n'(x_k)} + \frac{1}{n!}\frac{d}{dx}[f^{(n)}(\xi)]. \quad (3.6)$$

The fact that $f(x)/w_n(x)$ and the terms $1/(x-x_k)$ are differentiable implies the differentiability of $f^{(n)}(\xi)$.

From equation (3.5) a simple calculation shows that

$$\frac{f(x)/w_n(x) - f(x_{n+1})/w_n(x_{n+1})}{x - x_{n+1}} = \sum_{k=1}^{n} \frac{f(x_k)}{(x-x_k)(x_k - x_{n+1})w_n'(x_k)} +$$
$$+ \frac{f^{(n+1)}(\eta)}{(n+1)!}. \quad (3.7)$$

If $x_{n+1} \rightarrow x$, then the left-hand side of equation (3.7) tends to $(d/dx)[f(x)/w_n(x)]$. For each particular value of x_{n+1}, there is a corresponding value of η. For any sequence of values of x_{n+1} that converges to x, the corresponding values of η will remain in the closed interval I, so that by the Bolzano–Weierstrass theorem (see Apostol, 1957, p. 43) a convergent subsequence of values of η can be chosen with some limit point $\bar{\eta}$. Since interval I is closed, we conclude that $\bar{\eta} \in I$. If we take the limit of the right-hand side of equation (3.7) for this convergent subsequence, we have

$$\frac{d}{dx}\left[\frac{f(x)}{w_n(x)}\right] = \sum_{k=1}^{n} -\frac{f(x_k)}{(x-x_k)^2 w_n'(x_k)} + \frac{f^{(n+1)}(\bar{\eta})}{(n+1)!},$$

and this identity and equation (3.6) imply the statement of the theorem for $x \neq x_k$. But by recalling the assumed continuity of the derivative of f, one concludes that the statement must be true for any x. ∎

3.1. Numerical Differentiation

If we assume that function f is $n + j$ times continuously differentiable in an interval $[a, b]$ that contains points $x_1, \ldots, x_n, x$, equation (3.1) implies that

$$f^{(j)}(x) = \sum_{k=1}^{n} f(x_k) l_k^{(j)}(x) + \frac{d^j}{dx^j}\left[\frac{f^{(n)}(\xi)}{n!} w_n(x)\right].$$

By using Leibniz's formula and the statement of Theorem 3.1, we have

$$f^{(j)}(x) = \sum_{k=1}^{n} f_k l_k^{(j)}(x) + \sum_{i=0}^{j}\binom{j}{i}\left[\frac{d^i}{dx^i}\frac{f^{(n)}(\xi)}{n!}\right]\left[\frac{d^{j-i}}{dx^{j-i}} w_n(x)\right]$$
$$= \sum_{k=1}^{n} f_k l_k^{(j)}(x) + \sum_{i=0}^{j}\binom{j}{i}\frac{f^{(n+i)}(\eta_i)}{(n+i)!} w_n^{(j-i)}(x). \quad (3.8)$$

Thus $f^{(j)}(x)$ can be approximated by the first sum on the right-hand side and the second term gives the error of this approximation.

If the points $x_1, \ldots, x_n$ are equidistant, then identity (3.8) can be expressed in a less complicated form because we can employ relations obtained in Section 2.1.4 from the Fraser diagram. For instance, in the case of Newton's forward formula, we have

$$f^{(j)}(x) = f^{(j)}(x_0 + th) \approx \frac{d^j}{dx^j} \sum_{k=0}^{n} \binom{t}{k} \Delta^k f_0.$$

Since $t = (x - x_0)/h$ and therefore $dt/dx = 1/h$, the above relation can be rewritten as

$$f^{(j)}(x) \approx \frac{d^j}{dt^j} \sum_{k=0}^{n} \binom{t}{k} \Delta^k f_0 \frac{1}{h^j} = \frac{1}{h^j} \sum_{k=j}^{n} \frac{d^j}{dt^j} \binom{t}{k} \Delta^k f_0. \qquad (3.9)$$

In the last equation we used the fact that for $k < j$, the binomial coefficient $\binom{t}{k}$ is a polynomial in t of degree less than j, and so its jth derivative vanishes.

For $j = n$, we have

$$f^{(n)}(x) \approx \frac{1}{h^n} \frac{d^n}{dt^n} \binom{t}{n} \Delta^n f_0 = \frac{1}{h^n} \Delta^n f_0, \qquad (3.10)$$

since the leading coefficient of the polynomial representation of $\binom{t}{n}$ is equal to $(n!)^{-1}$.

If $n = 1$, formula (3.10) implies

$$f'(x) \approx \frac{\Delta f_0}{h} = \frac{f_1 - f_0}{h}, \qquad (3.11)$$

and from examination of the error term and Theorem 3.1, $f'(x) - \Delta f_0/h$ is $O(h)$ for $x \in [x_0, x_1]$. For $n = 2$, we have

$$f''(x) \approx \frac{\Delta^2 f_0}{h^2} = \frac{\Delta f_1 - \Delta f_0}{h^2} = \frac{f_2 - 2f_1 + f_0}{h^2}, \qquad (3.12)$$

and again the error is $O(h)$. By the notation $O(h)$, we mean that there is a constant C such that

$$| f'(x) - \Delta f_0/h | \leq Ch$$

for all h sufficiently small but positive. More generally, we say that an error term is $O(h^k)$ if the bound of the error has the form Ch^k.

Similar formulas can be obtained by using other interpolation polynomials, and in fact, formulas such as Gauss' [i.e., (2.13) and (2.14)], which tend to center the interpolation points about x_0, are generally more accurate for numerical differentiation. In particular, if the number of interpolation points is odd and if they are placed symmetrically about x_0, then the coefficient of the lowest power of h in the error term vanishes at $x = x_0$. The details of the verification of this technique, which is called the *centered difference approximation*, are worked out in Conte and de Boor (1972, Section 5.1). Thus a numerical differentiation formula we shall find useful in our discussion of differential equations is

$$f''(x_0) \approx (f_1 - 2f_0 + f_{-1})/h^2.$$

The error of this approximation is $O(h^2)$.

Example 3.1. We approximate the first derivative of the functions x^3, at the point $x = 0.5$. We let $n = 3$, $h = 1$, and use (3.9).

In our case $j = 1$; consequently, from (3.9),

$$f'(x) \approx \frac{1}{h} \sum_{k=1}^{n} \frac{d}{dt} \binom{t}{k} \Delta^k f_0.$$

Since $t = (0.5 - 0)/1 = 0.5$, and $\binom{t}{k} = t(t-1) \cdots (t-k+1)/k!$,

$$\frac{d}{dt}\binom{t}{1} = 1, \qquad \frac{d}{dt}\binom{t}{2} = \frac{2t-1}{2} = 0,$$

$$\frac{d}{dt}\binom{t}{3} = \frac{3t^2 - 6t + 2}{6} = \frac{-0.25}{6},$$

and consequently, since $\Delta f_0 = f_1 - f_0 = 1^3 - 0^3 = 1$, and, as the reader may check, $\Delta^2 f_0$ and $\Delta^3 f_0$ are similarly seen to be 6, we have

$$f'(0.5) \approx \frac{1}{1}[1 \cdot 1 + 0 \cdot 6 - \frac{0.25}{6} \cdot 6] = 0.75,$$

which is, of course, the true value.

An alternative expression for the error of approximation by differentiation of the interpolating polynomial is

$$f^{(j)}(x) - \sum_{k=1}^{n} f_k l_k^{(j)}(x) = \left(\prod_{k=1}^{n-j} (x - \xi_k) \right) \frac{f^{(n)}(\eta)}{(n-j)!}$$

for some η and distinct points $\xi_1, \ldots, \xi_n$ in the smallest interval containing $x_0, x_1, \ldots, x_n$, and x. See Isaacson and Keller (1966, Section 6.5) for a derivation of this bound and for a different approach to numerical differentiation.

We shall now bound the error arising from error in the functional values. Let $f_1, \ldots, f_n$ be the exact values, and let $\tilde{f}_1, \ldots, \tilde{f}_n$ be approximations of these values. Assume that for $k = 1, 2, \ldots, n$,

$$|f_k - \tilde{f}_k| \le \delta,$$

where δ is a given error bound. From considerations in Section 2.1.3, one may confirm that the errors of the kth-order differences are bounded by $2^k\delta$. As a result, the approximation error bound of formula (3.9) is

$$\delta_1 = \frac{1}{h^j}\left[\delta \sum_{k=j}^{n} \frac{d^j}{dt^j}\binom{t}{k} 2^k\right]. \tag{3.13}$$

It can be proven that the error of any numerical differentiation formula obtained from interpolation polynomials with equally spaced abscissas takes the form

$$\delta_1 = \frac{1}{h^j}\,\delta R(t),$$

where R does not depend on h. Such error arises from floating-point round-off, if nothing else.

If, in attempting to evaluate the derivative more accurately, one decreases the value of h, then as a consequence of the above relation as $h \to 0$, the rounding error δ_1 increases without bound. This fact implies that numerical differentiation formulas are generally ill conditioned and cannot provide arbitrary accuracy. In fact, numerical differentiation is notoriously risky and should be avoided if at all possible.

From equations (3.8) and (3.13) an optimal value of h can be determined as the minimizing value of the function

$$\sum_{i=0}^{n} \binom{j}{i} \frac{M_{n+i}}{(n+i)!}\,\Omega_n^{(j-i)}(h) + \frac{1}{h^j}\,\delta \sum_{k=j}^{n} T_k^{(j)} \cdot 2^k,$$

where

$$M_{n+i} = \max_{x\in[a,b]} |f^{(n+i)}(x)|, \qquad \Omega_n^{(j-i)}(h) = \max_{x\in[a,b]} |w_n^{(j-i)}(x)|,$$

and

$$T_k^{(j)} = \max_{x\in[a,b]} \left|\frac{d^j}{dt^j}\binom{t}{k}\right|.$$

Problems

3.1. Determine the numerical differentiation formulas corresponding to the interpolation formulas (2.11), (2.12), (2.13), (2.14), and (2.15).

3.2. Assume that errors of the estimates $\tilde{f}_1, \tilde{f}_2, \ldots, \tilde{f}_n$ of the functional values $f_1, f_2, \ldots, f_n$ are normally distributed random numbers with zero means and common standard deviations δ. Compute the expected square error of formula (3.9).

3.3. For $h = 0.1, 0.01, 0.001, 0.0001$, and 10^{-6}, approximate the derivative of $f(x) = \sin x$ at $x = 0.1, 0.2, 0.4$ by use of (3.9) with $n = 3$. Compare with exact value. Add an "error" of $\pm 10^{-4}$ to your function values and repeat the above calculations, letting $\tilde{f}_j = (-1)^j 10^{-4} + f_j$.

3.4. Prove in detail that for $x \in [x_0, x_1]$, $|f'(x) - (f_1 - f_0)/h|$ is $O(h)$ and $|f''(x_0) - (f_1 - 2f_0 + f_{-1})/h^2|$ is $O(h^2)$.

3.2. Numerical Quadrature

3.2.1. Interpolatory Quadrature Formulas

Let us take n interpolation points on the interval $[a, b]$ and let us approximate the function f with an interpolation polynomial constructed from these points. If we assume that the function f is n times differentiable on the interval $[a, b]$, it is clear that Theorem 2.3 implies equation (3.1). After integrating both sides of this equation, we obtain the identity

$$\int_a^b f(x)\, dx = \int_a^b p(x)\, dx + \int_a^b \frac{f^{(n)}(\xi)}{n!}\, w_n(x)\, dx. \qquad (3.14)$$

The first term on the right, which may be evaluated exactly by a simple analytic formula, can be considered to be an approximate integration formula. The second term gives the error of this formula.

When the interpolation is represented in the Lagrangian form, ([equation (2.5)], the preceding equation implies

$$\int_a^b f(x)\, dx \approx \int_a^b \sum_{k=1}^n f(x_k) l_k(x)\, dx = \sum_{k=1}^n f(x_k) \int_a^b l_k(x)\, dx = \sum_{k=1}^n f_k \cdot A_k, \qquad (3.15)$$

where the coefficients A_k depend only on the interpolation points.

The last formula is called an *interpolatory-type* formula. If the interpolation points x_i are evenly spaced, (3.15) is called a *Newton–Cotes* formula. In the following we shall describe the most important numerical integration rules of the Newton–Cotes and compound Newton–Cotes type.

Divide the interval $[a, b]$ into m equal subintervals, and let the points of the subintervals be denoted by $a = x_0 < x_1 < \cdots < x_{m-1} < x_m = b$. Let

$$h = \frac{b - a}{m}.$$

Apply formula (3.15) with $n = 2$ to each subinterval, and let the subinterval endpoints be the interpolating points. Then we have

$$\int_{x_k}^{x_{k+1}} f(x)\, dx \approx \sum_{i=k}^{k+1} f(x_i) A_i$$

$$= f(x_k) \int_{x_k}^{x_{k+1}} \frac{x - x_{k+1}}{x_k - x_{k+1}}\, dx + f(x_{k+1}) \int_{x_k}^{x_{k+1}} \frac{x - x_k}{x_{k+1} - x_k}\, dx$$

$$= \frac{h}{2} [f(x_k) + f(x_{k+1})]. \tag{3.16}$$

Adding the integrals (3.16) over each subinterval we obtain the approximation

$$\int_a^b f(x)\, dx \approx h[\tfrac{1}{2}f(x_0) + f(x_1) + \cdots + f(x_{m-1}) + \tfrac{1}{2}f(x_m)] \triangleq t_m, \tag{3.17}$$

which is called the *trapezoidal rule*. The trapezoidal rule is the compound second-order Newton–Cotes formula. Given a fixed quadrature formula such as a Newton–Cotes rule, a *compound* rule is a rule that divides the interval of integration into m $(m > 1)$ equal subintervals and applies that fixed quadrature formula to each of the m subintervals. The trapezoidal rule as well as the rectangular formula and Simpson's rule below are examples of compound rules.

Denoting the midpoints of the subintervals $[x_k, x_{k+1}]$ by y_{k+1} and using formula (3.15) for $n = 1$ and for the only interpolating point y_{k+1} and adding them, we have the *rectangular formula*:

$$\int_a^b f(x)\, dx \approx h[f(y_1) + f(y_2) + \cdots + f(y_m)] \triangleq r_m. \tag{3.18}$$

If m is an even number, and if the interval $[a, b]$ is partitioned into $m/2$ equal subintervals and formula (3.15) with $n = 3$ is used (the endpoints and midpoints of the subinterval serving as interpolating points), *Simpson's rule* is obtained:

$$\int_a^b f(x)\, dx \approx \frac{h}{3} [f(x_0) + 4f(x_1) + 2f(x_2) + \cdots + 2f(x_{m-2})$$

$$+ 4f(x_{m-1}) + f(x_m)] \triangleq s_m. \tag{3.19}$$

The rectangular formula and Simpson's rule are compound Newton–Cotes formulas of order 1 and 3, respectively. If an interpolation-type formula uses both endpoints of the interval of integration as points x_k in (3.15), then it is said to be a *closed* formula. The formula is said to be *open* if neither endpoint a nor b is among the function evaluation points x_i, $1 \leq i \leq n$. Thus the Newton–Cotes formula underlying the rectangular rule is open, and the Newton–Cotes rules yielding the trapezoidal and Simpson's rules are both closed. Open and *half-open* formulas (that is, interpolation rules using the lower but not the upper endpoint as an interpolation point) can be used as "predictors" in the numerical solution of differential equations.

In the full generality of this terminology, "interpolatory-type" formulas refer to quadrature rules of the form

$$\int_a^b w(x)f(x)\,dx \approx \sum_{i=1}^n A_i f_i, \qquad (3.20)$$

which are exact whenever the function $f(x)$ is a polynomial of degree less than n. In (3.20), in contrast to earlier usage, $w(x)$ denotes a *weight* function, a weight function being a nonnegative, integrable, but otherwise arbitrary function. In general, $w(x)$ is selected so that the integrals $\int_a^b w(x)x^k\,dx$ can be evaluated exactly from an analytic formula.

It is sometimes useful to represent the A_i as the solutions of the linear system of equations

$$\sum_{i=1}^n A_i x_i^k = \int_a^b w(x)x^k\,dx, \qquad 0 \leq k < n. \qquad (3.21)$$

We leave it to the reader to confirm that the A_i satisfying (3.21) give a rule that is equivalent to interpolating $f(x)$ at x_i, $1 \leq i \leq n$, and integrating the weighted interpolation polynomial in place of f.

3.2.2. Error Analysis and Richardson Extrapolation

The error of the trapezoidal rule can be determined by applying (3.14) to each subinterval and then adding the errors corresponding to each subinterval. Using equation (3.14), one finds that the error of formula (3.16) is given by

$$\int_{x_k}^{x_{k+1}} \frac{f''(\xi_k(x))}{2!}\,w_2(x)\,dx = \int_{x_k}^{x_{k+1}} \frac{f''(\xi_k(x))}{2}\,(x - x_k)(x - x_{k+1})\,dx$$

$$= h^3 \int_0^1 \frac{f''(\xi_k(x))}{2}\,t(t - 1)\,dt.$$

It can be shown that if $f''(x)$ is continuous, so is $f''(\xi_k(x))$. Since $t(t - 1)$ has no roots in $(0,1)$, the mean-value theorem of integral calculus implies that the right-hand side of the above equation is equal to

$$\frac{h^3 f''(\eta_k)}{2} \int_0^1 t(t-1)\, dt = -\frac{h^3 f''(\eta_k)}{12},$$

where $\eta_k \in [x_k, x_{k+1}]$. The error of the trapezoidal rule is equal to the sum of the errors of the corresponding formulas on the subintervals, and so using the intermediate value theorem for derivatives (see Apostol, 1957, p. 94) the error of the trapezoidal rule is given by

$$-\sum_{k=0}^{m-1} \frac{h^3 f''(\eta_k)}{12} = -\frac{h^3}{12} \sum_{k=0}^{m-1} f''(\eta_k) = \frac{-h^3 m f''(\eta)}{12}$$

$$= -\frac{(b-a)^3 m f''(\eta)}{12 m^3} = -\frac{(b-a)^3 f''(\eta)}{12 m^2}, \quad (3.22)$$

where $\eta \in [a, b]$.

Using the trapezoidal rule when m is even, we have

$$\int_a^b f(x)\, dx = t_m - \frac{(b-a)^3 f''(\eta)}{12 m^2},$$

and

$$\int_a^b f(x)\, dx = t_{m/2} - \frac{(b-a)^3 f''(\bar\eta)}{3 m^2}.$$

If $f''(\eta)$ and $f''(\bar\eta)$ are very close to each other, as is often the case for large m and smooth integrand, then by multiplying the first equation by 4 and subtracting the second one we have

$$\int_a^b f(x)\, dx \approx \frac{4 t_m - t_{m/2}}{3}, \quad (3.23)$$

which is known as the *corrected trapezoidal rule*.

The rationale behind the corrected trapezoidal rule is *Richardson's extrapolation* (otherwise known as the *deferred approach to the limit*), which we now discuss. Let $T(h)$ denote some numerical estimate of a quantity t, h being some measure of discretization. We suppose that the error in estimation has the form

$$t - T(h) = ch^r + dh^s + \text{higher-order terms in } h, \quad (3.24)$$

in which r is known and $0 < r < s$, but c and d are unknown constants. We may conclude from (3.24) that

$$T(h) = t + ch^r + dh^s,$$
$$T(2h) = t + 2^r ch^r + 2^s dh^s,$$

and the Richardson extrapolation rule for the model (3.24) is

$$T' = \frac{2^r T(h) - T(2h)}{2^r - 1} = t + d_1 h^s + \text{higher-order terms}, \quad (3.25)$$

where $d_1 = d(2^r - 2^s)/(2^r - 1)$. Thus T' has a higher order (namely s) of convergence than the basic rule T.

In Theorem 3.12 in Section 3.2.4, it will be seen that if f has a continuous nth derivative (n an even number), then for t_m the trapezoidal rule and t the exact integral, we have

$$t_m = t + c_2 h^2 + c_4 h^4 + c_6 h^6 + \cdots + c_{n-2} h^{n-2} + \text{higher-order terms}, \quad (3.26)$$

where $h = (b - a)/m$. Thus model (3.24) is valid and the corrected trapezoidal rule is an instance of Richardson's extrapolation. The order of the corrected trapezoidal rule is seen to be 4, as opposed to the uncorrected rule, which is 2. From examination of the above representation (3.26), one may consider using Richardson's extrapolation repeatedly to further increase the order of convergence. This is, in fact, the basis of Romberg integration, to be discussed in Section 3.2.5.

Example 3.2. Let $f_1(x) = \exp(x)$ and $f_2(x)$ be a sawtooth function having period 0.06 and amplitude 0.03, as illustrated in Figure 3.1. Let N_{20} stand for the 20-point Newton–Cotes formula, t_{20} the 20-point trapezoidal formula, and ct_{20} the 20-point corrected trapezoidal formula:

$$ct_{20} = (4t_{20} - t_{10})/3.$$

In Table 3.1, we present the relative errors associated with the above quadrature rules applied to f_1 and f_2. In all cases, the domain $[a, b]$ of integration was the unit interval $[0,1]$, and note that in all cases, the quadrature rules used 20 function values. The computations were done in double precision, giving an effective word

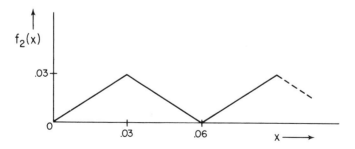

Fig. 3.1. A sawtooth function.

Table 3.1. A Comparison of Relative Errors Associated with Some Quadrature Formulas

Function	t_{20}	ct_{20}	N_{20}
f_1	2×10^{-3}	3×10^{-8}	$\approx 10^{-27}$
f_2	5×10^{-2}	6×10^{-3}	2.4

length of about 27 decimal places. Consequently, from developments in Chapter 1, one may confirm that roundoff error is negligible in comparison to truncation error, with the exception of N_{20} on f_1. The performance of N_{20} in Table 3.1 illustrates a rule of thumb that higher-order rules are only advantageous for smooth functions (and negligible roundoff error).

We shall recall from (3.18) that the rectangular formula is given by

$$\int_a^b f(x)\, dx \approx h[f(y_1) + f(y_2) + \cdots + f(y_m)] = r_m. \tag{3.27}$$

Using a development similar to the error analysis of the trapezoidal rule, one can prove that the error of the rectangular formula is given by

$$\int_a^b f(x)\, dx - r_m = -\frac{(b-a)^3 f''(\xi)}{24m^2},$$

where $\xi \in [a, b]$. The *corrected rectangular formula* is defined to be

$$\int_a^b f(x)\, dx \approx \frac{4r_m - r_{m/2}}{3}. \tag{3.28}$$

The corrected rectangular formula is, of course, the Richardson extrapolation of the elementary rectangular formula (3.27).

We recall that Simpson's rule [given in (3.19)] is the approximation

$$s_m = \frac{h}{3}\, [f(x_0) + 4f(x_1) + 2f(x_2) + \cdots + 2f(x_{m-2}) + 4f(x_{m-1}) + f(x_m)]. \tag{3.29}$$

Young and Gregory (1972, p. 396) show that the error of Simpson's rule is given by

$$\int_a^b f(x)\, dx - s_m = -\frac{(b-a)^5 f^{(4)}(\xi)}{180m^4}, \tag{3.30}$$

where $\xi \in [a, b]$. From this formula we see that the approximation error of Simpson's rule is approximately h^4, where $h = (b - a)/m$, as before. Consequently, Richardson's extrapolation with $r = 4$ yields the *corrected Simpson's rule*, which is

$$\int_a^b f(x)\, dx \approx \frac{16s_m - s_{m/2}}{15}.$$ (3.31)

Applying Theorem 3.1, one can prove that the error of a closed n-point Newton–Cotes formula can be written in the form

$$\frac{f^{(n)}(\xi)}{n!} \int_a^b w_n(x)\, dx, \qquad \text{if } n \text{ is even},$$ (3.32)

$$\frac{f^{(n+1)}(\xi)}{(n+1)!} \int_a^b x w_n(x)\, dx, \qquad \text{if } n \text{ is odd}.$$ (3.33)

See Young and Gregory (1972, Section 7.7) or Isaacson and Keller (1966, Section 7.1.1) for details.

Since the rectangular formula r_m is a Riemann sum, for integrable functions the value of r_m converges to the value of the integral for $m \to \infty$ regardless of existence of derivatives. The trapezoidal rule can be written in the form

$$t_m = h[f(x_0) + \cdots + f(x_{n-1})] + \frac{h}{2}[f(x_n) - f(x_0)],$$

where the first term is a Riemann sum and the second term tends to zero for $h \to 0$, and it can be easily proven that

$$s_m = \frac{4t_m - t_{m/2}}{3}.$$ (3.34)

Consequently, under the hypothesis that f is Riemann integrable, both the trapezoidal and Simpson's rules converge to the value of the integral for $m \to \infty$. But we generalize the preceding observations in the following assertion:

Theorem 3.2. Let $\sum_{i=1}^n A_i f(x_i)$ be some fixed quadrature rule that integrates constant functions exactly on the unit interval. If $I_m(f)$ denotes the compound rule obtained by applying the fixed rule to each of m equal intervals of $[a, b]$, then for any Riemann-integrable function f,

$$\lim_{m \to \infty} I_m(f) = \int_a^b f(x)\, dx.$$

Proof. For appropriately chosen points $\{x_{ij}\}$ with x_{ij} in the jth sub-interval of $[a, b]$, $1 \leq j \leq n$,

$$I_m(f) = \sum_{j=1}^{m} \left(\frac{b-a}{m} \right) \sum_{i=1}^{n} A_i f(x_{ij})$$

$$= \sum_{i=1}^{n} A_i \left[\sum_{j=1}^{m} f(x_{ij}) \frac{b-a}{m} \right].$$

The term enclosed in brackets is a Riemann sum for $\int_a^b f(x)\, dx$, and consequently

$$\lim_{m \to \infty} I_m(f) = \sum_{i=1}^{n} A_i \int_a^b f(x)\, dx.$$

$$= \int_a^b f(x)\, dx.$$

In the last equation we have used the assumption that the quadrature rule $\sum_{i=1}^{n} A_i f(x_i)$ is exact for constant functions on the unit interval to conclude that $\sum_{i=1}^{n} A_i = 1$. ∎

The theorem to follow is among the strongest statements concerning asymptotic convergence of quadrature approximations.

Theorem 3.3. Let $I(f)$ denote some fixed n-point quadrature rule evaluated at the function f, and let k be a fixed positive integer. Suppose that if f has a continuous kth derivative, the error of the rule I is of the form

$$E(f) = \int_a^b f(x)\, dx - I(f) = c(b-a)^{k+1} f^{(k)}(\xi) \tag{3.35}$$

for some ξ in $[a, b]$. If $I_m(f)$ denotes the compound rule obtained by applying I to each of the m equal subintervals of $[a, b]$, then

$$\lim_{m \to \infty} m^k I_m(f) = c(b-a)^k [f^{(k-1)}(b) - f^{(k-1)}(a)]. \tag{3.36}$$

Proof. From (3.35) applied to each of the m subintervals, we have the error of approximation

$$E_m(f) = \sum_{i=1}^{m} c \left(\frac{b-a}{m} \right)^{k+1} f^{(k)}(\xi_i),$$

where ξ_i is in the ith subinterval. We may use the distributive law to write

$$E_m(f) = c \left(\frac{b-a}{m} \right)^k \left[\sum_{i=1}^{m} f^{(k)}(\xi_i) \frac{b-a}{m} \right].$$

Observe that the term in brackets is a Riemann sum approximation of $\int_a^b f^{(k)}(x)\,dx$, and (3.36) follows. ∎

Note that from (3.32) and (3.33), all Newton–Cotes quadrature rules satisfy the hypothesis of the theorem.

If the functional values $f(x_k)$, $1 \le k \le n$, cannot be calculated exactly and approximations $\tilde{f}_k$ must be used, then of course the errors of the function values also have an effect on the error of the integrating formulas. Assume that for $k = 1, 2, \ldots, n$,

$$|f(x_k) - \tilde{f}_k| \le \delta,$$

where δ is a fixed positive number. Then the difference of formulas (3.15) with exact and approximating function values has the form

$$\left| \sum_{k=1}^{n} f(x_k)A_k - \sum_{k=1}^{n} \tilde{f}_k A_k \right| \le \delta \sum_{k=1}^{n} |A_k|. \qquad (3.37)$$

Since in the case of the trapezoidal, rectangular, and Simpson's formulas [equations (3.17), (3.18), and (3.19), respectively] the coefficients A_k are positive and their sum is equal to the length of interval $[a, b]$, the right-hand side of (3.37) is equal to $\delta(b - a)$, which tends to zero if $\delta \to 0$. Thus, in contrast to numerical differentiation, numerical integration is stable for quadrature rules with nonnegative weights, the stability being in the sense that the approximation error remains bounded as the discretization becomes finer. Unfortunately, for $n > 7$, n-point Newton–Cotes rules have some negative weights and can consequently be unstable. Gaussian integration (to be discussed) yields interpolatory formulas that, as we shall prove, always have positive weights.

Example 3.3. The values of a function are given in the following table:

x	$f(x)$	x	$f(x)$	x	$f(x)$
0	8.415	4	5.646	8	1.987
1	7.833	5	4.794	9	0.998
2	7.174	6	3.894	10	0.000
3	6.442	7	2.955		

From the trapezoidal rule we have

$$\int_0^{10} f(x)\,dx \approx t_{10} = 45.931.$$

The values in this example are points of the function $f(x) = 10 \sin(1 - 0.1x)$, and the exact value of the integral is equal to 45.970. Since $| f''(x) | \leq 0.1$, by virtue of (3.22) the upper bound for the error of the trapezoidal rule is given by $(10^3 \times 0.1)/(12 \times 10^2) \sim 0.083$, and the actual error is equal to 0.039.

Problems

3.5. Determine a step length h sufficient for computing the integral $\int_0^1 e^{x^2}\, dx$ to six decimal places by the trapezoidal rule.

3.6. Determine the step length h for computing the integral $\int_0^1 \sin{}^\alpha x\, dx$ to six significant places by the rectangular rule ($\alpha > 2$).

3.7. Show that (3.21) does yield a quadrature rule that is exact for $(n-1)$-degree polynomials. Show that if $w(x) = 1$, the solution of (3.21) is unique. HINT: The fact that an $(n-1)$-degree polynomial has at most $n-1$ roots can be used here.

3.8. Show that (3.32) and (3.33) imply that the error has the form $c(b-a)^{k+1}$ $f^{(k)}(\xi)$. What are c and k in these cases?

3.9. Find A_1, A_2, and A_3 so that the interpolatory-type rule

$$A_1 f(-1) + A_2 f(-\tfrac{1}{3}) + A_3 f(\tfrac{1}{3})$$

equals $\int_{-1}^1 f(x)\, dx$ for all polynomials of degree ≤ 2.

3.2.3. Gaussian Quadrature

Let $w(x)$ denote an integrable nonnegative weight function with at most finitely many zeros in a (perhaps unbounded) interval $[a, b]$.

Definition 3.1. An n-point quadrature formula $\sum_{i=1}^n A_i f(x_i)$ is a *Gaussian* quadrature formula with respect to the weight function $w(x)$ and the interval $[a, b]$ if it is exact for all polynomials p of degree less than $2n$; that is, if

$$\sum_{i=1}^n A_i p(x_i) = \int_a^b w(x) p(x)\, dx \tag{3.38}$$

for all polynomials p of degree less than $2n$.

We have only shown n-point interpolatory polynomials to be exact for degree less than n. It turns out, however, that Gaussian quadrature formulas exist, and that they are, in fact, interpolatory formulas with carefully selected quadrature points. These points are roots of orthogonal polynomials.

Definition 3.2. With respect to a given interval $[a, b]$ and weight function $w(x)$, an nth-degree polynomial $P_n(x)$ is *orthogonal* to polynomials of degree less than n if for any such polynomial p,

$$\int_a^b w(x)P_n(x)p(x)\, dx = 0. \tag{3.39}$$

For brevity, we shall refer to polynomials P_n as in this definition as "orthogonal." We assert the following two theorems about orthogonal polynomials:

Theorem 3.4. For any interval $[a, b]$, weight function $w(x)$, and positive integer n, there exists an orthogonal polynomial P_n that

(i) is unique up to a multiplicative constant, and

(ii) has n distinct real roots, all contained in $[a, b]$.

We omit the proof to this and the next theorem. The proofs may be found in, for example, Stroud (1974, Section 3.7). Problems 3.13 and 3.14 provide hints for a proof of Theorem 3.4.

Theorem 3.5. Let $\{P_j\}_{j\geq0}$ be a sequence of orthogonal polynomials, each having degree equal to its subscript. Assume each P_j is scaled so that its leading coefficient (i.e., the coefficient of x^j) is 1. Then the sequence of orthogonal polynomials satisfies the recurrence relationship

$$P_n(x) = (x - b_n)P_{n-1}(x) - c_n P_{n-2}(x), \tag{3.40}$$

where $P_1(x) = x - b_1$, $P_0(x) = 1$. The constants b_n and c_n are given by

$$b_n = \theta_{n,n-1}/\theta_{n-1,n-1} + A_{n-1,n-2},$$
$$c_n = \theta_{n-1,n-1}/\theta_{n-2,n-2},$$

where

$$\theta_{m,k} = \int_a^b w(x)x^m P_k(x)\, dx,$$

and $A_{n-1,n-2}$ is the coefficient of x^{n-2} in $P_{n-1}(x)$.

In Table 3.2 we give the characteristics of the orthogonal polynomials associated with some of the more popular weight functions. The reader may consult Chapter 1 of Davis and Rabinowitz (1975) for a more complete summary.

Table 3.2. Characteristics of Certain Orthogonal Polynomials

Interval	Weight function	Name of polynomial $P_n(x)$	b_n	c_n
$[-1,1]$	1	Legendre polynomial	0	$\dfrac{(n-1)^2}{(2n-1)(2n-3)}$
$[-1,1)$	$(1-x)^\alpha(1+x)^\beta$ $\alpha > -1, \beta > -1$	Jacobi polynomial	$\dfrac{(\alpha+\beta)(\beta-\alpha)}{(2n+\alpha+\beta)(2n+\alpha+\beta-2)}$	$\dfrac{4(n-1)(n+\alpha-1)(n+\beta-1)(n+\alpha+\beta-1)}{(2n+\alpha+\beta-1)(2n+\alpha+\beta-2)^2(2n+\alpha+\beta-3)}, \; n > 2$ $\dfrac{4(\alpha+1)(\beta+2)}{(\alpha+\beta+3)(\alpha+\beta+2)^2}, \; n = 2$
$[-\infty,\infty)$	e^{-x^2}	Hermite polynomial	0	$(n-1)/2$
$[0,\infty)$	e^{-x}	Laguerre polynomial	$2n-1$	$(n-1)^2$

We are now in a position to proceed with the main developments concerning Gaussian quadrature.

Theorem 3.6. If $\{x_1, x_2, \ldots, x_n\}$ are the roots of an orthogonal polynomial P_n and $I_n(f) = \sum_{i=1}^{n} A_i f(x_i)$ is the interpolatory-type quadrature formula associated with these points, then I_n is the unique n-point Gaussian quadrature formula for the interval $[a, b]$ and the weight function $w(x)$.

Proof. In order to establish that I_n is a Gaussian quadrature formula, we need only prove that it is exact for all polynomials p of degree less than $2n$. Let p be an arbitrary polynomial of degree less than $2n$. By a fundamental property (the Euclidean division ring property) of polynomials, p has a factorization

$$p(x) = q(x)P_n(x) + r(x),$$

where P_n is the nth-degree orthogonal polynomial and q and r are polynomials having degree less than n. Then we have, using the orthogonality property of P_n and the fact that q has degree less than n,

$$\int_a^b w(x)p(x)\,dx = \int_a^b w(x)q(x)P_n(x)\,dx + \int_a^b w(x)r(x)\,dx$$
$$= \int_a^b w(x)r(x)\,dx. \tag{3.41}$$

But since the quadrature points are the roots of P_n, we have, noting the linearity of I_n,

$$I_n(p(x)) = I_n(q(x)P_n(x) + r(x)) = I_n(q(x)P_n(x)) + I_n(r(x))$$
$$= I_n(r(x)).$$

Because I_n is an n-point interpolatory formula, it is exact for $r(x)$, which has degree less than n, and in summary,

$$I_n(p) = I_n(r) = \int_a^b w(x)r(x)\,dx = \int_a^b w(x)p(x)\,dx, \tag{3.42}$$

and we can conclude, therefore, that I_n is a Gaussian quadrature formula.

It remains to be seen that I_n is unique. Let I_n' be any other n-point Gaussian quadrature formula, and let $P_n'(x)$ be defined by

$$P_n'(x) = \prod_{k=1}^{n} (x - x_k'), \tag{3.43}$$

where $\{x_1', x_2', \ldots, x_n'\}$ is the set of quadrature points of I_n'. Since I_n' is a Gaussian formula, we have

$$I_n'(P_n'(x)q(x)) = \int_a^b w(x)P_n'(x)q(x)\,dx \qquad (3.44)$$

for all polynomials q of degree less than n. But from (3.43), it is evident that $I_n'(P_n'(x)q(x)) = 0$ for every function $q(x)$, and thus (3.44) implies that P_n' is an orthogonal polynomial. But from part (i) of Theorem 3.4 we know that P_n is unique up to a scale factor, and therefore that the quadrature points of I_n' and I_n are identical. Since they are both interpolatory-type formulas, the quadrature weights must also agree, and, in conclusion, $I_n = I_n'$.   ∎

Example 3.4. If a function f is "well behaved" one can usually anticipate that Gaussian quadrature will be as accurate as any other quadrature rule having a like number of points. For example, let $f_1(x) = e^x$ and $f_2(x) = \sin x$. We approximate the integrals of these functions over the interval [0,1] by the closed 10-point Newton-Cotes formula, designated N_{10}, and G_{10}, the 10-point Gaussian quadrature formula [for weight function $w(x) = 1$]. In Table 3.3, we tabulate the relative errors. The computations were done in double precision, and so from considerations in Chapter 1 it may be confirmed that the error is due to truncation rather than roundoff.

Observe that

$$I_n(P_n^2) = 0 \neq \int_a^b w(x)[P_n(x)]^2\,dx$$

which illustrates that n-point Gaussian quadrature need not be exact for polynomials of degree $2n$.

In addition to being exact for $(2n - 1)$-degree polynomials, Gaussian quadrature possesses another prominent feature not generally shared by

Table 3.3. Comparison of Relative Errors in Newton–Cotes and Gaussian Quadrature

Function	N_{10}	G_{10}
f_1	3×10^{-12}	6×10^{-22}
f_2	4×10^{-12}	8×10^{-23}

other systems of interpolation-type quadrature formulas. We state this feature as a theorem.

Theorem 3.7. Let I_n denote an n-point Gaussian quadrature formula with respect to some given weight function $w(x)$ and finite interval $[a, b]$. Then

$$\lim_{n \to \infty} I_n(f) = \int_a^b w(x) f(x) \, dx \tag{3.45}$$

for every continuous function $f(x)$ on $[a, b]$.

Proof. Our proof requires that we establish the fact that every weight A_i in the Gaussian quadrature formula I_n is positive. Let i be a fixed index and let $P(x)$ be the $(n-1)$-degree polynomial determined by

$$P(x) = \prod_{j \neq i} (x - x_j), \tag{3.46}$$

where the x_j are the quadrature points of I_n. Then $P(x)^2$ has degree $2n - 2 < 2n$, and I_n is exact for P^2. Consequently,

$$I_n(P^2) = \sum_{j=1}^n A_j P(x_j)^2 = A_i P(x_i)^2$$
$$= \int_a^b w(x) P(x)^2 \, dx. \tag{3.47}$$

From this equation, A_i is seen to be the ratio of two positive numbers and consequently it must be positive.

Also, since I_n integrates constant functions (which are polynomials of degree 0) exactly, we must have that

$$\sum_{j=1}^n |A_j| = \sum_{j=1}^n A_j = \int_a^b w(x) \, dx. \tag{3.48}$$

Let $f(x)$ be a continuous function and p_n be its nth-degree Bernstein approximation. Then by the Weierstrass theorem (Chapter 2), we have that as $n \to \infty$,

$$\max_{a \leq x \leq b} |f(x) - p_n(x)| \to 0. \tag{3.49}$$

Letting $g_n = f - p_n$ and

$$\|g_n\|_\infty = \max_{a \leq x \leq b} |g_n(x)|, \tag{3.50}$$

and noting that I_n integrates p_n exactly, we have

$$\int_a^b w(x)f(x)\,dx - I_n(f) = \left[\int_a^b w(x)p_n(x)\,dx - I_n(p_n)\right]$$
$$+ \left[\int_a^b w(x)g_n(x)\,dx - I_n(g_n)\right]. \quad (3.51)$$

We observe that the first term in brackets is zero because I_n is exact for nth-degree polynomials.

Thus

$$\int_a^b w(x)f(x) - I_n(f) = \int_a^b w(x)g_n(x)\,dx - I_n(g_n),$$

but since $w(x)$ is nonnegative,

$$\left|\int_a^b w(x)g_n(x)\,dx\right| \le \|\,g_n\,\|_\infty \int_a^b w(x)\,dx \quad (3.52)$$

and

$$|\,I_n(g_n)\,| \le \sum_{i=1}^n |\,A_i\,|\,\|\,g_n\,\|_\infty = \|\,g_n\,\|_\infty \int_a^b w(x)\,dx. \quad (3.53)$$

As a result,

$$\left|\int_a^b w(x)f(x)\,dx - I_n(f)\right| \le 2\,\|\,g_n\,\|_\infty \int_a^b w(x)\,dx. \quad (3.54)$$

Since (3.49) is equivalent to the assertion that $\|\,g_n\,\|_\infty \to 0$, the proof is now complete. ∎

One can see from this proof that we may anticipate convergence for any sequence of quadrature rules that are exact for increasingly high-degree polynomials and that have positive quadrature weights. Actually, all we really need for this proof is that the sums of the absolute values of the weights be uniformly bounded. However, the Newton–Cotes rules, for example, do not possess this property, and continuous functions exist whose integrals the Newton–Cotes rules fail to approximate arbitrarily accurately. Krylov (1962) contains the details supporting this assertion.

We shall now discuss the construction of the quadrature points for Gaussian integration. Once these points have been located, the quadrature weights can be found using either the linear equation method (3.21) or direct integration of the interpolation polynomial, as in equation (3.15). There is also a direct formula for the weights of an n-point Gaussian quadrature formula. Let $P_n^*(x)$ denote an orthonormal nth-degree polynomial that has been scaled so that

$$\int_a^b w(x)[P_n^*(x)]^2\, dx = 1. \tag{3.55}$$

The symbol k_n will denote the coefficient of x^n in P_n^*, and $P_n^{*\prime}$ the derivative of P_n^*. With this terminology, we have that the weight given to $f(x_k)$ in an n-point formula is

$$A_k = -(k_{n+1}/k_n)[P_{n+1}^*(x_k)P_n^{*\prime}(x_k)]^{-1}. \tag{3.56}$$

In view of Theorem 3.6, the natural way to find the interpolation points is to find the roots of the orthogonal polynomial. This, in turn, can be done in a straightforward manner by making use of the recursive relationship for orthogonal polynomials given by Theorem 3.5. For it is evident that this relation allows us to recursively compute the coefficients of the nth orthogonal polynomial in terms of the coefficients of the $(n-1)$th and $(n-2)$th orthogonal polynomials. Once we have the coefficients of P_n, the nth orthogonal polynomial, we can use methods to be described in Chapter 5 in order to find their roots. This is, in essence, the technique used in the program GRULE in Davis and Rabinowitz (1975) for finding the quadrature points and weights if the weight function $w(x) = 1$. In this important case, the orthogonal polynomials happen to be Legendre polynomials. The following formula provides a good approximation to the kth root of the nth Legendre polynomial:

$$x_k = \frac{4k-1}{4n+2}\,\pi + \frac{n-1}{8n^3}\cot\frac{4k-1}{4n+2}\,\pi + O(n^{-4}). \tag{3.57}$$

Davis and Rabinowitz (1975) use the approximation provided by this formula as a starting value in a Newton's method (Chapter 5) calculation for finding the roots of P_n.

An interesting alternative procedure for simultaneously computing the points and weights for Gaussian quadrature has been provided by Golub and Welsch (1969). Let $\{P_j^*\}$ be a system of orthonormal polynomials as in (3.55). An obvious consequence of Theorem 3.5 is that these polynomials satisfy the three-term recurrence relationship

$$P_j^*(x) = (\alpha_j x + \beta_j)P_{j-1}^*(x) - \gamma_j P_{j-2}^*(x), \qquad j > 2. \tag{3.58}$$

While the values of the constants α_j, β_j, γ_j in (3.58) are different constants from those in Theorem 3.5 because of different scaling, they may easily be computed from the relations given in that theorem. Relationship (3.58) may be expressed in matrix form as

$$
x\begin{bmatrix} P_0{}^*(x) \\ P_1{}^*(x) \\ \vdots \\ P_{N-1}^*(x) \end{bmatrix} = \begin{bmatrix} -\beta_1/\alpha_1 & 1/\alpha_1 & 0 & & & 0 \\ \gamma_2/\alpha_2 & -\beta_2/\alpha_2 & 1/\alpha_2 & & & \\ 0 & & & \ddots & & \vdots \\ \vdots & & & & & 1/\alpha_{N-1} \\ 0 & & & & \gamma_N/\alpha_N & -\beta_N/\alpha_N \end{bmatrix} \begin{bmatrix} P_0{}^*(x) \\ P_1{}^*(x) \\ \vdots \\ P_{N-1}^*(x) \end{bmatrix}
$$

$$
+ \begin{bmatrix} 0 \\ 0 \\ \vdots \\ 0 \\ P_N{}^*(x)/\alpha_N \end{bmatrix} \tag{3.59}
$$

which may be written as

$$
x\mathbf{P}(x) = \mathbf{T}\mathbf{P}(x) + (1/\alpha_N)\mathbf{e}_N P_N{}^*(x),
$$

where $\mathbf{T}$ is a tridiagonal matrix, $\mathbf{e}_N = (0, 0, \ldots, 1)^{\mathrm{T}}$, and $\mathbf{P}(x) = \big(P_0{}^*(x), \ldots, P_{N-1}^*(x)\big)^{\mathrm{T}}$. (Again, T as a superscript denotes "transpose.") It is clear that x_j is a root of $P_N{}^*(x)$ if and only if x_j is an eigenvalue of $\mathbf{T}$.

As we shall observe in Chapter 7, tridiagonal matrices are especially amenable to numerical computation of eigenvalues, and so this method has much to recommend it in comparison with the alternative of searching for the roots of the polynomial P_N by means of techniques of Chapter 5. Moreover, in terms of quantities defined above, Golub and Welsch (1969) give a rule for the quadrature weight at the point x_j:

$$
A_j = [\mathbf{P}(x_j)^{\mathrm{T}}\mathbf{P}(x_j)]^{-1}, \qquad j = 1, 2, \ldots, N. \tag{3.60}
$$

A related method was considered by Gautschi (1968). Stroud and Secrest (1966) give extensive tables of Gauss formulas.

From analysis of the Hermite approximation polynomial that agrees with $f(x_j)$ and $f'(x_j)$ at each quadrature point x_j, one may establish that the error of Gaussian quadrature is given by the following theorem:

Theorem 3.8. If $f(x)$ has a continuous derivative of order $2n$, then

$$
\int_{-1}^{1} w(x)f(x)\, dx - I_n(f) = \frac{f^{(2n)}(\eta)}{(2n)!k_n{}^2}, \tag{3.61}
$$

for some $\eta \in [-1, 1]$ and for k_n being the leading coefficient of the ortho-normal polynomial $P_n{}^*$ on the interval $[-1, 1]$. For Legendre polynomials [orthonormal if $w(x) = 1$], $k_n = \binom{2n}{n} 2^{-n}$.

The details of the proof are to be found in Davis and Rabinowitz (1975, p. 227).

Problems

3.10. Determine the interpolating point for the Gaussian quadrature for $n = 1$, $w(x) = 1$, and domain $[-1,1]$.

3.11. For 5-point Gaussian quadrature with weight and interval as above, the quadrature points turn out to be

$$x_3 = 0, \qquad x_4 = -x_2 = 0.53846931,$$
$$x_5 = -x_1 = 0.90617985,$$

and the weights are

$$A_3 = 0.56888889,$$
$$A_2 = A_4 = 0.47862867,$$
$$A_1 = A_5 = 0.23692689.$$

Compute the 5-point Gaussian approximations of x^6, e^x, $\sin^2 x \cos^3 x$, and $\sin(10x + 0.2)$.

3.12. (Continuation) Compare the accuracy of your result with that of the 6-point trapezoidal rule and the corrected 6-point trapezoidal rule.

3.13. Prove part (i) of Theorem 3.4. HINT: Scale so that the leading coefficients are the same and think about something clever to do with the square of the difference $r(x)^2$, so that you can show

$$\int_a^b w(x)r^2(x)\, dx = 0.$$

3.14. Prove part (ii) of Theorem 3.4. HINT: Let $x_1, \ldots, x_r$ be the roots of P_n that lie in $[a, b]$, and take $Q(x) = \prod_{i=1}^r (x - x_i)$. Show that if $r < n$, then

$$\int_a^b w(x)P_n(x)Q(x)\, dx \neq 0.$$

3.2.4. The Euler–Maclaurin Formula

The theory of Bernoulli numbers and Bernoulli polynomials provides a basis for the Euler–Maclaurin formula for improving the accuracy of the trapezoidal rule through use of derivatives at the interval endpoints, and provides the theoretical foundation of the popular Romberg integration

method, which is discussed in the next section. We begin the discussion with the definition and derivation of the salient properties of Bernoulli numbers.
   Consider the function

$$g(z) = \begin{cases} \dfrac{z}{e^z - 1}, & \text{if } z \neq 0, \ |z| < 2\pi, \\ 1, & \text{if } z = 0. \end{cases}$$

It can be proved that function g, viewed as a complex function, is analytic. Let the power series of g be denoted by

$$g(z) = \sum_{n=0}^{\infty} \frac{1}{n!} b_n z^n. \tag{3.62}$$

The numbers b_n are called *Bernoulli numbers*. The most important properties of Bernoulli numbers are summarized in the following theorem.

Theorem 3.9. (a) $b_0 = 1$, and for $n \geq 1$,

$$b_n = -\sum_{k=0}^{n-1} \frac{n!}{(n+1-k)!k!} b_k; \tag{3.63}$$

   (b) for n odd and $n \neq 1$, $b_n = 0$.

Proof. (a) Using the definition of Bernoulli numbers we have

$$z = (e^z - 1)g(z) = \sum_{j=1}^{\infty} \frac{1}{j!} z^j \sum_{k=0}^{\infty} \frac{1}{k!} b_k z^k = \sum_{m=1}^{\infty} z^m \sum_{k=0}^{m-1} \frac{b_k}{k!(m-k)!},$$

which implies $b_0 = 1$ and, for $m \geq 1$,

$$\sum_{k=0}^{m-1} \frac{b_k}{k!(m-k)!} = 0.$$

Assuming that $m - 1 = n$, and solving for b_n from this relation, we get recursive equation (3.63).
   (b) Since equation (3.62) implies that

$$g(-z) = \sum_{n=0}^{\infty} (-1)^n \frac{1}{n!} b_n z^n, \tag{3.64}$$

and from the definition of function g we have

$$g(-z) = \frac{-z}{e^{-z} - 1} = \frac{ze^z}{e^z - 1} = z + \frac{z}{e^z - 1} = z + \sum_{n=0}^{\infty} \frac{1}{n!} b_n z^n. \tag{3.65}$$

The coefficients of z^n on the right-hand side of equations (3.64) and (3.65) must be equal to each other, which implies statement (b). ∎

Assume that the value of x is fixed and consider the function

$$
G(z, x) = \begin{cases} e^{xz} \dfrac{z}{e^z - 1}, & \text{if } z \neq 0,\ |z| < 2\pi, \\ 1, & \text{if } z = 0. \end{cases}
$$

It is easy to show that function G, as a complex function, is an analytic function of z for all fixed values of x. Let the power series of G be denoted by

$$
G(z, x) = \sum_{n=0}^{\infty} \frac{1}{n!} B_n(x) z^n. \tag{3.66}
$$

The functions B_n are called *Bernoulli polynomials*.

Theorem 3.10. The functions B_n have the following properties:

(a) $B_n(x) = \displaystyle\sum_{i=0}^{n} \frac{n!}{i!(n-i)!} b_{n-i} x^i$ $(n = 0, 1, 2, \ldots)$.

Consequently the function B_n is a polynomial of degree n. Also

(b) $B_n(x) = b_n + n \displaystyle\int_0^x B_{n-1}(t)\, dt.$

(c) $B_n(1 - x) = (-1)^n B_n(x).$

(d) $B_n(1) = B_n(0) = b_n,$ for even n.

(e) $B_n(x) = (-1)^{n/2}\, n!\, 2^{1-n} \pi^{-n} \displaystyle\sum_{j=1}^{\infty} j^{-n} \cos 2\pi j x,$ $0 \le x \le 1$, n even.

(f) $b_n \sim (-1)^{(n/2-1)} \dfrac{n!}{2^{n-1}\pi^n}$ for even n.

(g) $B_n(x) - b_n$ has the same sign for all x in $[0,1]$, the sign being equal to the sign of b_n and

$$
|B_n(x) - b_n| \le 2\,|b_n| \text{for even } n.
$$

Proof. (a) By the definition of function G_n we have

$$
G(z, x) = e^{xz} g(z) = \sum_{i=0}^{\infty} x^i z^i \frac{1}{i!} \sum_{k=0}^{\infty} \frac{1}{k!} b_k z^k
$$

$$
= \sum_{n=0}^{\infty} z^n \sum_{i=0}^{n} \frac{b_{n-i}}{i!(n-i)!} x^i,
$$

and comparing it with equation (3.66) we see that the uniqueness of the coefficients of power series in the variable z implies the statement.

(b) Statement (a) implies

$$B_n'(x) = \sum_{i=1}^{n} \frac{n!}{(i-1)!(n-i)!} b_{n-i} x^{i-1}$$

$$= n \sum_{j=0}^{n-1} \frac{(n-1)!}{j!(n-1-j)!} b_{n-1-j} x^j = n B_{n-1}(x).$$

After integrating both sides over interval $[0, x]$, we have

$$B_n(x) = B_n(0) + \int_0^x n B_{n-1}(x) \, dx = b_n + n \int_0^x B_{n-1}(x) \, dx.$$

(c) Using definition (3.66) we obtain

$$G(z, 1 - x) = \sum_{n=0}^{\infty} \frac{1}{n!} B_n(1 - x) z^n,$$

and simple calculations show that

$$G(z, 1 - x) = e^{(1-x)z} \frac{z}{e^z - 1} = e^{x(-z)} \frac{-z}{e^{-z} - 1}$$

$$= G(-z, x) = \sum_{n=0}^{\infty} (-1)^n \frac{1}{n!} B_n(x) z^n.$$

The statement follows from the uniqueness of the coefficients of the two above power series.

(d) Let n be even; then using (a) and (c),

$$B_n(1) = B_n(1 - 1) = B_n(0) = b_n,$$

and for $n > 1$ and odd

$$B_n(1) = -B_n(1 - 1) = -B_n(0) = 0.$$

(e) Since for even n the values of function B_n are the same at the endpoints of interval $[0,1]$, the Fourier series of B_n according to the orthogonal system $\{1, \sin 2\pi j x, \cos 2\pi j x, j \geq 1\}$ converges to B_n. For a discussion of the relevant properties of the Fourier series, we refer the reader to Apostol (1957, Section 15.3).

(f) We prove that

$$\lim_{k \to \infty} \sum_{j=1}^{\infty} \frac{1}{j^k} = 1,$$

which immediately implies the assertion.

Using the convexity of function $1/x^k$ we have

$$1 + \int_2^\infty \frac{1}{x^k}\, dx \leq \sum_{j=1}^\infty \frac{1}{j^k} \leq 1 + \frac{1}{2^k} + \int_2^\infty \frac{1}{x^k}\, dx. \qquad (3.67)$$

Since for $k \geq 2$

$$\int_2^\infty \frac{1}{x^k}\, dx = \int_2^\infty x^{-k}\, dx = \left[\frac{x^{-k+1}}{-k+1} \right]_2^\infty = \frac{1}{(k-1)2^{k-1}},$$

both the left- and right-hand sides of inequality (3.67) tend to 1 for $k \to \infty$, which implies the limit relation (f).

(g) For even n using statements (e) and (f) of the theorem we have

$$B_n(x) - b_n = B_n(x) - B_n(0)$$
$$= (-1)^{n/2}\, n!\, 2^{1-n}\pi^{-n} \sum_{j=1}^\infty \frac{1}{j^n}\, (1 - \cos 2\pi j x),$$

from which the statement follows immediately. ∎

By use of statement (a) of Theorem 3.9 and statement (b) of Theorem 3.10, the Bernoulli numbers and Bernoulli polynomials can be determined successively:

$$b_0 = 1, \quad b_1 = -\tfrac{1}{2}, \quad b_2 = \tfrac{1}{6}, \quad b_3 = 0, \quad b_4 = -\tfrac{1}{30}, \ldots$$
$$B_0(x) = 1, \quad B_1(x) = x - \tfrac{1}{2}, \quad B_2(x) = x^2 - x + \tfrac{1}{6},$$
$$B_3(x) = x^3 - \tfrac{3}{2}x^2 + \tfrac{1}{2}x, \ldots$$

Theorem 3.11. If function y is m times differentiable on the interval $[0,1]$, then

$$\int_0^1 y(t)\, dt - \frac{1}{2}\, [y(0) + y(1)] + \sum_{k=2}^m \frac{1}{k!}\, b_k [y^{(k-1)}(1) - y^{(k-1)}(0)]$$
$$= \frac{1}{m!} \int_0^1 y^{(m)}(t) B_m(1 - t)\, dt. \qquad (3.68)$$

Proof. The theorem will be proved by induction. For $m = 1$, using the method of integration by parts, we have

$$\int_0^1 y'(t) B_1(1 - t)\, dt = \int_0^1 y'(t)\, (\tfrac{1}{2} - t)\, dt$$
$$= [y(t)(\tfrac{1}{2} - t)]_0^1 + \int_0^1 y(t)\, dt$$
$$= \int_0^1 y(t)\, dt - \tfrac{1}{2}\, [y(0) + y(1)];$$

thus identity (3.68) is proved for $m = 1$.

Let us assume that equation (3.68) is true for m. Applying statements (b) and (d) of Theorem 3.10, we obtain

$$\frac{1}{(m+1)!} \int_0^1 y^{(m+1)}(t) B_{m+1}(1-t)\, dt$$

$$= \frac{1}{(m+1)!} [y^{(m)}(t) B_{m+1}(1-t)]_0^1$$

$$+ \frac{1}{(m+1)!} \int_0^1 y^{(m)}(t) B'_{m+1}(1-t)\, dt$$

$$= \frac{1}{m!} \int_0^1 y^{(m)}(t) B_m(1-t)\, dt + \frac{1}{(m+1)!} b_{m+1}[y^{(m)}(0) + y^{(m)}(1)].$$

By replacing the first term of the right-hand side with the left-hand side of equation (3.68), we obtain identity (3.68) for $m+1$.

Equation (3.68) is called the *Euler identity*. The second term of the left-hand side is equal to the trapezoidal rule for $n = 1$.

Theorem 3.12. If m is even and function f is m times differentiable in the interval $[a, b]$, then

$$\int_a^b f(x)\, dx = t_n + \sum_{k=1}^{(m/2)-1} \frac{1}{(2k)!} b_{2k}[f^{(2k-1)}(a) - f^{(2k-1)}(b)]h^{2k}$$

$$+ \frac{1}{m!} h^{m+1} \int_0^1 [B_m(t) - b_m] \sum_{i=0}^{n-1} f^{(m)}(a + ih + th)\, dt, \quad (3.69)$$

where $h = (b-a)/n$ and t_n is the trapezoidal rule for $n+1$ points.

Proof. Divide the interval $[a, b]$ into n uniform subintervals and let $a = x_0, x_1, \ldots, x_n = b$ denote the endpoints of the subintervals. Consider now subinterval $[x_i, x_{i+1}]$ and apply equation (3.68) to conclude that

$$\int_{x_i}^{x_{i+1}} f(x)\, dx = h \int_0^1 f(x_i + th)\, dt = \frac{h}{2} [f(x_i) + f(x_{i+1})]$$

$$+ \sum_{k=1}^{m/2} \frac{1}{(2k)!} b_{2k}[f^{(2k-1)}(x_i) - f^{(2k-1)}(x_{i+1})]h^{2k}$$

$$+ \frac{1}{m!} h^{m+1} \int_0^1 f^{(m)}(x_i + ht) B_m(1-t)\, dt.$$

In the above calculation we used the fact that the odd Bernoulli numbers are equal to zero.

Now we use the fact (Theorem 3.10, part c) that since m is even,

$B_m(1 - x) = B_m(x)$ and the observation that the last summand in the above equation can be rewritten

$$\frac{h^m b_m}{m!} [f^{(m-1)}(x_i) - f^{(m-1)}(x_{i+1})] = \frac{h^{m+1}}{m!} \int_0^1 f^{(m)}(x_i + ht)(-b_m) \, dt$$

to restate the earlier expression for $\int_{x_i}^{x_{i+1}} f(x) \, dx$ as

$$\int_{x_i}^{x_{i+1}} f(x) \, dx = h \int_0^1 f(x_i + th) \, dt$$

$$= \frac{h}{2} [f(x_i) + f(x_{i+1})]$$

$$+ \sum_{k=1}^{(m/2-1)} \frac{1}{(2k)!} b_{2k} [f^{(2k-1)}(x_i) - f^{(2k-1)}(x_{i+1})] h^{2k}$$

$$+ \frac{1}{m!} h^{m+1} \int_0^1 f^{(m)}(x_i + ht)[B_m(t) - b_m] \, dt.$$

If we add the above equation for $i = 0, 1, 2, \ldots$, equation (3.69) is obtained. ∎

Equation (3.69) is called the *Euler–Maclaurin formula*, and it can be considered to be a corrected formula generalizing the formula in Section 3.2.1, for the trapezoidal rule. The error of the Euler–Maclaurin formula is

$$R_m = \frac{1}{m!} h^{m+1} \int_0^1 [B_m(t) - b_m] \sum_{i=0}^{n-1} f^{(m)}(a + ih + th) \, dt.$$

Since in view of part (g) of Theorem 3.10, the function $B_m(t) - b_m$ does not change the sign in the interval $[0,1]$, after applying the second mean value theorem of integral calculus we have

$$R_m = \frac{1}{m!} h^{m+1} \sum_{i=0}^{n-1} f^{(m)}(a + ih + \xi h) \int_0^1 [B_m(t) - b_m] \, dt,$$

where $\xi \in [0,1]$. Since

$$\int_0^1 [B_m(t) - b_m] \, dt = \left[\frac{B_{m+1}(t)}{m + 1} - b_m t \right]_0^1 = -b_m,$$

the theorem of Darboux (see Apostol, 1957, p. 94) implies that

$$R_m = -\frac{1}{m!} b_m h^{m+1} \sum_{i=0}^{n-1} f^{(m)}(a + ih + \xi h) = -\frac{1}{m!} b_m h^{m+1} n f^{(m)}(\zeta),$$
$$(3.70)$$

where ζ is a number in $[a, b]$.

Example 3.5. We determine the sum of the pth power of the first n positive integers. Let $f(x) = x^p$, $a = 0$, $b = n$, $h = 1$, and $m = 2[p/2] + 2$. Then $m \geq p + 1$, so that $f^{(m)}(x) \equiv 0$, and consequently $R_m = 0$. Using equation (3.69), we have

$$\int_0^n x^p \, dx = \sum_{i=1}^n i^p - \frac{1}{2} n^p - \sum_{k=1}^{[p/2]} \frac{1}{(2k)!} b_{2k} p(p-1) \cdots (p - 2k + 2) n^{p-2k+1}$$

$$= \sum_{i=1}^n i^p - \frac{1}{2} n^p - \sum_{k=1}^{[p/2]} \frac{1}{2k} \binom{p}{2k-1} b_{2k} n^{p-2k+1},$$

which implies

$$\sum_{i=1}^n i^p = \frac{n^{p+1}}{p+1} + \frac{n^p}{2} + \sum_{k=1}^{[p/2]} \frac{1}{2k} \binom{p}{2k-1} b_{2k} n^{p-2k+1}.$$

In the special case of $p = 1$,

$$\sum_{i=1}^n i = \frac{n^2}{2} + \frac{n}{2} + 0 = \frac{n(n+1)}{2},$$

and in the case of $p = 2$,

$$\sum_{i=1}^n i^2 = \frac{n^3}{3} + \frac{n^2}{2} + \frac{1}{2} \binom{2}{1} b_2 n = \frac{n^3}{3} + \frac{n^2}{2} + \frac{n}{6} = \frac{n(n+1)(2n+1)}{6},$$

and if $p = 3$, then

$$\sum_{i=1}^n i^3 = \frac{n^4}{4} + \frac{n^3}{3} + \frac{1}{2} \binom{3}{1} b_2 n^2 = \frac{n^4}{4} + \frac{n^3}{2} + \frac{n^2}{4} = \left[\frac{n(n+1)}{2} \right]^2.$$

Approximating the derivatives in the summation of equation (3.69) by formulas (3.9), we obtain *Gregory's formula*:

$$\int_a^b f(x) \, dx = t_n + \frac{h}{12} [\Delta f_0 - \Delta f_{n-1}] - \frac{h}{24} [\Delta^2 f_0 + \Delta^2 f_{n-2}]$$

$$+ \frac{19h}{720} [\Delta^3 f_0 - \Delta^3 f_{n-3}] - \frac{3h}{160} [\Delta^4 f_0 + \Delta^4 f_{n-4}] + \cdots,$$

where $x_0 = a$. The use of this formula is much more convenient than the use of formula (3.69) because the differences can be computed much more easily than the derivatives.

Problems

3.15. Show that

$$\frac{t}{e^t - 1} + \frac{t}{2} = \frac{t}{2} \coth \frac{t}{2},$$

and use this formula for proving the properties of Bernoulli numbers.

3.16. Evaluate

$$\int_0^\pi \frac{\sin x}{x}\, dx \quad \text{and} \quad \int_0^1 \exp x^2\, dx$$

with an error less than 10^{-8}.

3.2.5. Romberg Integration

In the previous section a corrected form of the trapezoidal rule was discussed. Let us examine the error term R_m given by (3.70). It can be written as

$$R_m = \left[\frac{f^{(m)}(\zeta)}{m!}\, h^m\right][(-1)hb_m n].$$

Assuming that the function f is analytic and the points $a - h$, $b + h$, belong to the domain of convergence of Taylor's series of functions f, we conclude that the first factor of the right-hand side tends to zero for $m \to \infty$. Using statement (f) of Theorem 3.10, we also see that $b_m \to 0$ for $m \to \infty$, which implies that for $m \to \infty$ and fixed n, $R_m \to 0$. Thus under the above condition,

$$I = \int_a^b f(x)\, dx = t_n + \sum_{=1}^\infty a_j h^{2j}, \tag{3.71}$$

where

$$a_j = \frac{1}{(2j)!}\, b_{2j}[f^{(2j-1)}(a) - f^{(2j-1)}(b)].$$

Let

$$T_{0,k} = t_{2^k} \quad (k = 0, 1, 2, \ldots),$$

and recursively for $m = 1, 2, \ldots$, let

$$T_{m,k} = \frac{1}{4^m - 1}\,(4^m T_{m-1,k+1} - T_{m-1,k}). \tag{3.72}$$

The process of successively constructing these corrected trapezoidal formulas is known as *Romberg's method*. The formulas $T_{m,k}$ are *Romberg formulas*.

We shall first prove that for $k = 0, 1, 2, \ldots$ and $m = 1, 2, \ldots$, and some double sequence $\{b_{i,j}\}$, I can be represented in the form

$$I = \int_a^b f(x)\, dx = T_{m,k} + \sum_{j=m+1}^\infty b_{m,j}\left(\frac{b-a}{2^k}\right)^{2j}. \tag{3.73}$$

The proof will be made by induction on m. For $m = 0$, equation (3.71) for $b_{mj} = a_j$ gives (3.73). Assume that (3.73) is true for $m - 1$ and $k = 0, 1, 2, \ldots$. Then using (3.72) we have

$$
\begin{aligned}
T_{m,k} &= \frac{1}{4^m - 1} \left\{ 4^m \left[I - \sum_{j=m}^{\infty} b_{m-1,j} \left(\frac{b - a}{2^{k+1}} \right)^{2j} \right] \right. \\
&\qquad \left. - \left[I - \sum_{j=m}^{\infty} b_{m-1,j} \left(\frac{b - a}{2^k} \right)^{2j} \right] \right\} \\
&= I - \sum_{j=m+1}^{\infty} b_{m-1,j} \left(\frac{4^m}{2^{2j}} - 1 \right) \frac{1}{4^m - 1} \left(\frac{b - a}{2^k} \right)^{2j}, \quad (3.74)
\end{aligned}
$$

which proves (3.73). From (3.74) a recursive relation for the coefficients $b_{m,j}$ can also be obtained; for $m \geq 1$, $j = m + 1, m + 2, \ldots$,

$$
b_{m,j} = b_{m-1,j} \left(\frac{4^m}{2^{2j}} - 1 \right) \frac{1}{4^m - 1}. \quad (3.75)
$$

Equations (3.71), (3.73), and (3.75) show that $T_{m,k}$ converges to I, and the leading term in its error is the order of

$$
\left(\frac{b - a}{2^k} \right)^{2(m+1)},
$$

while the leading term of the error of $T_{0,k}$ is $[(b - a)/2^k]^2$. Consequently, the convergence of approximations $T_{m,k}$ for large m is much faster than the convergence of $T_{0,k}$ for the trapezoidal rule. Romberg's method can be regarded as resulting from iteratively applying Richardson's extrapolation to the trapezoidal rule.

Problems

   3.17. Compute the integral $\int_{-4}^{4} dx/(1 + x^2)$ using the Romberg method for order $k = 1, 2, \ldots, 8$. For each pair m, k, print out the error.

   3.18. Show that the formula $T_{1,k}$ is the $(2^k + 1)$-point compound Simpson's formula.

   3.19. Write out the sums that $T_{1,3}$ and $T_{2,3}$ represent and demonstrate whether or not they are Newton–Cotes formulas.

3.3. Supplementary Notes and Discussion

Common sense would seem to suggest that equally spaced quadrature points (as used in Newton–Cotes rules) should give the most representative sampling of the function values and thus generally lead to the most accurate quadrature formulas. It is thus a tribute to analysis when we find that intuition is wrong and that unequally spaced Gaussian quadrature points typically give dramatic improvement in accuracy over equally spaced quadrature points in Newton–Cotes formulas, as is seen in the representative Table 3.3. In fact, as we have referenced, it is known that in contrast to Gaussian quadrature, Newton–Cotes formulas do not generally converge, with increasing number of points, to the exact integral of a continuous function.

The strongest statement we made concerning asymptotic convergence rates was Theorem 3.3, which asserts that if a function f has a continuous kth derivative, the error of compound rules is asymptotically C/m^k, m being the number of subintervals. Despite this strong result, experience has shown that usually one should look first to Gaussian quadrature for the most accurate estimate. Stroud (1974, Section 3.14) amplifies this remark, asserting, "Often a Gauss formula will be much superior to any other formula with the same number of points."

For n-point Newton–Cotes formulas, we presented in equation (3.33) error formulas that depended on $f^{(n+1)}$ in case n is odd and $f^{(n)}$ otherwise. Also, in equation (3.61) we see that the error for n-point Gauss formulas depends on $f^{(2n)}$. One may naturally desire error bounds that are determined by lower-order derivatives, for use in cases in which the requisite high-order derivatives are difficult to compute. The theory of Peano kernels leads to error bounds using lower derivatives. Let a quadrature formula I be exact for all polynomials of order $m - 1$. Then if d is any nonnegative integer not exceeding m, there exists a function $K_d(x)$ such that

$$\int_a^b w(x)f(x)\,dx - I(f) = \int_a^b f(x)K_d(x)\,dx. \qquad (3.76)$$

The function K_d is called *Peano's kernel function*. The construction and use of this function and the representation (3.76) is given in Stroud (1974, Section 3.12) and Davis and Rabinowitz (1975, Section 4.3). Expression (3.76) is the basis of the standard way for deriving the error bound for Simpson's rule.

The effect of roundoff error may be studied and bounded by use of developments given toward the end of Section 3.2.2. A detailed "worst-case" analysis of roundoff error is undertaken in Section 4.2 of Davis and Rabinowitz (1975). From formulas given there, it is seen that for n-point formulas $\sum_{i=1}^{n} A_i f(x_i)$ for $\int_a^b f(x)\,dx$, the roundoff bound is given by the sum $T_1 + T_2 + T_3$, where

$$T_1 = qM \sum_{i=1}^{n} |A_i|,$$

$$T_2 = (1 + q)M\alpha_2 \sum_{i=1}^{n} |A_i|,$$

$$T_3 = (1.06)(1 + q)M\alpha_1(1 + \alpha_2) \sum_{i=1}^{n} |A_i|(n - i + 1).$$

In the above equations, q, α_1, and α_2 are small multiples of 2^{-t}, t being the floating-point binary word length of the mantissa and M being the least upper bound for the absolute value of the integrand f. Davis and Rabinowitz (1975) conclude that if, as in the usual case, the quadrature points number between 10 and 100, "in ordinary circumstances ... no great damage is done [*by rounding*]." But the authors go on to warn that if the number of points is very large, as is often the case for multivariate quadrature formulas, then the effect of rounding must be considered.

Numerical analysis for multivariate integration is a more difficult and incomplete subject than the univariate case studied in this chapter. Two primary sources of difficulty are: (1) not all domains in two (or more) dimensions are equivalent under scaling and translation (whereas all intervals are equivalent to the unit interval) and so in two dimensions, quadrature rules must be fundamentally dependent on the domain of integration, and (2) orthogonal polynomials in several variables are complicated and computationally burdensome if not intractable. Yet as we have seen, efficient quadrature is closely tied to methods using orthogonal polynomials. The reader is referred to the book by Stroud (1971) on the subject of computational multivariate integration.

There is an interesting link between integration and probability theory. The discipline known as the *Monte Carlo method* exploits this link to provide probabilistic simulation techniques for numerical approximation of integrals. These methods and their relatives, the *number-theoretic methods*, are competitive for integrands of high dimension. One may refer to Chapter 6 of Yakowitz (1977) for a discussion of this subject, and to Chapter 6 of Stroud (1971) for a development of quasi-Monte Carlo multivariate integration.

Recently, Yakowitz *et al.* (1978) have provided a Monte Carlo integration method such that if $\{U_i\}_{i=1}^n$ are random numbers and the integrand f has a continuous second derivative on the unit interval (which is taken as the domain of integration), and if we let E denote "expectation", I the exact integral, and I_n the Monte Carlo approximation, we have

$$E[(I - I_n)^2] = O(1/n^4).$$

Previously, the best bound for Monte Carlo formulas was $O(1/n)$.

4

General Theory for Iteration Methods

A large portion of numerical methods is based on the determination of the unknown numbers, n-tuples, and functions by recursively applying some operation or formula. Such methods are known collectively as *iteration methods*. It is known that certain numerical problems (such as finding roots of polynomials or eigenvalues of matrices) are amenable *only* to iteration methods. The zeros of univariate functions are real or complex numbers; solutions of linear or nonlinear systems of algebraic equations are vectors; and solutions of differential and integral equations are functions. The common property of these solution sets is that these are sets on which distances of elements may be naturally defined. A function giving a distance measure between every pair of points is called a metric.

In essence, this chapter is a compendium of those standard results of mathematical analysis and functional analysis that are most germane to deriving properties of iteration methods for numerical analysis problems. In isolating the mathematical underpinnings for numerical iteration schemes in a separate chapter, our intention is to attempt to make transparent the common structure of these iterations methods and to avoid unnecessary repetition.

We have however tried to provide for the reader who insists on having his mathematical abstractions motivated. Such a reader may skip to later chapters and wait until developments in this chapter are specifically required in our analysis of numerical methods. He will then be led back to a particular result or example of this chapter, and from that point, he can read "backward" to pick up missing concepts and terminology.

4.1. Metric Spaces

For any sets A and B, $A \times B$ denotes the *Cartesian product* of A and B, that is, the set of pairs $\{(x, y) : x \in A, y \in B\}$.

Definition 4.1. A pair (M, ϱ) is called a *metric space* if M is an abstract set and ϱ is a real-valued function defined on the set $M \times M$ with the following properties:

1. for arbitrary $x, y \in M$, $\varrho(x, y) \geq 0$, and $\varrho(x, y) = 0$ if and only if $x = y$;
2. for all $x, y \in M$, $\varrho(x, y) = \varrho(y, x)$;
3. for all $x, y, z \in M$, $\varrho(x, z) \leq \varrho(x, y) + \varrho(y, z)$.

The function ϱ is called the *metric*.

It is obvious that any subset of M with the metric ϱ is also metric space.

For any $x \in M$, the set

$$G(x, r) = \{y \mid y \in M, \varrho(x, y) < r\}$$

is called an *open sphere* with center x and radius $r > 0$, and the set

$$\bar{G}(x, r) = \{y \mid y \in M, \varrho(x, y) \leq r\}$$

is called a *closed sphere* with center x and radius $r > 0$.

Let M_1 be a subset of M. A point $x \in M_1$ is called an *interior point* of M_1 if for some positive r the sphere $G(x, r)$ is a subset of M_1. The point $x \in M_1$ is called the *boundary point* of M_1 if for arbitrary $r > 0$ the set $G(x, r)$ contains points that belong to M_1 and points that do not belong to M_1.

The set M_1 is called an *open set* if each point of M_1 is an interior point of M_1. The set M_1 is called a *closed set* if the set difference $M - M_1$ is an open set.

The empty set is considered to be both open and closed.

Definition 4.2. A sequence $\{x_n\}_{n=1}^{\infty}$ of elements of M is said to be *convergent* and to have the *limit point* $x \in M$, if $\varrho(x_n, x) \to 0$ as $n \to \infty$. We shall write $x_n \to x$ if x is the limit point of $\{x_n\}$.

It is easy to prove that the limit point of a convergent sequence is unique. For assume that there are two limit points x, x' of the sequence

$\{x_n\}_{n=1}^{\infty}$. Then for all values of n,

$$0 \leq \varrho(x, x') \leq \varrho(x, x_n) + \varrho(x_n, x') = \varrho(x_n, x) + \varrho(x_n, x'),$$

where both terms of the right-hand side tend to zero for $n \to \infty$. Consequently $\varrho(x, x') = 0$, which implies $x = x'$.

Through use of the above definition, the convergence of a sequence can be established only in the case that the limit point exists. On the other hand, the Cauchy convergence criterion does not use the limit point.

Definition 4.3. A sequence $\{x_n\}_{n=1}^{\infty}$ of elements of M is called a *Cauchy sequence*, if $\varrho(x_n, x_m) \to 0$ as $n, m \to \infty$.

In arbitrary metric spaces, a Cauchy sequence is not necessarily convergent, as the following example shows:

Example 4.1. Let M be the positive real numbers, and for $x, y \in M$, let $\varrho(x, y) = |x - y|$. It is easy to verify that (M, ϱ) is a metric space. Consider the sequence $\{1/n\}_{n=1}^{\infty}$, which has no limit point in M, but this sequence is nevertheless a Cauchy sequence since

$$\varrho(x_n, y_n) = \left| \frac{1}{n} - \frac{1}{m} \right| \to 0,$$

for $m, n \to \infty$.

On the other hand, the following converse is true:

Theorem 4.1. If the sequence $\{x_n\}_{n=1}^{\infty}$ is convergent, then it is also a Cauchy sequence.

Proof. Let the limit point of sequence $\{x_n\}_{n=1}^{\infty}$ be denoted by x. Then

$$0 \leq \varrho(x_n, x_m) \leq \varrho(x_n, x) + \varrho(x, x_m) = \varrho(x_n, x) + \varrho(x_m, x).$$

If $m, n \to \infty$, then both terms of the right-hand side tend to zero, which also implies that $\varrho(x_n, x_m) \to 0$. ∎

Definition 4.4. A metric space (M, ϱ) is called a *complete* metric space if all Cauchy sequences of elements in M have limit points in M.

We now discuss an important property of metric functions.

Theorem 4.2. If $x_n \to x$ and $y_n \to y$ for $n \to \infty$, then also $\varrho(x_n, y_n) \to \varrho(x, y)$; that is, the metric ϱ regarded as a bivariate function is continuous.

Proof. Using property 3 of metric spaces, we have

$$\varrho(x_n, y_n) \le \varrho(x_n, x) + \varrho(x, y) + \varrho(y, y_n),$$

which implies

$$\varrho(x_n, y_n) - \varrho(x, y) \le \varrho(x_n, x) + \varrho(y_n, y). \tag{4.1}$$

Similarly we can obtain the inequality

$$\varrho(x, y) - \varrho(x_n, y_n) \le \varrho(x_n, x) + \varrho(y_n, y). \tag{4.2}$$

The right-hand sides of inequalities (4.1) and (4.2) are the same, and also the absolute values of the left-hand sides are equal to each other, but their signs are different. As a consequence we have

$$0 \le |\varrho(x_n, y_n) - \varrho(x, y)| < \varrho(x_n, x) + \varrho(y_n, y).$$

Since both terms of the right-hand side tend to zero, for $n \to \infty$ $\varrho(x_n, y_n) - \varrho(x, y) \to 0$, which completes the proof. ∎

4.2. Examples of Metric Spaces

Example 4.2. Let M be the set of real or complex numbers and let $\varrho(x, y) = |x - y|$ for all $x, y \in M$. We now show that the pair (M, ϱ) satisfies the axioms of Definition 4.1.

1. $\varrho(x, y) \ge 0$, and $\varrho(x, y) = 0$ if and only if $|x - y| = 0$, which is equivalent to $x = y$;

2. for all $x, y \in M$,

$$\varrho(x, y) = |x - y| = |y - x| = \varrho(y, x);$$

3. for all $x, y, z \in M$

$$\varrho(x, z) = |x - z| \le |x - y| + |y - z| = \varrho(x, y) + \varrho(y, z), \tag{4.3}$$

because, with the notation

$$a = x - y, \ b = y - z,$$

inequality (4.3) is equivalent to the well-known triangle inequality

$$|a + b| \le |a| + |b|,$$

which is true for real and complex numbers.

It is well known from real and complex analysis that the Cauchy convergence criterion is a necessary and sufficient condition for the convergence of real or complex sequences; consequently, (M, ϱ) is a complete metric space.

Example 4.3. Let M be the set of n-dimensional vectors with real or complex elements.

(a) Let the metric ϱ_∞ be defined by

$$\varrho_\infty(\mathbf{x}, \mathbf{y}) = \max_i |x_i - y_i|,$$

for all $\mathbf{x} = (x_1, \ldots, x_n)^\mathrm{T}$, $\mathbf{y} = (y_1, \ldots, y_n)^\mathrm{T} \in M$.

Conditions 1 and 2 of Definition 4.1 are obviously satisfied, and so only condition 3 needs to be proven. For $i = 1, 2, \ldots, n$,

$$|x_i - y_i| + |y_i - z_i| \geq |x_i - z_i|, \tag{4.4}$$

which follows from the previous example. Let

$$|x_{i_0} - z_{i_0}| = \max_i |x_i - z_i|.$$

Then inequality (4.4) implies

$$|x_{i_0} - y_{i_0}| + |y_{i_0} - z_{i_0}| \geq |x_{i_0} - z_{i_0}| = \varrho_\infty(\mathbf{x}, \mathbf{z}).$$

But clearly $\varrho_\infty(\mathbf{x}, \mathbf{y}) \geq |x_{i_0} - y_{i_0}|$ and $\varrho_\infty(\mathbf{y}, \mathbf{z}) \geq |y_{i_0} - z_{i_0}|$, and so

$$\varrho_\infty(\mathbf{x}, \mathbf{y}) + \varrho_\infty(\mathbf{y}, \mathbf{z}) \geq \varrho_\infty(\mathbf{x}, \mathbf{z}).$$

(b) Let the function ϱ_1 be defined by $\varrho_1(\mathbf{x}, \mathbf{y}) = \sum_{i=1}^n |x_i - y_i|$. The first two axioms of Definition 4.1 are obviously satisfied; the third condition can be proven by adding inequality (4.4) for $i = 1, 2, \ldots, n$.

(c) Let the distance of vectors $\mathbf{x}$, $\mathbf{y}$ be the *Euclidean distance*:

$$\varrho_2(\mathbf{x}, \mathbf{y}) = \left\{ \sum_{i=1}^n |x_i - y_i|^2 \right\}^{1/2}.$$

The first two properties of metrics are also satisfied in this case; the third condition follows immediately from (4.4) and from the Cauchy–Schwarz (Taylor, 1964, p. 6) inequality

$$\varrho_2(\mathbf{x}, \mathbf{z})^2 = \sum_{i=1}^n |x_i - z_i|^2 \leq \sum_{i=1}^n (|x_i - y_i| + |y_i - z_i|)^2$$

$$= \sum_{i=1}^n |x_i - y_i|^2 + 2 \sum_{i=1}^n |x_i - y_i| |y_i - z_i| + \sum_{i=1}^n |y_i - z_i|^2$$

$$\leq \sum_{i=1}^n |x_i - y_i|^2 + 2 \left\{ \sum_{i=1}^n |x_i - y_i|^2 \right\}^{1/2} \left\{ \sum_{i=1}^n |y_i - z_i|^2 \right\}^{1/2} + \sum_{i=1}^n |y_i - z_i|^2$$

$$= \left[\left\{ \sum_{i=1}^n |x_i - y_i|^2 \right\}^{1/2} + \left\{ \sum_{i=1}^n |y_i - z_i|^2 \right\}^{1/2} \right]^2 = [\varrho_2(\mathbf{x}, \mathbf{y}) + \varrho_2(\mathbf{y}, \mathbf{z})]^2.$$

We remark that from the completeness of real numbers it follows that the metric spaces (M, ϱ) for $\varrho = \varrho_1$, $\varrho = \varrho_2$, and $\varrho = \varrho_\infty$ are all complete metric spaces.

The metrics defined in this example are special cases of the more general distance

$$\varrho_p(\mathbf{x}, \mathbf{y}) = \left\{ \sum_{i=1}^{n} |x_i - y_i|^p \right\}^{1/p},$$

where $p \geq 1$ is a real number. The first and second properties of the metrics are obviously satisfied; the triangle inequality follows immediately from the Minkowski inequality (see Taylor, 1964, p. 563). It can be verified that $\varrho_\infty(\mathbf{x}, \mathbf{y}) = \lim_{p \to \infty} \varrho_p(\mathbf{x}, \mathbf{y})$.

We remark that the metric ϱ_1 defined in Example 4.2 is the special case of this distance for $n = 1$.

Example 4.4. Let $[a, b]$ be a closed, finite interval and let $C[a, b]$ be the set of continuous functions defined on $[a, b]$. If $M = C[a, b]$ and for all functions $f, g \in M$

$$\varrho_\infty(f, g) = \max_{a \leq x \leq b} |f(x) - g(x)|, \tag{4.5}$$

then (M, ϱ_∞) is a complete metric space.

First we remark that the function $|f(x) - g(x)|$ is continuous and consequently reaches a finite maximal value in the interval $[a, b]$ (see Apostol, 1957, p. 73). The axioms of metric spaces can be verified by the argument given in part (a) of the previous example; the details are not discussed here.

The metric (4.5) is the limiting case for $p = \infty$ of the metric

$$\varrho_p(f, g) = \left\{ \int_a^b |f(x) - g(x)|^p \, dx \right\}^{1/p},$$

where $p \geq 1$. As in Example 4.3, the first two properties in Definition 4.1 are obviously satisfied; the triangle inequality follows from the Minkowski inequality.

A sequence $\{f_n\}_{n=1}^{\infty}$ is said to converge uniformly to a function f if

$$\varrho_\infty(f_n, f) \to 0,$$

for $n \to \infty$. It is known from mathematical analysis (see Apostol, 1957, p. 394) that the limit function of a uniformly convergent sequence of continuous functions defined on a closed, finite interval $[a, b]$ is also continuous, which implies that the metric space (M, ϱ_∞) is complete.

Problems

4.1. Let M be an abstract set and for all $x, y \in M$ let

$$\varrho(x, y) = \begin{cases} 1, & \text{if } x \neq y, \\ 0, & \text{if } x = y. \end{cases}$$

Prove that (M, ϱ) is a metric space. Is it a complete metric space?

4.2. Construct a metric space (M, ϱ) that contains spheres $G(x_1, r_1)$ and $G(x_2, r_2)$ with the properties

$$G(x_1, r_1) \subset G(x_2, r_2), \qquad r_1 > r_2 > 0.$$

4.3. Let M be the set of real numbers and for all $x, y \in M$ let $\varrho(x, y) = |x - y|^2$. Is (M, ϱ) a metric space?

4.3. Operators on Metric Spaces

Let (M, ϱ), (M', ϱ') be two (not necessarily different) metric spaces.

Definition 4.5. A single-valued mapping (in other words, a function) from the set M into the set M' is called an *operator* from metric space (M, ϱ) into metric space (M', ϱ').

Let A be an operator. Let us introduce the following notation:

$D(A) = \{x \mid x \in M \text{ and } A(x) \text{ is defined}\},$

$R(A) = \{x' \mid x' \in M' \text{ and there exists } x \in D(A) \text{ such that } x' = A(x)\}.$

The set $D(A)$ is called the *domain* of A, and set $R(A)$ is called the *range* of A. Obviously $D(A) \subset M$ and $R(A) \subset M'$.

Definition 4.6. The operator A is *continuous* at a point $x \in D(A)$ if for every sequence $\{x_n\}_{n=1}^{\infty}$ from $D(A)$ converging to x, $A(x_n) \to A(x)$ as $n \to \infty$.

This definition is the generalization of the idea of continuity of real functions.

Definition 4.7. The operator A is called a *bounded operator* if there exists a nonnegative constant K (independent of x and y) such that for all $x, y \in D(A)$,

$$\varrho'(A(x), A(y)) \leq K \cdot \varrho(x, y). \tag{4.6}$$

The interpretation of inequality (4.6) is that the distance of the images must not be greater than the distance of x and y multiplied by the fixed constant K.

First we remark that usually the continuity of operators does not imply their boundedness, as the following example shows.

Example 4.5. Let $M = M' = (-\infty, \infty)$ and let $\varrho(x, y) = \varrho'(x, y) = |x - y|$ for all $x, y \in M$. Consider the function $A(x) = x^2$ with domain $D(A) = (-\infty, \infty)$. Obviously operator A is continuous, but it is not bounded. For assume that A is bounded; that is,

$$|x^2 - y^2| \le K|x - y|,$$

for all $x, y \in (-\infty, \infty)$. Assuming that $x \ne y$, we have $|x + y| \le K$, which is a contradiction, because for $x = K$, $y = 1$, this inequality does not hold.

Theorem 4.3. If operator A is bounded, then it is continuous.

Proof. Let $\{x_n\}_{n=1}^{\infty}$ be a sequence in $D(A)$ with limit point $x \in D(A)$. We have to prove that $A(x_n) \to A(x)$.

Since operator A is bounded, we have

$$0 \le \varrho'\big(A(x_n), A(x)\big) \le K\varrho(x_n, x),$$

where the right-hand side tends to zero for $n \to \infty$, which implies that $A(x_n) \to A(x)$. ∎

In our numerical analysis applications of metric space theory, a special class of bounded operators, the set of contraction mappings, has a very important role. Contraction mappings (or contractions, for short) are defined next.

Definition 4.8. An operator A is called a *contraction* if there exists a nonnegative number $0 \le q < 1$ such that for all $x, y \in D(A)$,

$$\varrho'\big(A(x), A(y)\big) \le q \cdot \varrho(x, y).$$

Since contractions are bounded operators, Theorem 4.3 implies that contractions are also continuous operators.

4.4. Examples of Bounded Operators

Example 4.6. Let $M = M' = R^1$, with the metric defined in Example 4.2., and let M_1 be an interval in R^1. Let g denote a differentiable function on the interval M_1. If the value $K = \sup_{x \in M_1} |g'(x)|$ is finite, then recalling Lagrange's mean value theorem, we have

$$\varrho'(g(x), g(y)) = |g(x) - g(y)| = |g'(\xi)| \cdot |x - y|$$
$$\le K \cdot |x - y| = K \cdot \varrho(x, y).$$

Example 4.7. Let $M = M' = R^n$ and let M_1 be a convex subset of R^n. Assume that $g_1, \ldots, g_n$ are n-variate, continuous, differentiable real functions defined on the subset M_1, and assume further that the partial derivatives of these functions are bounded on M_1. Consider the operator

$$A(\mathbf{x}) = \begin{pmatrix} g_1(\mathbf{x}) \\ g_2(\mathbf{x}) \\ . \\ . \\ . \\ g_n(\mathbf{x}) \end{pmatrix} = \mathbf{g}(\mathbf{x}).$$

with domain M_1 and range in R^n. We shall prove that operator A is bounded for all three metrics defined in Example 4.3.

(a) In the case of distance ϱ_∞, by using Lagrange's mean value theorem, we have

$$\varrho_\infty(A(\mathbf{x}), A(\mathbf{y})) = \max_i |g_i(\mathbf{x}) - g_i(\mathbf{y})|$$

$$= \max_i \left| \sum_{j=1}^n \frac{\partial g_i(\xi_i)}{\partial x_j} (x_j - y_j) \right|$$

$$\leq \max_i \sum_{j=1}^n \sup_{x \in M_1} \left| \frac{\partial g_i(\mathbf{x})}{\partial x_j} \right| \times |x_j - y_j|.$$

Since $|x_j - y_j| \leq \max_i |x_i - y_i| = \varrho_\infty(\mathbf{x}, \mathbf{y})$,

$$\varrho_\infty(A(\mathbf{x}), A(\mathbf{y})) \leq K_\infty \varrho_\infty(\mathbf{x}, \mathbf{y}), \tag{4.7}$$

where

$$K_\infty = \max_i \sum_{j=1}^n \sup_{x \in M_1} \left| \frac{\partial g_i(\mathbf{x})}{\partial x_j} \right|.$$

(b) Using metric ϱ_1 as in the above procedure, we see that

$$\varrho_1(A(\mathbf{x}), A(\mathbf{y})) = \sum_{i=1}^n |g_i(\mathbf{x}) - g_i(\mathbf{y})|$$

$$= \sum_{i=1}^n \left| \sum_{j=1}^n \frac{\partial g_i(\xi_i)}{\partial x_j} (x_j - y_j) \right|$$

$$\leq \sum_{j=1}^n \sum_{i=1}^n \sup_{x \in M_1} \left| \frac{\partial g_i(\mathbf{x})}{\partial x_j} \right| \times |x_j - y_j|,$$

which implies

$$\varrho_1(A(\mathbf{x}), A(\mathbf{y})) \leq K_1 \varrho_1(\mathbf{x}, \mathbf{y}), \tag{4.8}$$

where

$$K_1 = \max_j \sum_{i=1}^n \sup_{x \in M_1} \left| \frac{\partial g_i(\mathbf{x})}{\partial x_j} \right|.$$

(c) By using the Cauchy–Schwarz inequality one can prove that

$$\varrho_2(A(\mathbf{x}), A(\mathbf{y})) \leq K_2\varrho_2(\mathbf{x}, \mathbf{y}), \tag{4.9}$$

where

$$K_2 = \left\{ \sum_{i=1}^{n} \sum_{j=1}^{n} \sup_{x \in M_1} \left| \frac{\partial g_i(\mathbf{x})}{\partial x_j} \right|^2 \right\}^{1/2}.$$

Example 4.8. Let M and M' be the set of real or complex vectors of dimension n, let **B** be an nth-order real or complex matrix, and let **c** be a fixed real or complex vector in M. Consider the operator

$$A(\mathbf{x}) = \mathbf{Bx} + \mathbf{c}$$

with domain $D(A) = M$. We can prove that operator A is bounded in case of all the three metrics defined in Example 4.3. Since the proof is analogous to the proof of the previous example, the details of calculations are not discussed here. The final formulas are also the same as equations (4.7), (4.8), and (4.9), with

$$K_\infty = \max_i \sum_{j=1}^{n} |b_{ij}|,$$

$$K_1 = \max_j \sum_{i=1}^{n} |b_{ij}|,$$

$$K_2 = \left\{ \sum_{i=1}^{n} \sum_{j=1}^{n} |b_{ij}|^2 \right\}^{1/2},$$

where the numbers b_{ij} denote the ijth element of matrix **B**.

Example 4.9. Let $M = M' = C[a, b]$ with the metric as defined in Example 4.4. Assume that function k is defined and continuous on the square $[a, b] \times [a, b]$, and the function f is continuous on $[a, b]$. Consider the operator

$$A(y)(x) = \int_a^b k(x, s)y(s)\, ds + f(x), \tag{4.10}$$

with the domain $D(A) = C[a, b]$. Since A denotes the operator, $A(y)$ denotes the function that is the image of function y under the mapping A. The symbol $A(y)(x)$ is used to denote the value of function $A(y)$ at number x. We shall prove that operator A is bounded.

It is obvious that $R(A) \subset M' = C[a, b]$. A simple calculation shows the boundedness of operator A. Let y_1, y_2 be two functions from $C[a, b]$; then

$$\varrho'(A(y_1), A(y_2)) = \max_{x \in [a,b]} \left| \int_a^b k(x, s)\, y_1(s)\, ds + f(x) - \int_a^b k(x, s)\, y_2(s)\, ds - f(x) \right|$$

$$\leq \max_{x \in [a,b]} \int_a^b |k(x, s)| \cdot |y_1(s) - y_2(s)|\, ds$$

$$\leq \max_{s \in [a,b]} |y_1(s) - y_2(s)| \times \max_{x \in [a,b]} \int_a^b |k(x, s)|\, ds$$

$$= K_\infty \cdot \varrho(y_1, y_2),$$

where

$$K_\infty = \max_{x \in [a,b]} \int_a^b | k(x, s) | \, ds.$$

Example 4.10. Let M, M', k, f, and ϱ be as in Example 4.9. Consider the operator

$$A(y)(x) = \int_a^x k(x, s) \, y(s) \, ds + f(x). \tag{4.11}$$

It can be proved, using the method employed in Example 4.9, that operator (4.11) is also bounded by the same constant as in Example 4.9.

We remark that operator (4.10) is called the *Fredholm integral operator* and operator (4.11) is called the *Volterra integral operator*.

4.5. Iterations of Operators

Let (M, ϱ) be a metric space, and let A be an operator with domain $D(A) \subset M$ and assume that its range $R(A)$ is a subset of $D(A)$. Then for arbitrary $x \in D(A)$ the point $A(A(x))$ is uniquely determined. The mapping $x \to A(A(x))$ is called the square of operator A and is denoted by A^2. The powers of A with a higher exponent can be defined recursively by the equation

$$A^m(x) = A(A^{m-1}(x)), \quad m \geq 2.$$

We shall make the easy verification that the boundedness of operator A implies the boundedness of operator A^m for all m. Consider the following inequality:

$$\varrho(A^m(x), A^m(y)) = \varrho(A(A^{m-1}(x)), A(A^{m-1}(y)))$$
$$\leq K\varrho(A^{m-1}(x), A^{m-1}(y)) \leq \cdots \leq K^m \varrho(x, y). \tag{4.12}$$

In many cases, a more accurate estimation can be given by using special properties of the specific operator being used.

Example 4.11. Consider the generalized version of operator (4.11):

$$A(y)(x) = \int_a^x k(x, s, y(s)) \, ds + f(x),$$

with domain

$$D(A) = \{y \mid y \in C[a, b] \text{ and for } x \in [a, b], Q_1 \leq y(x) \leq Q_2\},$$

where Q_1, Q_2 may be infinite, function k is continuous on the set $[a,b] \times [a,b] \times$

$[Q_1, Q_2]$, and function f is continuous on the interval $[a, b]$. Assume that for all $x, s \in [a, b]$, and $y_1, y_2 \in [Q_1, Q_2]$,

$$| k(x, s, y_1) - k(x, s, y_2) | \le L | y_1 - y_2 |, \qquad (4.13)$$

where L is a fixed constant.

Let us consider the operator A^m for $m \ge 1$. For $y_1, y_2 \in D(A)$ we have

$$\max_{a \le t \le x} | A^m(y_1)(t) - A^m(y_2)(t) |$$

$$\le \int_a^x \max_{\substack{a \le t \le s \\ a \le u \le b}} | k(u, t, A^{m-1}(y_1)(t)) - k(u, t, A^{m-1}(y_2)(t)) | \, ds$$

$$\le L \int_a^x \max_{a \le t \le s} | A^{m-1}(y_1)(t) - A^{m-1}(y_2)(t) | \, ds.$$

Introducing the notation

$$E_{y_1, y_2}^{(m)}(x) = \max_{a \le t \le x} | A^m(y_1)(t) - A^m(y_2)(t) |, \qquad m \ge 1,$$

we obtain the recursive inequality

$$E_{y_1, y_2}^{(m)}(x) \le L \int_a^x E_{y_1, y_2}^{(m-1)}(s) \, ds,$$

and since

$$E_{y_1, y_2}^{(1)}(x) \le L(x - a)\varrho(y_1, y_2),$$

by induction one can verify that for $m \ge 1$,

$$E_{y_1, y_2}^{(m)}(x) \le \frac{L^m(x - a)^m}{m!} \varrho(y_1, y_2),$$

which implies

$$\varrho(A^m(y_1), A^m(y_2)) \le \frac{L^m(b - a)^m}{m!} \varrho(y_1, y_2). \qquad (4.14)$$

Note that the expression on the right side is the mth term in the Taylor's series for $e^{L(b-a)}$. Thus, since this series converges, the mth term must tend to zero as $m \to \infty$, and so A^m is a contraction for m large enough that $L^m(b - a)^m/m!$ is less than 1.

Example 4.12. Let **B** be an $n \times n$ matrix with real or complex elements and assume that the magnitudes of the eigenvalues of **B** are less than 1. Observe that a sequence of matrices converges to the zero matrix if and only if the corresponding sequences of bounds K_∞, K_1, K_2 defined in Example 4.8 tend to zero. Let operator A be defined by $A(\mathbf{x}) = \mathbf{Bx}$; then obviously $A^m(\mathbf{x}) = \mathbf{B}^m\mathbf{x}$ for $m \ge 1$. Assuming that matrix **B** has the above eigenvalue property, none of the values K_∞, K_1, K_2 need necessarily be less than one, and consequently, in general, inequality (4.12) does not imply that $\mathbf{B}^m \to \mathbf{0}$ for $m \to \infty$. But we will prove that the above eigenvalue property of **B** is a necessary and sufficient condition for the convergence relation $\mathbf{B}^m \to \mathbf{0}$.

The proof will be accomplished in several stages.

(a) Let **T** be arbitrary invertible matrix of appropriate dimension, and let $\mathbf{C} = \mathbf{TBT}^{-1}$. The identities (which the reader may easily establish)

$$\mathbf{C}^m = \mathbf{TB}^m\mathbf{T}^{-1} \text{ and } \mathbf{B}^m = \mathbf{T}^{-1}\mathbf{C}^m\mathbf{T}$$

imply that $\mathbf{B}^m \to \mathbf{0}$ if and only if $\mathbf{C}^m \to \mathbf{0}$.

(b) Assume that matrix **B** is a hypermatrix of the form

$$\mathbf{B} = \begin{pmatrix} \mathbf{B}_1 & & & & 0 \\ & \mathbf{B}_2 & & & \\ & & \ddots & & \\ 0 & & & & \mathbf{B}_s \end{pmatrix}$$

Since

$$\mathbf{B}^m = \begin{pmatrix} \mathbf{B}_1{}^m & & & & 0 \\ & \mathbf{B}_2{}^m & & & \\ & & \ddots & & \\ 0 & & & & \mathbf{B}_s{}^m \end{pmatrix},$$

$\mathbf{B}^m \to \mathbf{0}$ if and only if $\mathbf{B}_i{}^m \to \mathbf{0}$ for $i = 1, 2, \ldots, s$.

(c) It is known from theory of matrices that arbitrary real or complex square matrix **B** can be transformed into the hypermatrix form

$$\mathbf{B} = \begin{pmatrix} \mathbf{J}_1 & & & & 0 \\ & \mathbf{J}_2 & & & \\ & & \ddots & & \\ 0 & & & & \mathbf{J}_s \end{pmatrix},$$

where the submatrices $\mathbf{J}_k$ are called *Jordan blocks* and are subdiagonal matrices of the form (see Wilkinson, 1965, pp. 9–12)

$$\mathbf{J}_j = \begin{pmatrix} \lambda_j & & & & 0 \\ 1 & \lambda_j & & & \\ & \ddots & \ddots & & \\ 0 & & & 1 & \lambda_j \end{pmatrix}$$

where the repeated diagonal element λ_j is an eigenvalue of **B**, and each eigenvalue can be found in at least one Jordan block. By induction, one can verify that

$$
\mathbf{J}_j{}^m =
\begin{pmatrix}
\lambda_j{}^m & & & & \\
\binom{m}{1}\lambda_j{}^{m-1} & \lambda_j{}^m & & & \\
\vdots & \vdots & \ddots & & \\
\binom{m}{k-1}\lambda_j{}^{m-k+1} & \binom{m}{k-2}\lambda_j{}^{m-k+2} & \cdots & \lambda_j{}^m
\end{pmatrix},
$$

where k is the order of submatrix $\mathbf{J}_j$. From examination of this form it is evident that the sequence of matrices $\{\mathbf{J}_j{}^m\}_m$ tends to $\mathbf{0}$ if and only if $|\lambda_j| < 1$.

By combining steps (a), (b), and (c) the proof is completed. ∎

Problems

4.4. Let $M - [0, \pi]$ with the metric $\varrho(x, y) = |x - y|$, and let $A(x) = \sin x$ with domain $[0, \pi]$. Is operator A a bounded operator?

4.5. Let $M = (-\infty, \infty)$ with the metric $\varrho = \varrho'$ given by

$$
\varrho(x, y) =
\begin{cases}
0, & \text{if } x = y, \\
1, & \text{if } x \neq y,
\end{cases}
$$

and let f be a function with domain $D(f) = M$ having the property

$$
|f(x)| \leq Q, \qquad \text{for all } x \in M,
$$

where Q is a fixed positive constant. Show that operator $A(x) = f(x)$ is bounded.

4.6. Let M be the set of differentiable functions on the interval $[a, b]$ and let $A(y)(x) = y'(x)$ for all $y \in M$. Show that operator A is not bounded if the metric is defined as in Example 4.4.

4.6. Fixed-Point Theorems

As we shall see in the following chapters, a great many equations to be solved by numerical techniques can be written in the form $x = A(x)$, where A is an operator with domain and range in the same metric space. Solutions of equations of this type are called *fixed points* for the corresponding operator A. The following theorem guarantees the existence and uniqueness of fixed points of certain operators defined on metric spaces and also gives a constructive algorithm for computing them. For that reason the following theorem will be very useful for the analysis of iterative algorithms.

Theorem 4.4. Assume that metric space (M, ϱ) is complete, and let M_1 be a closed subset of M. Assume that the operator A satisfies the following conditions:

(a) $\text{Domain}(A) = M_1$ and $\text{Range}(A) \subset M_1$;

(b) for $k = 1, 2, \ldots$, the operators A^k are bounded and majorized by the corresponding bounds K_k, such that the series $\sum_{k=1}^{\infty} K_k$ is convergent.

Then operator A has exactly one fixed point in M_1, and it can be found as the limit point of the sequence $\{x_i\}$ where,

$$
\begin{aligned}
x_1 &= A(x_0), \qquad x_0 \in M_1 \\
x_2 &= A(x_1) = A^2(x_0), \\
&\vdots \\
x_{n+1} &= A(x_n) = A^{n+1}(x_0), \\
&\vdots
\end{aligned}
\tag{4.15}
$$

Proof. The proof contains several steps. First we prove that sequence (4.15) is a Cauchy sequence. Using Definition 4.7. and equations (4.15), we have, for $m > n$,

$$
\begin{aligned}
\varrho(x_n, x_m) &\leq \varrho(x_n, x_{n+1}) + \varrho(x_{n+1}, x_{n+2}) + \cdots + \varrho(x_{m-1}, x_m) \\
&= \varrho\big(A^n(x_0), A^n(x_1)\big) + \varrho\big(A^{n+1}(x_0), A^{n+1}(x_1)\big) \\
&\quad + \cdots + \varrho\big(A^{m-1}(x_0), A^{m-1}(x_1)\big) \\
&\leq \varrho(x_0, x_1)[K_n + K_{n+1} + \cdots + K_{m-1}].
\end{aligned}
\tag{4.16}
$$

Since the numbers $K_k, k \geq 1$, are nonnegative, inequality (4.16) implies that, for every $m \geq n$,

$$
\varrho(x_n, x_m) \leq \varrho(x_0, x_1)[K_n + K_{n+1} + \cdots + K_{m-1} + \cdots].
\tag{4.17}
$$

Using the convergence of the infinite series $\sum_{k=1}^{\infty} K_k$ we conclude that the right-hand side of (4.17) tends to zero as $n \to \infty$.

Since metric space (M, ϱ) is complete and subset M_1 is closed, the fact that sequence $\{x_n\}_{n=0}^{\infty}$ is a Cauchy sequence implies that sequence $\{x_n\}_{n=0}^{\infty}$ is also convergent. Let the limit point be denoted by x^*.

Next we show that point x^* is a fixed point of A. Operator A is bounded; as a result it is also continuous and therefore both sides of the equation

$$
x_{n+1} = A(x_n)
$$

converge to a limit point. Thus for $n \to \infty$, we have

$$x^* = \lim x_{n+1} = \lim A(x_n) = A(x^*),$$

which proves that point x^* is a fixed point.

Toward proving the uniqueness of the fixed point, assume that both x^* and x^{**} are fixed points. Using the boundedness of the operators A^n, $n \geq 1$, we have

$$0 \leq \varrho(x^*, x^{**}) = \varrho\big(A(x^*), A(x^{**})\big) = \varrho\big(A^2(x^*), A^2(x^{**})\big)$$
$$= \cdots = \varrho\big(A^n(x^*), A^n(x^{**})\big) \leq K_n \varrho(x^*, x^{**})$$

for all values of n. Since the series $\sum_{n=1}^{\infty} K_n$ is convergent, $K_n \to 0$ for $n \to \infty$, which implies that $\varrho(x^*, x^{**}) = 0$; that is, $x^* = x^{**}$. ∎

Through inequality (4.17), the rate of the convergence $x_n \to x^*$ can be bounded. For $m \to \infty$ and fixed value of n, the inequality

$$\varrho(x_n, x^*) \leq \varrho(x_0, x_1) \sum_{k=n}^{\infty} K_k \qquad (4.18)$$

can be obtained. If point x_1 is known, inequality (4.18) implies that x_n will approximate x^* with error not exceeding ε, if n is large enough to satisfy

$$\sum_{k=n}^{\infty} K_k \leq \frac{\varepsilon}{\varrho(x_0, x_1)}. \qquad (4.19)$$

We remark that the value of n depends on the speed of convergence of the series $\sum_{k=1}^{\infty} K_k$. This inequality can be used to bound in advance the number of iterative calculations required to obtain a specified accuracy. Therefore, inequality (4.18) is called an *a priori error bound*.

Often the value of n obtained from (4.19) is much larger than the minimum number of steps until the error is actually less than ε. This "overestimation" supports the use of the *a posteriori error bounds*, which can be applied during the course computations. Thus by calculating the distances between successive terms, we can estimate the distance of the actual approximation and the fixed point. An *a posteriori upper bound* for the sequence (4.15) can be obtained from (4.18) at the nth stage by replacing x_0 and x_1 with x_{n-1} and x_n, respectively. Thus

$$\varrho(x_n, x^*) \leq \varrho(x_{n-1}, x_n) \sum_{k=1}^{\infty} K_k. \qquad (4.20)$$

This implies that $\varrho(x_n, x^*)$ will not be greater than ε if

$$\varrho(x_{n-1}, x_n) \leq \frac{\varepsilon}{\sum_{k=1}^{\infty} K_k}. \qquad (4.21)$$

In practical applications the points $A(x)$, $x \in M_1$, can only be approximated. For this reason the terms of the sequence (4.15) are approximate. We now estimate the error that arises due to inexact computation of functional values $A(x)$.

In place of the sequence determined by

$$x_{n+1} = A(x_n),$$

in the approximate calculations we compute the sequence

$$z_{n+1} = \tilde{A}(z_n),$$

where $z_0 = x_0$ and operator $\tilde{A}$ is an approximation of operator A. Assume that $D(\tilde{A}) = M_1$, $R(\tilde{A}) \subset M_1$, and for all $x \in M_1$,

$$\varrho\big(A(x), \tilde{A}(x)\big) \leq \delta, \qquad (4.22)$$

where δ is a fixed positive number.

We prove by induction that for $n = 1, 2, \ldots,$

$$\varrho(x_n, z_n) \leq \delta[1 + K_1 + K_1^2 + \cdots + K_1^{n-1}], \qquad (4.23)$$

where K_1 is the bound for operator A.

For $n = 1$, (4.23) is true because using inequality (4.22) we have

$$\varrho(x_1, z_1) = \varrho\big(A(x_0), \tilde{A}(z_0)\big) = \varrho\big(A(x_0), \tilde{A}(x_0)\big) \leq \delta = \delta \times 1.$$

Assuming that (4.23) is true for n, we have

$$\begin{aligned}
\varrho(x_{n+1}, z_{n+1}) &= \varrho\big(A(x_n), \tilde{A}(z_n)\big) \\
&\leq \varrho\big(A(x_n), A(z_n)\big) + \varrho\big(A(z_n), \tilde{A}(z_n)\big) \\
&\leq K_1 \varrho(x_n, z_n) + \delta \\
&\leq K_1 \delta[1 + K_1 + K_1^2 + \cdots + K_1^{n-1}] + \delta \\
&= \delta[1 + K_1 + K_1^2 + \cdots + K_1^n],
\end{aligned}$$

which completes the proof.

Note that (4.23) may be rewritten

$$\varrho(x_n, z_n) \leq \begin{cases} \delta(1 - K_1^n)/(1 - K_1), & K_1 \neq 1, \\ n\delta, & K_1 = 1. \end{cases}$$

Inequality (4.23) intimates that the error arising from the use of an approximate operator $\tilde{A}$ can be arbitrary large for $K_1 \geq 1$. In other words, the effect of a small perturbation in the computation of the terms of the iteration sequence can be arbitrary large at later times. Such behavior is called *numerical instability*.

As a special case of Theorem 4.4 we can prove the following useful fact.

Theorem 4.5. (Contraction Mapping Theorem) If metric space (M, ϱ) is complete, M_1 is a closed subset in M, and operator A, having domain $D(A) = M_1$ and range $R(A) \subset M_1$, is a contraction, then there is a unique fixed point of A in M_1, and it can be obtained as the limit point of the sequence with an arbitrary initial point $x_0 \in M_1$:

$$
\begin{aligned}
x_1 &= A(x_0), \\
x_2 &= A(x_1), \\
&\;\;\vdots \\
x_{n+1} &= A(x_n), \\
&\;\;\vdots
\end{aligned}
\tag{4.24}
$$

Proof. It is enough to prove that the conditions of Theorem 4.4. are satisfied. For $k = 1, 2, 3, \ldots$ and $x, y \in M_1$, inequality (4.12) implies

$$
\varrho\big(A^k(x), A^k(y)\big) \leq q^k \varrho(x, y),
$$

where q is the bound of A, which is less than one. Thus we can choose $K_k = q^k$, and consequently

$$
\sum_{k=1}^{\infty} K_k = \sum_{k=1}^{\infty} q^k = \frac{q}{1-q}.
$$

The other conditions are obviously satisfied. ∎

In this special but important case, the error estimates (4.18), (4.20), (4.21), and (4.23) have simple forms. The a priori error bound for approximation x_n is given by

$$
\varrho(x_n, x^*) \leq \varrho(x_0, x_1) \frac{q^n}{1-q},
\tag{4.25}
$$

and the number of steps n needed to ensure error less than or equal to ε can be calculated from

$$
\varrho(x_0, x_1) \frac{q^n}{1-q} \leq \varepsilon,
$$

which is equivalent to the inequality

$$n \geq \log \left[\frac{\varepsilon(1-q)}{\varrho(x_0, x_1)} \right] \Big/ \log q. \tag{4.26}$$

The a posteriori error bound for x_n is determined by

$$\varrho(x_n, x^*) \leq \varrho(x_{n-1}, x_n) \frac{q}{1-q}, \tag{4.27}$$

and the condition for assuring that the error does not exceed ε is the inequality

$$\varrho(x_{n-1}, x_n) \leq \frac{\varepsilon(1-q)}{q}.$$

Finally, the estimation (4.23) in this special case has the form

$$\varrho(x_n, z_n) \leq \frac{\delta}{1-q}, \tag{4.28}$$

which shows the numerical stability of algorithm (4.24).

Example 4.13. Let $M = M_1 = R^1$ for $x, y \in M$, let $\varrho(x, y) = |x - y|$, and for $x \in M_1$, let $A(x) = qx$, where q is a constant satisfying $0 \leq q < 1$. In this case $x^* = 0$ and for arbitrary $x_0 \in M_1$, we have

$$x_n = qx_{n-1} = q^2 x_{n-2} = \cdots = q^n x_0,$$

which implies

$$\varrho(x_n, x^*) = |x_n - x^*| = |q^n x_0| = \frac{q^n}{1-q} |x_0 - x_0 q|$$
$$= \frac{q^n}{1-q} \varrho(x_0, x_1), \tag{4.29}$$

and

$$\varrho(x_n, x^*) = |x_n - x^*| = |q^n x_0| = \frac{q}{1-q} |q^{n-1} x_0 - q^n x_0|$$
$$= \frac{q}{1-q} \varrho(x_{n-1}, x_n). \tag{4.30}$$

For $x \in M_1$ let $\tilde{A}(x) = A(x) + \delta = qx + \delta$, where δ is a fixed positive constant; then

$$z_0 = x_0,$$
$$z_1 = qz_0 + \delta = qx_0 + \delta,$$
$$z_2 = qz_1 + \delta = q^2 x_0 + \delta(1 + q),$$
$$z_3 = qz_2 + \delta = q^3 x_0 + \delta(1 + q + q^2),$$

and generally it is easy to verify that for arbitrary $n \geq 1$,

$$z_n = q^n x_0 + \delta(1 + q + q^2 + \cdots + q^{n-1}),$$

which implies

$$\varrho(x_n, z_n) = \delta(1 + q + q^2 + \cdots + q^{n-1}). \tag{4.31}$$

Equations (4.29)–(4.31) show that the error bounds for algorithm (4.15) cannot be improved.

The following theorem is a simple consequence of Theorem 4.5.

Theorem 4.6. Let (M, ϱ) be a complete metric space. Assume that operator A with domain $D(A) \subset M$ and range $R(A) \subset M$ has a fixed point $x^* \in M$. Assume that operator A is a contraction in the set $\bar{G}(x^*, r) \subset D(A)$, where $r > 0$ is a fixed positive number. Then x^* is the only fixed point in $\bar{G}(x^*, r)$ and it can be obtained as the limit point of the sequence $x_{n+1} = A(x_n)$, where x_0 is an arbitrary point in $\bar{G}(x^*, r)$.

Proof. We prove that for $M_1 = \bar{G}(x^*, r)$ the conditions of Theorem 4.5 are satisfied. Since (M, ϱ) is a complete metric space, M_1 is a closed subset of M, $M_1 \subset D(A)$, and A is a contraction on M_1, it is sufficient to prove that for $x \in M_1$, $A(x) \in M_1$. This follows immediately from the inequality

$$\varrho(A(x), x^*) = \varrho(A(x), A(x^*)) \leq q\varrho(x, x^*) \leq qr < r. \qquad \blacksquare$$

For applying Theorem 4.4, all the operators A^n, $n \geq 1$, must be examined. In this regard, we show that if operator A is bounded, and for some integer $n \geq 1$ the operator A^n is a contraction, then the constants K_v, $v \geq 1$, can be chosen so that the series $\sum_{v=1}^{\infty} K_v$ is convergent.

Observe that for $m = nk + r$ and $x, y \in D(A)$,

$$\varrho(A^m(x), A^m(y)) = \varrho((A^n)^k(A^r(x)), (A^n)^k(A^r(y)))$$
$$\leq q^k \varrho(A^r(x), A^r(y)) \leq q^k K_1^r \varrho(x, y),$$

where K_1 is the bound for A and q is the bound for A^n, which is hypothesized to be less than 1. Consequently K_m can be chosen as $q^k K_1^r$. This observation implies that

$$\sum_{k=1}^{\infty} K_k = \sum_{k=1}^{\infty} \left(q^k \sum_{r=0}^{n-1} K_1^r \right) = \left(\sum_{r=0}^{n-1} K_1^r \right) \sum_{k=1}^{\infty} q^k$$
$$= \left(\sum_{r=0}^{n-1} K_1^r \right) \frac{q}{1 - q}.$$

Problems

4.7. Dropping the condition of completeness of metric space (M, ϱ) and keeping the other conditions of Theorem 4.5, construct an example for which no fixed point exists.

4.8. Dropping the condition that subset M_1 is closed and keeping the other conditions of Theorem 4.5, give an example for which no fixed point exists.

4.9. Construct an operator that is bounded but its bound is equal to one and, keeping all other properties of Theorem 4.5, give examples for which (a) no fixed point exists, (b) infinitely many fixed points exist.

4.7. Systems of Operator Equations

Consider the metric spaces (M_{i0}, ϱ_i) for $i = 1, 2, \ldots, n$, and assume that these metric spaces are complete. Let M be the set of vectors $\mathbf{x} = (x_1, \ldots, x_n)$, where $x_i \in M_{i0}$, $1 \leq i \leq n$, and define a metric for M by the equation

$$\varrho(\mathbf{x}, \mathbf{y}) = \max_{1 \leq i \leq n} \varrho_i(x_i, y_i), \qquad (4.32)$$

where the numbers y_i denote the components of vector $\mathbf{y}$. It is easy to verify that (M, ϱ) is a complete metric space.

For $i = 1, 2, \ldots, n$, let M_{i1} be closed subset of M_{i0} and let

$$M_1 = \{\mathbf{x} \mid \mathbf{x} \in M, \text{ and } x_i \in M_{i1}\}, \qquad \text{for } 1 \leq i \leq n,$$

which is obviously a closed subset of M. A sequence of vectors $\mathbf{x}_m = (x_1^{(m)}, \ldots, x_n^{(m)})$ converges to a vector $\mathbf{x} = (x_1, \ldots, x_n)$ if and only if for $i = 1, 2, \ldots, n$ the sequences $x_i^{(m)}$ tend to x_i.

Let A_i, $1 \leq i \leq n$, be an operator with domain $D(A_i) = M_1$ and range $R(A_i) \subset M_{i1}$, and introduce the following operator:

$$A(\mathbf{x}) = \big(A_1(\mathbf{x}), \ldots, A_n(\mathbf{x})\big),$$

which has domain $D(A) = M_1$ and range $R(A) \subset M_1$.

Assume furthermore that the operators A_i satisfy the inequalities

$$\varrho_i\big(A_i(\mathbf{x}), A_i(\mathbf{y})\big) \leq \sum_{k=1}^{n} a_{ik}\varrho_k(x_k, y_k), \qquad \mathbf{x}, \mathbf{y} \in M_1, \qquad (4.33)$$

where the numbers a_{ik} are fixed positive constants. Under the above condition, operator A is bounded, as the following shows:

$$\varrho(A(\mathbf{x}), A(\mathbf{y})) = \max_i \varrho_i(A_i(\mathbf{x}), A_i(\mathbf{y}))$$

$$\leq \max_i \sum_{k=1}^{n} a_{ik} \varrho_k(x_k, y_k)$$

$$\leq \varrho(\mathbf{x}, \mathbf{y}) \max_i \sum_{k=1}^{n} a_{ik}$$

$$= \varrho(\mathbf{x}, \mathbf{y}) K, \quad \text{for } K = \max_i \sum_{k=1}^{n} a_{ik}.$$

Assume that $K < 1$. Then metric space (M, ϱ) and operator A satisfy the conditions of the contraction mapping theorem (Theorem 4.5). Consequently operator A has a unique fixed point $\mathbf{x}^*$ in M_1 that can be determined as the limit point of sequence $\mathbf{x}^{(k+1)} = A(\mathbf{x}^{(k)})$, where $\mathbf{x}^{(0)} \in M_1$ is arbitrary. This sequence can be written also in the form

$$x_i^{(k+1)} = A_i(x_1^{(k)}, x_2^{(k)}, \ldots, x_n^{(k)}), \quad 1 \leq i \leq n, \quad (4.34)$$

where $x_i^{(k)}$ and $x_i^{(k+1)}$ denote the ith components of vectors $\mathbf{x}^{(k)}$ and $\mathbf{x}^{(k+1)}$.

In representation (4.34), for computation of $x_i^{(k+1)}$ we use the components of the previous approximation $\mathbf{x}^{(k)}$. But observe that the new components $x_1^{(k+1)}, \ldots, x_{i-1}^{(k+1)}$ are also known. Replacing the coordinates $x_1^{(k)}, \ldots, x_{i-1}^{(k)}$ by the terms $x_1^{(k+1)}, \ldots, x_{i-1}^{(k+1)}$, we have the iteration scheme, for $i = 1, 2, \ldots, n$, given by

$$x_i^{(k+1)} = A_i(x_1^{(k+1)}, \ldots, x_{i-1}^{(k+1)}, x_i^{(k)}, \ldots, x_n^{(k)}). \quad (4.35)$$

This method is called *Seidel-type* iteration. We indicate that iteration (4.35) usually converges to the fixed point faster than method (4.34).

First we observe that iteration (4.35) is identical with the iteration of the form (4.24) applied to the operator $\tilde{A}$ corresponding to the mapping $\mathbf{x} \to \mathbf{y}$ defined as follows:

$$y_1 = A_1(x_1, x_2, \ldots, x_n),$$

$$y_2 = A_2(y_1, x_2, \ldots, x_n),$$

$$\vdots$$

$$y_i = A_i(y_1, \ldots, y_{i-1}, x_i, \ldots, x_n), \quad (4.36)$$

$$\vdots$$

$$y_n = A_n(y_1, y_2, \ldots, y_{n-1}, x_n).$$

It is sufficient to prove that the operator corresponding to the mapping (4.36) is a contraction and the bound of the operator is not greater than K, the bound for A. Consider the ith equation of (4.36). The definition of metric ϱ implies that

$$\varrho\big(\tilde{A}(\mathbf{x}^{(1)}), \tilde{A}(\mathbf{x}^{(2)})\big) = \max_i \varrho_i(y_i^{(1)}, y_i^{(2)}), \qquad (4.37)$$

for all $\mathbf{x}^{(1)}, \mathbf{x}^{(2)} \in M_1$, where $\mathbf{y}^{(1)} = \tilde{A}(\mathbf{x}^{(1)})$ and $\mathbf{y}^{(2)} = \tilde{A}(\mathbf{x}^{(2)})$. Let us now assume that vectors $\mathbf{x}^{(1)}$ and $\mathbf{x}^{(2)}$ are fixed.

Inequality (4.33) implies that for all i,

$$\varrho_i(y_i^{(1)}, y_i^{(2)}) \leq \sum_{j=1}^{i-1} a_{ij}\varrho_j(y_j^{(1)}, y_j^{(2)}) + \sum_{j=i}^{n} a_{ij}\varrho_j(x_j^{(1)}, x_j^{(2)}). \qquad (4.38)$$

Let

$$\alpha_i = \sum_{j=1}^{i-1} a_{ij}, \qquad \beta_i = \sum_{j=i}^{n} a_{ij},$$

and assume that the maximizing index of formula (4.37) is at $i = i_0$. Then from (4.37) and (4.38) we have

$$\varrho\big(\tilde{A}(\mathbf{x}^{(1)}), \tilde{A}(\mathbf{x}^{(2)})\big) = \varrho(\mathbf{y}^{(1)}, \mathbf{y}^{(2)}) = \varrho_{i_0}(y_{i_0}^{(1)}, y_{i_0}^{(2)})$$
$$\leq \alpha_{i_0}\varrho(\mathbf{y}^{(1)}, \mathbf{y}^{(2)}) + \beta_{i_0}\varrho(\mathbf{x}^{(1)}, \mathbf{x}^{(2)}).$$

Since $1 - \alpha_{i_0} \geq 1 - K > 0$, then if we introduce the notation

$$K' = \max_i \frac{\beta_i}{1 - \alpha_i} = \frac{\beta_{i^*}}{1 - \alpha_{i^*}},$$

where i^* is the maximizing index, we have

$$\varrho\big(\tilde{A}(\mathbf{x}^{(1)}), \tilde{A}(\mathbf{x}^{(2)})\big) \leq \frac{\beta_{i_0}}{1 - \alpha_{i_0}} \varrho(\mathbf{x}^{(1)}, \mathbf{x}^{(2)})$$
$$\leq K'\varrho(\mathbf{x}^{(1)}, \mathbf{x}^{(2)}).$$

Finally we prove that $K' \leq K$. Simple calculation shows that

$$K - K' = \max_i (\alpha_i + \beta_i) - \max_i \left\{ \frac{\beta_i}{1 - \alpha_i} \right\}$$
$$\geq \alpha_{i^*} + \beta_{i^*} - \frac{\beta_{i^*}}{1 - \alpha_{i^*}} = \frac{\alpha_{i^*}(1 - \alpha_{i^*} - \beta_{i^*})}{1 - \alpha_{i^*}}$$
$$\geq \frac{\alpha_{i^*}(1 - K)}{1 - \alpha_{i^*}} \geq 0,$$

which completes the proof.

4.8. Norms of Vectors and Matrices

Distances of vectors are defined in Example 4.3. The metric measures the distance between two vectors, and so it is not explicitly the measure of the magnitude of a single vector. However, in analogy to the definition of the absolute values of real numbers being their distance from zero, the length of a vector $\mathbf{x}$, which is termed the norm of $\mathbf{x}$, can sometimes be defined to be the distance $\varrho(\mathbf{x}, \mathbf{0})$ of $\mathbf{x}$ from the $\mathbf{0}$ vector.

However, norms are defined in such a way as to preserve the scaling property and the triangle inequality of Euclidean distance, and not all metric functions $\varrho(\mathbf{x}, \mathbf{0})$ satisfy this definition.

Definition 4.9. A *norm* for a vector space M is any real-valued function on M, denoted by $\| \mathbf{x} \|$, $\mathbf{x} \in M$, which satisfies the following three properties:

(a) $\| \mathbf{x} \| \geq 0$ and $\| \mathbf{x} \| = 0$ if and only if $\mathbf{x} = \mathbf{0}$;

(b) for an arbitrary real or complex number c,

$$\| c\mathbf{x} \| = | c | \, \| \mathbf{x} \|;$$

(c) $\| \mathbf{x} + \mathbf{y} \| \leq \| \mathbf{x} \| + \| \mathbf{y} \|.$

We now show that the vector metrics introduced in Example 4.3 determine norms.

Theorem 4.7. Let

$$\| \mathbf{x} \|_\infty = \varrho_\infty(\mathbf{x}, \mathbf{0}) = \max_{1 \leq i \leq n} | x_i |,$$

$$\| \mathbf{x} \|_1 = \varrho_1(\mathbf{x}, \mathbf{0}) = \sum_{i=1}^{n} | x_i |,$$

$$\| \mathbf{x} \|_2 = \varrho_2(\mathbf{x}, \mathbf{0}) = \left\{ \sum_{i=1}^{n} | x_i |^2 \right\}^{1/2}, \tag{4.39}$$

where $\mathbf{x}$ is a real or complex vector of the dimension n.

The functions $\| \mathbf{x} \|_\infty$, $\| \mathbf{x} \|_1$, and $\| \mathbf{x} \|_2$ are vector norms.

Proof. Properties (a) and (b) follow immediately from the corresponding properties of distances in metric spaces. Property (c) can be proved as follows:

The reader may easily verify that if ϱ is either ϱ_∞, ϱ_1, or ϱ_2 from Example 4.3, then

$$\varrho(\mathbf{x} + \mathbf{y}, \mathbf{0}) = \varrho(\mathbf{x}_1, -\mathbf{y}).$$

Consequently we have that

$$
\begin{aligned}
\| \mathbf{x} + \mathbf{y} \| = \varrho(\mathbf{x} + \mathbf{y}, \mathbf{0}) = \varrho(\mathbf{x}, -\mathbf{y}) &\leq \varrho(\mathbf{x}, \mathbf{0}) + \varrho(\mathbf{0}, -\mathbf{y}) \\
&= \varrho(\mathbf{x}, \mathbf{0}) + \varrho(-\mathbf{y}, \mathbf{0}) = \| \mathbf{x} \| + \| -\mathbf{y} \| \\
&= \| \mathbf{x} \| + \| \mathbf{y} \|. \qquad \blacksquare
\end{aligned}
$$

Consider, however, the discrete metric $\varrho(x, y) = 0$ if $x = y$ and 1 otherwise, introduced in Problem 4.1. It is clear that the function $\varrho(x, 0)$ fails to possess property (b) of the definition of "norm."

It is useful to define norms for matrices. The definition is closely connected to the idea of bounded operators.

Definition 4.10. A *norm* for the set $\mathscr{S}$ of nth-order matrices with real or complex coefficients is any real-valued function on $\mathscr{S}$, denoted by $\| \mathbf{A} \|$, $\mathbf{A} \in \mathscr{S}$, having the following four properties:

(a) $\| \mathbf{B} \| \geq 0$ and $\| \mathbf{B} \| = 0$ if and only if $\mathbf{B} = \mathbf{0}$;

(b) for arbitrary real or complex number c,

$$\| c\mathbf{B} \| = | c | \, \| \mathbf{B} \|;$$

(c) $\| \mathbf{B} + \mathbf{C} \| \leq \| \mathbf{B} \| + \| \mathbf{C} \|$;

(d) $\| \mathbf{BC} \| \leq \| \mathbf{B} \| \, \| \mathbf{C} \|$.

Note that this definition differs from that of vector norms only in property (d). The next theorem gives us some important matrix norms.

Theorem 4.8. Let $\mathbf{B}$ denote an nth-order matrix with real or complex coefficients $\{b_{ij}\}$. The functions $\| \mathbf{B} \|_\infty$, $\| \mathbf{B} \|_1$, and $\| \mathbf{B} \|_2$ determined by the formulas

$$\| \mathbf{B} \|_\infty = \max_{1 \leq i \leq n} \sum_{j=1}^{n} | b_{ij} |,$$

$$\| \mathbf{B} \|_1 = \max_{1 \leq j \leq n} \sum_{i=1}^{n} | b_{ij} |, \qquad (4.40)$$

$$\| \mathbf{B} \|_2 = \left\{ \sum_{i=1}^{n} \sum_{j=1}^{n} | b_{ij} |^2 \right\}^{1/2}.$$

are all matrix norms.

Proof. It is obvious that properties (a) and (b) are satisfied. Properties (c) and (d) can be proved by simple calculations. For example, property (d) will be proved for the norm $\| \cdot \|_2$.

Since the jth element of the ith row of matrix **BC** is equal to $\sum_{k=1}^{n} b_{ik} c_{kj}$, using the Cauchy–Schwarz (Taylor, 1966, p. 5) inequality we have

$$\| \mathbf{BC} \|_2^2 = \sum_{i=1}^{n} \sum_{j=1}^{n} \left| \sum_{k=1}^{n} b_{ik} c_{kj} \right|^2$$

$$\leq \sum_{i=1}^{n} \sum_{j=1}^{n} \left(\sum_{k=1}^{n} | b_{ik} | \, | c_{kj} | \right)^2$$

$$\leq \sum_{i=1}^{n} \sum_{j=1}^{n} \left(\sum_{k=1}^{n} | b_{ik} |^2 \sum_{l=1}^{n} | c_{lj} |^2 \right)$$

$$= \sum_{i=1}^{n} \sum_{j=1}^{n} \sum_{k=1}^{n} \sum_{l=1}^{n} | b_{ik} |^2 | c_{lj} |^2$$

$$= \left(\sum_{i=1}^{n} \sum_{k=1}^{n} | b_{ik} |^2 \right) \left(\sum_{j=1}^{n} \sum_{l=1}^{n} | c_{lj} |^2 \right)$$

$$= \| \mathbf{B} \|_2^2 \, \| \mathbf{C} \|_2^2. \qquad \blacksquare$$

Corollary 4.1. The successive application of property (d) implies that $\| \mathbf{B}^k \| \leq \| \mathbf{B} \|^k$ for arbitrary square matrix **B** and positive integer k.

In studying iterative processes, it is useful to use vector and matrix norms jointly. A vector norm $\| \cdot \|_v$ and a matrix norm $\| \cdot \|_m$ are said to be *compatible* if for all vectors $\mathbf{x} \in M$, and matrices $\mathbf{A} \in \mathscr{S}$,

$$\| \mathbf{Ax} \|_v \leq \| \mathbf{A} \|_m \| \mathbf{x} \|_v.$$

If we have compatible norms, then from property (d) of Definition 4.10, it is evident that the useful inequalities

$$\| \mathbf{ABx} \|_v \leq \| \mathbf{A} \|_m \| \mathbf{B} \|_m \| \mathbf{x} \|_v,$$

$$\| \mathbf{A}^k \mathbf{x} \|_v \leq \| \mathbf{A} \|_m^k \| \mathbf{x} \|_v, \qquad k = 1, 2, \ldots,$$

hold. With respect to the vector and matrix norms defined in Theorems 4.7 and 4.8, we have the result to follow.

Theorem 4.9. The vector norms $\| \cdot \|_\infty$, $\| \cdot \|_1$, $\| \cdot \|_2$ are compatible with the matrix norms $\| \cdot \|_\infty$, $\| \cdot \|_1$, $\| \cdot \|_2$, respectively.

Proof. The assertion of the theorem follows from the boundedness of the operator $A(\mathbf{x}) = \mathbf{Bx}$, as discussed in Example 4.8. Using the fact that the matrix norms in each of the three cases are bounds of the operators, we have

$$\| \mathbf{Bx} \| = \varrho(\mathbf{Bx}, \mathbf{0}) = \varrho(\mathbf{Bx}, \mathbf{B0}) = \varrho\big(A(\mathbf{x}), A(\mathbf{0})\big)$$
$$\leq \| \mathbf{B} \| \varrho(\mathbf{x}, \mathbf{0}) = \| \mathbf{B} \| \| \mathbf{x} \|.$$ ∎

We note that the matrix norm $\| \cdot \|_2$ is termed the *Frobenius* norm.

Problems

4.10. Apply the result of Section 4.7 to find a condition or the convergence of the sequence

$$x_n = f(x_{n-1}, \ldots, x_{n-k}),$$

where $x_0, \ldots, x_{k-1}$ are suitable initial approximations and f is a k-variate real function.

4.11. Prove property (c) of Theorem 4.7 and properties (c) and (d) of Theorem 4.8 for all three norms.

4.12. Prove that all eigenvalues λ_i of matrix $\mathbf{B}$ satisfy the inequality $| \lambda_i | \leq \| \mathbf{B} \|$, where $\| \mathbf{B} \|$ is an arbitrary norm of the matrix $\mathbf{B}$.

4.13. Prove that the property $\| \mathbf{B} \| < 1$ implies that

$$\mathbf{B}^m \to \mathbf{0} \text{ for } m \to \infty.$$

4.9. The Order of Convergence of an Iteration Process

Let (M, ϱ) be a metric space. An operator A for (M, ϱ) with range $R(A) \subset M$ and an arbitrary element $x_0 \in M$ having been specified, the sequence $\{x_i\}$ determined recursively by $x_{i+1} = A(x_i)$ is called an *iteration process*. If $\{x_n\}$ converges to a point $x^* \in M$ and if there exists a maximal positive number t such that for some number K and for $n = 1, 2, \ldots$

$$\varrho(x_n, x^*) \leq K\varrho(x_{n-1}, x^*)^t, \tag{4.41}$$

where the constant K may depend on x_0, then we say that the *order of the iteration method* A is equal to t.

For $t = 1$, the convergence is called *linear*, and in the special case $t = 2$ the convergence is called *quadratic*. The number K is called the *coefficient of convergence*. Its value in the linear convergence case is of particular significance in judging how fast convergence occurs.

Observe that the order of the method (4.15) satisfying the conditions of Theorem 4.4 is at least one. If the iteration process obtained by application of a contraction is convergent, then the order of convergence is at least linear, and the bound q suffices as a coefficient of convergence.

In connection with convergence, if f and g are real-valued functions defined on the same metric space (M, ϱ), one writes that

$$f(x) = A + O(g(x))$$

at x^* if there exist constants K_1 and K_2 such that if $\varrho(x, x^*) < K_1$, then

$$| f(x) - A | < K_2 | g(x) |.$$

Thus, for example, if (M, ϱ) is as in Example 4.2, then we may confirm that

$$\cos x = 1 + O(x^2), \qquad \sin x = O(x)$$

at $x^* = 0$. Furthermore, in this notation, if operator A is an iteration method of order t at a fixed point x^*, then we may write that

$$\varrho(A(x), x^*) = O(\varrho(x, x^*)^t).$$

4.10. Inner Products

In Chapter 9, we shall repeatedly use the defining property of inner products. Their use elsewhere will be more self-evident, and so this section is only necessary reading in preparation for Chapter 9.

Let M denote a vector space (such as the set of real-valued n-tuples) over the field of real numbers.

Definition 4.11. An *inner product* is a real-valued function, denoted by $(\mathbf{x}, \mathbf{y})$, $\mathbf{x}, \mathbf{y} \in M$, defined on $M \times M$ such that if a_1, a_2, b_1, b_2 denote the scalars and if $\mathbf{x}, \mathbf{y}, \mathbf{w}, \mathbf{t} \in M$, then $(\mathbf{x}, \mathbf{x}) > 0$ if $\mathbf{x} \neq \mathbf{0}$, $(\mathbf{0}, \mathbf{0}) = 0$, $(\mathbf{x}, \mathbf{y}) = (\mathbf{y}, \mathbf{x})$, and

$$(a_1\mathbf{x} + a_2\mathbf{y}, b_1\mathbf{w} + b_2\mathbf{t}) = a_1b_1(\mathbf{x}, \mathbf{w}) + a_1b_2(\mathbf{x}, \mathbf{t}) + a_2b_1(\mathbf{y}, \mathbf{w}) + a_2b_2(\mathbf{y}, \mathbf{t}).$$

The most well-known inner product is the *Cartesian* inner product. Let M be the vector space of n-tuples. Take

$$(\mathbf{x}, \mathbf{y}) = \sum_{i=1}^{n} x_i y_i, \tag{4.42}$$

where $\mathbf{x}, \mathbf{y} \in M$ and x_j, y_j denote the jth coordinates of $\mathbf{x}$ and $\mathbf{y}$, respectively.

An inner product needed for developments in Chapter 9 is

$$(f, g) = \int_a^b f(x)g(x)\, dx, \qquad (4.43)$$

where M is $C[a, b]$.

It is an easy matter to prove that if $(\cdot, \cdot)$ is an inner product, then $\| \mathbf{x} \| = [(\mathbf{x}, \mathbf{x})]^{1/2}$ is a vector norm.

Problems

4.14. Show that the formulas (4.42) and (4.43) are inner products.

4.15. Prove that if $(\cdot, \cdot)$ is an inner product, then $\| \mathbf{x} \| = [(\mathbf{x}, \mathbf{x})]^{1/2}$ is a vector norm.

4.11. Supplementary Notes and Discussion

Ideally, those wishing to do advanced work on numerical analysis subjects should have a solid grounding in the mathematical discipline known as functional analysis (e.g., Taylor, 1964; Lang, 1969). Functional analysis, in turn, has as prerequisite, the basic subject matter of abstract linear algebra as well as measure theory. The fact of the matter is that functional analysis is the language and medium for much of the research-level discourse on iteration methods for numerical analysis.

Realistically, however, one cannot expect that the many engineers and applied and industrial mathematicians who need to use and develop numerical methods at an advanced level have the time or inclination necessary to obtain a solid grounding in functional analysis. We believe that with only a modicum of functional analysis concepts, readers with some energy and a solid background in undergraduate level mathematics subjects will be in a position to make worthwhile and respectable contributions to the practice and literature of numerical analysis.

The purpose of this chapter has been to supply this minimal collection of functional analysis concepts. Additionally, the material covered here is basic to our exposition. In the remaining chapters, we shall continually be referring to results and examples of this chapter, particularly to those examples that involve operations on spaces of functions and to the contraction mapping theorem.

Among the comprehensive numerical analysis books that make extensive use of functional analysis developments, we single out Blum (1972) as being particularly successful. This book is largely self-contained, but it is not easy reading for a person not already exposed to functional analysis. The book by Collatz (1966) entitled *Functional Analysis and Numerical Mathematics* is devoted to exploiting the connection between functional analysis and numerical analysis. The book edited by Swaminathan (1976) gives recent developments in the application of the theory of fixed points to numerical analysis. The research monograph by Varga (1976) is a systematic application of functional analysis and recent developments in the theory of splines to various numerical analysis subjects.

5

Solution of Nonlinear Equations

This chapter is devoted to solving equations of the form

$$f(x) = 0.$$

To begin with, f will be a real function. In later sections, we shall generalize to the vector-valued, multivariable case.

Only in a few very special cases are there direct formulas (formulas giving an exact answer, in the absence of roundoff, after finitely many iterations) for solving equations of the above type. For even in the case of the lowly polynomial, a famous result based on the work of Galois (Herstein, 1964, Section 5.6) is that direct methods for finding roots of polynomials of degree greater than four do not exist. In view of the limitations of direct methods, therefore, all the techniques given in this chapter are iterative, and the methods and results of Chapter 4 will serve us well.

5.1. Equations in One Variable

5.1.1. The Bisection Method

Assume that function f is continuous on an interval $[a, b]$ and that $f(a) > 0$ and $f(b) < 0$. Then the theorem of Bolzano (see Apostol, 1957, pp. 73–74) implies that there is at least one value of x in (a, b) such that $f(x) = 0$. Such a point x is said to be a *zero* (or *root*) of the function f.

Let s be the midpoint of interval $[a, b]$. If $f(s) = 0$, then s is a zero of the function f, and if $f(s) \neq 0$, then one of the inequalities holds: $f(s)f(a) < 0$ or $f(s)f(b) < 0$. In the first case the interval (a, s) contains at least one zero of function f. In the second case, at least one zero is in (s, b). By repeating the above step successively with $[a, s]$ or $[s, b]$ now playing the role of $[a, b]$ we may approximate a zero of f as closely as we wish. This approximation process is called the *bisection method*. We illustrate its application in Figure 5.1, in which x_2, x_3, x_4 denote successive midpoints.

After n steps, one zero of f will be located within an interval with length

$$\frac{b - a}{2^n}$$

and so both endpoints of the interval constructed at the nth step approximate at least one zero of function f with error not exceeding $(b - a)/2^n$. In the terminology of Section 4.9, the convergence of the bisection method is linear and the coefficient of convergence is $1/2$. If an a priori error bound $\varepsilon > 0$ is given, the number of steps n necessary for ensuring that the error is less than ε can be expressed by the inequality

$$\frac{b - a}{2^n} < \varepsilon,$$

which is equivalent to the condition

$$n > \log_2 \frac{b - a}{\varepsilon}. \tag{5.1}$$

Let r be a positive integer. A root x^* of a function f is said to have *multiplicity* r if $\lim_{x \to x^*} f(x)/(x - x^*)^r$ exists and is unequal to zero. We remark that roots with even multiplicity usually cannot be determined by using bisection or the next technique to be described because the sign of the function does not change in the neighborhood of a root of even multiplicity.

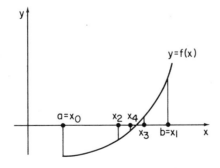

Fig. 5.1. Illustration of the bisection method.

Table 5.1. Application of the Bisection Method

n	Midpoints	Value of $f(s)$
1	-1.75	-3.796875
2	-1.625	-1.587891
3	-1.5625	-0.576416
4	-1.53125	-0.093292
5	-1.515625	$+0.142696$
6	-1.523438	$+0.025157$
$\vdots$	$\vdots$	$\vdots$
20	-1.525102	$+0.000004$

Example 5.1. The polynomial

$$f(x) = x^3 - 3x^2 - x + 9$$

has one root in the interval $(-2, -1.5)$. We compute the root to six decimal places.

Let $a = -2$ and $b = -1.5$; then $f(a) = -9 < 0$ and $f(b) = 0.375 > 0$. The number of iterative steps required can be determined by using (5.1) for $\varepsilon = 0.5 \times 10^{-6}$:

$$n > \log_2 \frac{0.5}{0.5 \times 10^{-6}} = \log_2 10^6 \simeq 19.93,$$

which implies $n = 20$. The midpoints of successive intervals and the values of corresponding functional values are shown in Table 5.1. Thus the value -1.525102 gives a root of f accurate to six decimals.

Problems

5.1. The equation $x^4 - 3x + 1 = 0$ has a zero in the interval $(0.3, 0.4)$. Compute it to five decimal places of accuracy using the bisection method.

5.2. Compute the only positive root of the equation $\sin x - x/2 = 0$ by the bisection method, to five decimal places of accuracy.

5.1.2. The Method of False Position (Regula Falsi)

In performing the bisection method we divide the interval into two equal parts at each step. The method of *regula falsi* can be viewed as a modified version of the bisection algorithm in which the intervals are divided according to the ratio of the functional values at the endpoints of the intervals. This procedure would seem sensible in the case that f is differentiable, for then on short intervals f is approximately a linear polynomial.

Let $[a, b]$ be the initial interval with the property that $f(a)f(b) < 0$; then s is defined by the intersection of the x axis and the chord joining the points $(a, f(a))$ and $(b, f(b))$. If $f(s) = 0$, then a zero of f has been discovered, and the algorithm terminates. If $f(s)f(a) < 0$, then the interval (a, s) contains a root of f, and if $f(s)f(b) < 0$, then the interval (s, b) contains at least one zero of f. By successively repeating the above procedure for the reduced intervals, one obtains a sequence x_k $(k = 0, 1, 2, \ldots)$ from the initial terms $x_0 = a$, $x_1 = b$. Consider the general step for the above method. Assume that the interval obtained in step k has as endpoints the points x_j and x_k. Then the equation of the chord joining points $(x_k, f(x_k))$ and $(x_j, f(x_j))$ has the form

$$y - f(x_k) = \frac{f(x_j) - f(x_k)}{x_j - x_k}(x - x_k),$$

which implies that the intersection of this chord and the x axis is x_{k+1}, where

$$x_{k+1} = x_k - \frac{f(x_k)(x_k - x_j)}{f(x_k) - f(x_j)}. \tag{5.2}$$

The application of regula falsi is illustrated in Figure 5.2.

It can be proved (Young and Gregory, 1972, Section 4.5) that for continuous functions f the sequence $\{x_k\}_{k=0}^{\infty}$ is convergent and the limit of the sequence is equal to a zero of function f.

For f sufficiently smooth, we can find a lower bound for the speed of convergence of the sequence $\{x_k\}$. Assume that f is twice continuously differentiable and that $|f'(x)|$ is bounded away from 0 in the interval $[a, b]$. Let x^* be a zero of function f in the interval (a, b). Using (5.2)

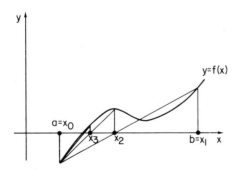

Fig. 5.2. Illustration of the regula falsi method.

we have

$$x_{k+1} - x^* = \frac{(x_k - x^*)f(x_j) - (x_j - x^*)f(x_k)}{f(x_j) - f(x_k)},$$

which implies

$$\frac{x_{k+1} - x^*}{(x_k - x^*)(x_j - x^*)} = \frac{f(x_j)/(x_j - x^*) - f(x_k)/(x_k - x^*)}{f(x_j) - f(x_k)}. \qquad (5.3)$$

Cauchy's mean value theorem (see Cooper, 1966, p. 42) asserts that if $g(x)$ and $h(x)$ are continuously differentiable real functions on the interval $[a, b]$, there exists a number ξ in that interval such that

$$\frac{g(b) - g(a)}{h(b) - h(a)} = \frac{g'(\xi)}{h'(\xi)}.$$

Applying this result to equation (5.3) and observing that

$$\frac{d}{dx}\left[\frac{f(x)}{x - x^*}\right] = \frac{f'(x)(x - x^*) - f(x)}{(x - x^*)^2},$$

we obtain the result that

$$\frac{x_{k+1} - x^*}{(x_k - x^*)(x_j - x^*)} = \frac{f'(\xi)(\xi - x^*) - f(\xi)}{(\xi - x^*)^2 f'(\xi)},$$

where ξ is a point between x_j and x_k. From Taylor's formula, we have

$$0 = f(x^*) = f(\xi) + f'(\xi)(x^* - \xi) + \frac{f''(\alpha)}{2}(x^* - \xi)^2,$$

where α is a point between x^* and ξ. Combining this equation with the one above, we see that

$$x_{k+1} - x^* = \frac{f''(\alpha)}{2f'(\xi)}(x_k - x^*)(x_j - x^*). \qquad (5.4)$$

Upon introducing the notation $\varepsilon_k = x_k - x^*$, $k \geq 0$, equation (5.4) implies

$$\varepsilon_{k+1} = \frac{f''(\alpha)}{2f'(\xi)}\varepsilon_k\varepsilon_j.$$

From our differentiability assumptions, there exist positive numbers m and M such that

$$|f'(x)| \geq m \quad \text{and} \quad |f''(x)| \leq M \qquad \text{for all } x \in [a, b].$$

Let us assume that $m > 0$. Then

$$| \varepsilon_{k+1} | \leq \frac{M}{2m} | \varepsilon_k | | \varepsilon_j |.$$

After multiplying both sides of this inequality by $K = M/2m$ we have

$$K | \varepsilon_{k+1} | \leq K^2 | \varepsilon_k | \| \varepsilon_j |,$$

which is equivalent to the inequality

$$d_{k+1} \leq d_k d_j, \qquad (5.5)$$

in which $d_l = K\varepsilon_l$, $l \geq 0$.

Assume that the initial approximations x_0, x_1 satisfy

$$d = \max\{d_0, d_1\} < 1.$$

Then from (5.5) we conclude that for $j \geq 1$, $d_j \leq d$, and so

$$d_{k+1} \leq d^{k+1}, \qquad k \geq 0. \qquad (5.6)$$

Inequality (5.6) shows the linear convergence of the regula falsi method. The coefficient of convergence is d. For $d < \frac{1}{2}$ this method converges faster than the bisection method.

Example 5.2. The root of polynomial $f(x) = x^3 - 3x^2 - x + 9$ of Example 5.1 can be approximated using the method of regula falsi. The sequence $\{x_n\}$ and the corresponding function values are shown in Table 5.2.

Table 5.2. Application of the Regula Falsi Method

n	x_n	$f(x_n)$
0	-2	-9
1	-1.5	0.3750
2	-1.52	0.076992
3	-1.524	0.016666
4	-1.524880	0.003147
5	-1.525057	0.000679
6	-1.525093	0.000136
7	-1.525100	0.000028
8	-1.525102	0.000006
9	-1.525102	0.000001

Problems

5.3. Write a computer program to check Tables 5.1 and 5.2. Continue these tables in double-precision iterations to 100, printing out all digits in the double-precision word.

5.4. Show that the regula falsi method makes use of inverse interpolation (Section 2.1.7).

5.5. The equation $x^4 - 3x + 1 = 0$ has a unique solution in $[0.3, 0.4]$. Determine this zero using the regula falsi method to five decimal places of accuracy.

5.6. Compute the only positive root of the equation $\sin x - x/2 = 0$ by regula falsi, to five decimal places of accuracy.

5.7. Prove that in the regula falsi method, the points $\{x_i\}$ converge to a solution of $f(x)$ if f is continuous on the initial interval, regardless of the existence of derivatives. HINT: Use the fact that a continuous function is uniformly continuous on a closed, bounded interval.

5.8. Prove that the convergence of the regula falsi method is faster than that of the bisection method if d in equation (5.6) is less than 0.5.

5.1.3. The Secant Method

In the method of regula falsi, at each step the signs of the corresponding function values determine the choice of reduced intervals. In the case of convex or concave functions, one endpoint of each reduced interval coincides with an endpoint of the initial interval $[a, b]$, and so the lengths of the reduced intervals do not tend to zero, because in (5.2) j will be either 1 or 0. Consequently upper bound (5.6) cannot be made smaller. This fact is the underlying reason for the slower convergence of the method of false position in comparison to the methods of the present and next sections.

We shall prove that a method giving faster convergence can be obtained if in each step we choose $j = k - 1$. The formula of this revised version of the method of regula falsi can be written as follows:

$$x_{k+1} = x_k - \frac{f(x_k)(x_k - x_{k-1})}{f(x_k) - f(x_{k-1})}, \tag{5.7}$$

and error estimation (5.5) takes the form

$$d_{k+1} \leq d_k d_{k-1}. \tag{5.8}$$

Rule (5.7) is called the *secant method*.

By induction we see that, for $k = 0, 1, 2, \ldots,$

$$d_k \leq d^{\gamma_k}, \tag{5.9}$$

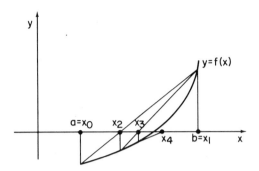

Fig. 5.3. Illustration of the secant method.

where γ_k is the kth term of the *Fibonacci sequence* defined by the recursive relation $\gamma_k = \gamma_{k-1} + \gamma_{k-2}, \gamma_0 = \gamma_1 = 1$, and $d = \max\{d_0, d_1\}$ as in (5.6). It can be verified by induction that

$$\gamma_k - \frac{1}{5^{1/2}} \left[\left(\frac{1 + 5^{1/2}}{2} \right)^{k+1} - \left(\frac{1 - 5^{1/2}}{2} \right)^{k+1} \right],$$

which implies that

$$\lim_{k \to \infty} \frac{\gamma_{k+1}}{\gamma_k} = \frac{1 + 5^{1/2}}{2} \simeq 1.618.$$

Thus the order of convergence of the secant method is equal to $(1 + 5^{1/2})/2$, which is faster than the linear convergence (order 1) of the regula falsi rule.

The geometric interpretation of the secant method is shown in Figure 5.3, which illustrates the construction of the first three steps of the secant method.

Example 5.3. In using the secant method for the problem of Example 5.1, we take as starting values $x_0 = -2$ and $x_1 = -1$ and obtain Table 5.3.

Table 5.3. Application of the Secant Method

n	x_n	$f(x_n)$
0	-2	-9
1	-1	6
2	-1.4	1.776000
3	-1.499	0.389743
4	-1.526841	-0.026330
5	-1.525079	0.000348
6	-1.525102	0.000000

Problems

5.9. Letting $x_0 = 0.3$ and $x_1 = 0.4$, solve the equation $f(x) = x^4 - 3x + 1 = 0$ to an accuracy of five decimal places by the secant method.

5.10. Find the positive root of $\sin x - x/2$ to five significant places by the secant method. Take $x_0 = \pi/2$ and $x_1 = 4\pi/5$.

5.1.4. Newton's Method

The method described in this section can be applied when a close initial approximation is available, and f has a continuous first derivative. It achieves a faster convergence rate than the previous algorithms, but like the secant method it may diverge if the initial approximation is not sufficiently close.

Let x_0 be an approximation of a zero x^* of function f. Approximate the function by its tangent at the point x_0; that is, the function $f(x)$ is approximated by the linear polynomial

$$\hat{f}(x) = f(x_0) + (x - x_0)f'(x_0),$$

and the next approximation of the zero of f is determined by the zero of this linear approximation, which is

$$x_1 = x_0 - \frac{f(x_0)}{f'(x_0)}.$$

By repeating the above step, one obtains the general formula of *Newton's method* (otherwise known as the *Newton–Raphson* method), which is

$$x_{k+1} = x_k - \frac{f(x_k)}{f'(x_k)}. \tag{5.10}$$

The graphical interpretation of Newton's method is given in Figure 5.4.

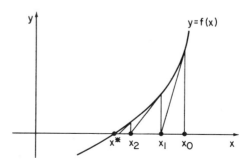

Fig. 5.4. Illustration of Newton's method.

Assume that between x^* and x_0 the following conditions hold: (1) f'' exists and is continuous, (2) both f' and f'' differ from zero, and (3) $f(x_0)f''(x_0) > 0$. We prove that the sequence $\{x_k\}_{k=0}^{\infty}$ converges to x^*. Assume that $x_0 > x^*$, $f(x_0) > 0$, $f''(x_0) > 0$. (All of the other cases can be handled by the analysis to follow). Under these assumptions, function f is strictly monotone increasing and convex, which implies $x_{k+1} \geq x^*$, $k \geq 1$, and $x_{k+1} \leq x_k$ [for the definition and main properties of convex functions see Hadley (1969), pp. 83–90)]. Thus sequence $\{x_k\}$ is bounded from below and nonincreasing, which implies the convergence of the sequence. Let the limit point be denoted by x^{**}; then for $k \to \infty$, equation (5.10) has the form

$$x^{**} = x^{**} - \frac{f(x^{**})}{f'(x^{**})};$$

that is, $f(x^{**}) = 0$. Since function f is strictly monotone increasing, there is no root other than x^* between x^* and x_0. Consequently, $x^{**} = x^*$.

Under certain regularity conditions, the speed of the convergence can be determined by employing devices used in the previous convergence results. From Taylor's formula for the function f we have

$$0 = f(x^*) = f(x_k) + f'(x_k)(x^* - x_k) + \tfrac{1}{2} f''(\xi)(x^* - x_k)^2,$$

where ξ is a point between x_k and x^*. Combining this equation with (5.10) we have

$$x_{k+1} - x^* = \frac{f''(\xi)}{2f'(x_k)} (x_k - x^*)^2. \tag{5.11}$$

If we assume that in an interval containing the sequence $\{x_k\}$ and x^*, the function f is twice differentiable and that for all x in this interval

$$0 < m \leq |f'(x)|, \qquad |f''(x)| \leq M,$$

and if, as before, we let $\varepsilon_j = x_j - x^*$, then (5.11) implies

$$|\varepsilon_{k+1}| \leq \frac{M}{2m} |\varepsilon_k|^2.$$

After multiplying both sides by $K = M/2m$ we have

$$d_{k+1} \leq d_k^2, \tag{5.12}$$

where $d_k = K\varepsilon_k$. Assume that $d = d_0 < 1$. Then by using the method of

Table 5.4. Application of Newton's Method

n	x_n	$f(x_n)$	$f'(x_n)$
0	-2	-9	23
1	-1.608695	-1.318155	16.415879
2	-1.528398	-0.049942	15.178390
3	-1.525108	-0.000082	15.128506
4	-1.525102	$+0.000000$	

finite induction we see that

$$d_{k+1} \leq d^{2^{k+1}}.$$

Inequality (5.12) shows the quadratic convergence of Newton's method, and the condition that $d_0 < 1$ gives us a sufficient condition on the starting point to ensure convergence.

Example 5.4. Choosing $x_0 = -2$ in the case of the polynomial $p(x) = x^3 - 3x^2 - x + 9$ of Example 5.1, we obtain Table 5.4.

In the above demonstration we assumed that $f'(x^*) \neq 0$, which follows from our assumption of the existence of a positive constant m, such that $0 < m \leq |f'(x)|$ in the neighborhood of x^*. Now we shall study the application of Newton's method for the case of roots having higher multiplicity. Assume that x^* is a zero of f with multiplicity $r > 1$. In this case, $f'(x^*) = 0$ and the preceding analysis does not apply. Nevertheless, we can prove that Newton's method is convergent when started from a sufficiently accurate initial approximation x_0. In the multiplicity r case, function f can be written in the form $f(x) = (x - x^*)^r g(x) \neq 0$. Let us introduce the function

$$h_t(x) = x - \frac{f(x)}{f'(x)} t,$$

where t is a real number. Simple calculations show that

$$\frac{d}{dx} [h_t(x^*)] = h_t'(x^*) = 1 - \frac{t}{r}. \tag{5.13}$$

Observe that for $0 < t < 2r$, $|h_t'(x)| < 1$. Assuming the continuity of

h_t' we conclude that for x in a sufficiently small neighborhood of x^*

$$| h_t'(x) | \leq q < 1,$$

where q is a fixed constant. Theorem 5.1 of the next section guarantees the convergence of the method. The convergence rate will be seen to be linear.

For $t = r$, (5.13) implies that $h_t'(x^*) = 0$. Consider the following modified version of Newton's method:

$$x_{k+1} = h_r(x_k) = x_k - r \frac{f(x_k)}{f'(x_k)}. \tag{5.14}$$

Theorem 5.2 of Section 5.1.5 will guarantee the quadratic convergence of method (5.14).

Example 5.5. Consider the polynomial $x^3 - 2x^2 + x$, for which the root $x^* = 1$ has multiplicity $r = 2$. Let $x_0 = 2$ and compute three steps by both the Newton method [equation (5.10)] and the modified version, equation (5.14). The results are shown in Table 5.5. The much faster convergence of method (5.14) is obvious.

The application of Newton's method requires computation of the values $f(x_k)$ and $f'(x_k)$ at each step. Consequently the number of operations used in Newton's method can be much larger than those required for methods discussed in the previous sections. If function f' is sufficiently smooth, the

Table 5.5. Example Showing Faster Convergence of the Modified Newton's Method

n	x_n	$f(x_n)$	$f'(x_n)$
		Newton's method	
0	2	2	5
1	1.6	0.576	2.28
2	1.347368	0.162580	1.056731
3	1.193517	0.044696	0.499379
		Method (5.14) with $r = 2$	
0	2	2	5
1	1.2	0.048	0.52
2	1.015385	0.000240	0.031479
3	1.000116	0.0000000	0.000231

Table 5.6. Application of an Approximation of Newton's Method

n	x_n	$f(x_n)$
0	-2	-9
1	-1.609695	-1.318156
2	-1.551384	-0.402860
3	-1.533869	-0.133207
4	-1.528077	-0.045074
5	-1.526117	-0.015366
6	-1.525449	-0.005252
7	-1.525221	-0.001797
8	-1.525143	-0.000615
9	-1.525116	-0.000210
10	-1.525107	-0.000072
11	-1.525104	-0.000025
12	-1.525103	-0.000008
13	-1.525102	$+0.000002$

values of f' do not change much in a neighborhood of x_0 and the effect of replacing $f'(x_k)$ by $f'(x_0)$ is small. In this case instead of method (5.10) or (5.14) we can use the formula

$$x_{k+1} = x_k - \frac{f(x_k)}{f'(x_0)} \quad \text{or} \quad x_{k+1} = x_k - r\,\frac{f(x_k)}{f'(x_0)}. \qquad (5.15)$$

It can be proven (on the basis of the fixed-point theorem) that starting from a sufficiently good initial approximation x_0, method (5.15) converges to x^* and the speed of convergence is linear.

Example 5.6. Consider the polynomial $f(x) = x^3 - 3x^2 - x + 9$ of Example 5.1 and let $x_0 = -2$. Then $f'(x_0) = 23$ and the results of applying (5.15) are shown in Table 5.6.

Problems

5.11. Apply Newton's method to approximate the value of $A^{1/k}$ where $A > 1$ and k is a positive integer.

5.12. Determine the value of $1/A$ (where $A > 1$) without using division operations.

5.13. Compute the zero of the equation

$$1 + \arctan x - x = 0$$

to six decimal places using Newton's method.

5.14. Determine the positive root of the function $\sin x - x/2$ to five significant places by the Newton method.

5.1.5. Application of Fixed-Point Theory

The methods described in the previous section are special cases of the general iteration scheme having the form

$$x_{k+1} = F_k(x_0, x_1, \ldots, x_k). \tag{5.16}$$

Method (5.16) is called an *i-point iteration algorithm* if function F_k depends on only the variables $x_{k-i+1}, \ldots, x_{k-1}, x_k$. An *i*-point iteration algorithm is called *stationary* if the functions $F_k, k = i, i + 1, \ldots,$ are identical to the same function F. If a stationary iteration process converges to a point x^*, then, assuming the continuity of F, we must have

$$x^* = F(x^*, x^*, \ldots, x^*).$$

The method of false position is not generally a stationary iteration method, because the value of j in (5.2) may or may not change from step to step. Newton's method and the secant method are obviously stationary iterations. Newton's method is a one-point algorithm, and the secant method a two-point algorithm.

Let us now consider a stationary one-point version of method (5.16), which we write as

$$x_{k+1} = g(x_k), \tag{5.17}$$

for some real function g.

Theorem 5.1. Let x^* denote a solution of $x = g(x)$. Assume that g' exists, and for some number $a > 0$, its magnitude is bounded by $q < 1$ on an interval $I = [x^* - a, x^* + a]$. Then in I, x^* is the only solution of equation $x = g(x)$, and the iteration process (5.17) converges to x^* from an arbitrary initial approximation $x_0 \in I$.

Proof. As in Example 4.6, we have that for $x, y \in I$.

$$\varrho'\big(g(x), g(y)\big) = |g(x) - g(y)|$$
$$= |g'(\xi)||x - y| \leq q|x - y|. \tag{5.18}$$

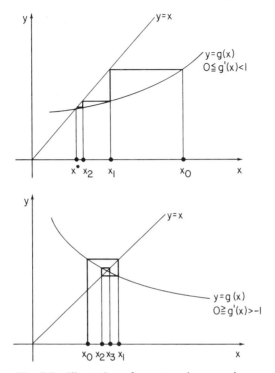

Fig. 5.5. Illustration of a contraction mapping.

Since q is hypothesized to be less than 1, is it evident that g, as a function on domain I, is a contraction mapping having a bound $\leq q$ and having range contained in I. The conclusion in Theorem 5.1 follows from our contraction mapping theorem (Theorem 4.5). In fact, from equation (4.27) following that theorem, we have the useful error bounds

$$|x_k - x^*| \leq \frac{q^k}{1-q} |g(x_0) - x_0|,$$

$$|x_k - x^*| \leq \frac{q}{1-q} |x_k - x_{k-1}|. \qquad \blacksquare$$

The application of the theorem in two different instances is illustrated in Figure 5.5.

Example 5.7. The polynomial $x^5 - x + 0.46$ has at least one root in the interval $I = [0.46, 0.5]$. The zero of this polynomial is obviously equal to the solution of equation

$$x = x^5 + 0.46.$$

Table 5.7. Application of an Iteration Formula

n	x_n	$f(x_n)$
0	0.5	−0.008750
1	0.491250	−0.002640
2	0.488610	−0.000761
3	0.487849	−0.000216
4	0.487633	−0.000061
5	0.487572	−0.000017
6	0.487554	−0.000004
7	0.487550	−0.000001
8	0.487548	−0.000000
9	0.487548	

Since in interval I

$$| (x^5 + 0.46)' | = | 5x^4 | \leq 5 \times 0.5^4 = 0.3125 = q,$$

and $q^2 \simeq 0.1$, method (5.17) is convergent and each two steps give at least a new decimal place of accuracy of the solution. Thus in order to obtain an error not exceeding 10^{-6}, at most $2 \cdot 6 = 12$ steps are sufficient. From the actual calculations in Table 5.7, it is seen that eight steps suffice.

The error bounds in Theorem 5.1 imply that the speed of iteration process (5.17) is at least linear. We now show that under certain assumptions the speed of convergence can be much faster.

Theorem 5.2. Suppose g is p times differentiable in an interval I centered at x^*. Suppose further that $g^{(j)}(x^*) = 0$ for $j = 1, 2, \ldots, p - 1$ and $g^{(p)}(x^*) \neq 0$. If $g^{(p)}$ is bounded in I, then iteration process (5.17) is convergent from a sufficiently good initial approximation x_0 in I and the convergence has order p.

Proof. Using Taylor's formula we have

$$g(x_k) = g(x^*) + \frac{g'(x^*)}{1!} (x_k - x^*) + \cdots + \frac{g^{(p-1)}(x^*)}{(p-1)!} (x_k - x^*)^{p-1}$$
$$+ \frac{g^{(p)}(\xi)}{p!} (x_k - x^*)^p,$$

for ξ between x_k and x^*. This implies that

$$x_{k+1} = x^* + \frac{g^{(p)}(\xi)}{p!}(x_k - x^*)^p. \tag{5.19}$$

Let

$$\varepsilon_l = x_l - x^*, \qquad K = \frac{1}{p!}\left[\sup_{x \in I}|f^{(p)}(x)|\right].$$

Then from equation (5.19) we have

$$|\varepsilon_{k+1}| \leq K|\varepsilon_k|^p.$$

Multiplying both sides by the constant $K_1 = K^{1/(p-1)}$ we have

$$K_1|\varepsilon_{k+1}| \leq [K_1|\varepsilon_k|]^p,$$

which implies that for $d_k = K_1\varepsilon_k$,

$$d_{k+1} \leq d_k^p. \tag{5.20}$$

Assume that $d = d_0 < 1$. Then, using the method of finite induction, one can prove that

$$d_{k+1} \leq d^{p^{k+1}}$$

which completes the proof. ∎

Finally we show a common fixed-point theory interpretation of the methods discussed in the previous sections.

If the function h differs from zero, then equations $f(x) = 0$ and

$$x = x + h(x)f(x) \tag{5.21}$$

have the same solutions. If h can be chosen in a way such that the derivative of the right-hand side of (5.21) at x^* is zero, then by Theorem 5.2 the corresponding iteration scheme has at least quadratic convergence. Simple calculation shows that this condition holds if

$$1 + f'(x^*)h(x^*) + f(x^*)h'(x^*) = 0.$$

Since $f(x^*) = 0$, this is equivalent to

$$h(x^*) = -1/f'(x^*).$$

If the values of f' differ from zero in the neighborhood of x^*, then we can choose

$$h(x) = -1/f'(x)$$

for all x. In this case the iteration method corresponding to (5.21) has the form

$$x_{k+1} = x_k - \frac{f(x_k)}{f'(x_k)},$$

which is Newton's method. One may conclude from this reasoning that in the general case of one-point stationary methods of form (5.21), which depend only on f and its first derivative at x^*, only quadratic convergence can be realized. Newton's method is consequently optimal in the sense that it achieves this upper bound.

If the value of $f'(x_k)$ is approximated by $f'(x_0)$, then method (5.15) is obtained. In approximating the derivatives $f'(x_k)$ by divided differences, we can obtain the method of regula falsi and the secant method.

Problems

5.15. Compute the solution of the equation

$$x = 1 + \arctan x$$

to an accuracy of six decimal places using the iteration method (5.17).

5.16. Show that the order of convergence of the method

$$x_{k+1} = x_k - \frac{f(x_k)}{f'(x_k)} - \frac{[f(x_k)]^2 f''(x_k)}{2[f'(x_k)]^3}$$

is at least 3, assuming that f is sufficiently many times differentiable and the derivatives are bounded away from zero in a neighborhood of the zero.

5.1.6. Acceleration of Convergence, and Aitken's δ^2-Method

Given that x^* is the fixed point for a given operator $g(x)$, in many instances it is possible to construct an operator $G(x)$ from $g(x)$ such that the fixed points of G coincide with those of g but such that the iteration process $\{x_n\}$ obtained from iterative application of G converge more rapidly than iteration process $\{x_n'\}$ obtained by iterative application of g. Such constructs G are known collectively as *acceleration* procedures.

We discuss a very prominent acceleration procedure, *Aitken's δ^2-method*, which is applicable in the case that g is a real function. Define

$$e(x) = g(x) - x.$$

Then fixed points of g are zeros of $e(x)$. A point x_n having been specified,

Aitken's δ^2-method prescribes choosing its successor x_{n+1} by a secant iteration of the form (5.7) in which we set $x_k = x_n$, $x_{k-1} = g(x_n)$, and replace f by e. Specifically,

$$x_{n+1} = G(x_n) \triangleq x_n - e(x_n)\left[\frac{x_n - g(x_n)}{e(x_n) - e(g(x_n))}\right]$$

$$= x_n - \frac{[g(x_n) - x_n]^2}{g(g(x_n)) - 2g(x_n) + x_n}$$

$$= \frac{x_n g(g(x_n)) - g(x_n)^2}{g(g(x_n)) - 2g(x_n) + x_n}.$$

Under the assumption that g is sufficiently many times differentiable, the reader will be able to confirm, by use of L'Hôpital's rule, that x^* is a fixed point of G if and only if it is a fixed point of g.

Let us now assume that g is continuously $p + 1$ times differentiable. We now show that if iterative application of g yields a sequence having order of convergence equal to p, where p is greater than 1, then for an initial point x_0 sufficiently close to x^*, G yields a sequence having order of convergence equal to $2p - 1$. For the fact that g has order of convergence p implies that the Taylor's series expansion of g about x^* has the form

$$g(x^* + \varepsilon) = x^* + A\varepsilon^p + B(\varepsilon)\varepsilon^{p+1},$$

where A is a fixed coefficient and $B(\varepsilon)$ depends on the deviation ε. Writing our earlier expression for G explicitly as a function of the deviation from the fixed point, we have

$$G(x^* + \varepsilon) = \frac{(x^* + \varepsilon)[g(g(x^* + \varepsilon))] - [g(x^* + \varepsilon)]^2}{g(g(x^* + \varepsilon)) - 2g(x^* + \varepsilon) + (x^* + \varepsilon)}.$$

If we define $\delta = g(x^* + \varepsilon) - x^* = A\varepsilon^p + B(\varepsilon)\varepsilon^{p+1}$, then we may write

$$g(g(x^* + \varepsilon)) = g(x^* + \delta) = x^* + A\delta^p + B(\delta)\delta^{p+1}.$$

Thus

$$G(x^* + \varepsilon) = \frac{(x^* + \varepsilon)[x^* + A\delta^p + B(\delta)\delta^{p+1}] - (x^* + \delta)^2}{[x^* + A\delta^p + B(\delta)\delta^{p+1}] - 2(x^* + \delta) + x^* + \varepsilon}$$

$$= \frac{x^*[A\delta^p + B(\delta)\delta^{p+1} - 2\delta + \varepsilon]}{A\delta^p + B(\delta)\delta^{p+1} - 2\delta + \varepsilon} - \frac{\delta^2 - \varepsilon A\delta^p - B(\delta)^{p+1}}{A\delta^p + B(\delta)\delta^{p+1} - 2\delta + \varepsilon}.$$

Assuming that $p \geq 2$, we have, upon substituting $A\varepsilon^p + B(\varepsilon)\varepsilon^{p+1}$ for δ, that

$$G(x^* + \varepsilon) = x^* - \frac{[A + B(\varepsilon)\varepsilon]^2\varepsilon^{2p} + O(\varepsilon^{(p+1)p})}{\varepsilon + O(\varepsilon^p)}$$

$$= x^* - A^2\varepsilon^{2p-1} + O(\varepsilon^{2p}),$$

which in view of Theorem 5.2 establishes the asserted convergence for the $p \geq 2$ case.

We leave it to the reader to modify the preceding developments in order to show that if x^* is a simple root of $e(x)$, and g gives first-order convergence (that is, $p = 1$), then for an initial point close enough to x^*, iterative application of G gives quadratic convergence.

Aitken's δ^2-method should preferably be executed in double precision, since differencing of very close elements occurs in both the numerator and denominator of the formula, and we know from Chapter 1 that the relative error can be arbitrarily large from such a subtraction.

Problems

5.17. Prove, under our differentiability assumptions, that x^* is a fixed point for g if and only if it is also a fixed point for G.

5.18. Establish quadratic convergence of G, starting from a point sufficiently close to x^*, under the assumptions that g yields linear convergence and that x^* is not a multiple root of g.

5.19. Suppose $f(x^* + \varepsilon)$ is $O(\varepsilon^j)$ and $g(x^* + \varepsilon)$ is $O(\varepsilon^k)$, where $j \geq k > 0$. Show that the sum $f + g$ is $O(\varepsilon^k)$ at x^*. What is the order of convergence of the product?

5.20. With f as above, show that $1/[1 + f(x)]$ is $1 + O(\varepsilon^j)$.

5.2. Solution of Polynomial Equations

In this section we shall deal with the numerical solution of *polynomial equations*, which are equations of the form

$$f(z) = a_0 z^n + a_1 z^{n-1} + \cdots + a_{n-1} z + a_n = 0,$$

the coefficients $a_0, a_1, \ldots, a_n$ being real or complex numbers. In the special case in which the coefficients are real numbers and a real root is to be computed, the methods described in the previous sections can be directly applied.

The bisection and regula falsi algorithms are not applicable to finding complex roots because their implementation requires ordering the values of f, which must therefore be real. The secant and Newton's methods, however, may be applied directly, with the understanding that the values of x and f are complex. In fact, our proof of the quadratic convergence of Newton's method holds for $f(x)$ an analytic complex function. As in the real case, both the secant and Newton's methods may fail to converge if the initial estimate is not close enough to the root. The secant method has the additional disadvantage that it may fail near a multiple root. Its use, therefore, is recommended only for locating a simple real root.

The main purpose of this section is to describe some special facts and techniques that are entirely motivated by the polynomial equation case.

5.2.1. Sturm Sequences

In this section we present a method for effectively determining whether a given interval $[a, b]$ contains a root of a given polynomial $f(x)$. In principle, one could use this method iteratively for approximating real roots, but this plan is not desirable because the Sturm sequence procedure is subject to mistakes arising from roundoff error in the immediate vicinity of a root and because faster convergence can be achieved, once the root has been localized to a fairly short interval, by employing the secant method or Newton's method, for example.

Before proceeding with the Sturm sequence method, it may be useful to use Descartes' rule of signs to verify whether the polynomial $f(x)$ has a positive root. The following is a statement of *Descartes' rule of signs*:

> The number of positive real roots p of a polynomial with real coefficients does not exceed v, the number of sign variations of the coefficients; moreover, $v - p$ is a nonnegative even integer.

A sign variation occurs when the coefficient of x^{j+1} is positive and the coefficient of x^j is negative, or vice versa. In counting sign variations, zero coefficients are ignored. For example, the coefficients of the polynomial $f(x) = 3x^5 - 2x^4 - x^3 + x + 1$ have *two* sign variations, and so the number of positive real roots is either zero or two.

One can test for negative real roots by applying Descartes' rule of signs to the polynomial $f_1(x) = f(-x)$. The coefficients of odd powers of x of f_1 will differ from the signs of those of $f(x)$. One can similarly test for roots of $f(x)$ greater than a given real number r by applying the rule of signs to $f_2(x) = f(x - r)$.

Descartes' rule of signs, therefore, provides a quick and easy test for the existence and very rough approximate location of real roots. The method that we now describe, based on Sturm sequences, is far more effective for computer tests for existence of roots in a given interval. Let $f(x)$ be a given nth-degree polynomial with real coefficients. We construct a sequence of polynomials of strictly decreasing degree by the following recursive rule:

$$-g_k(x) = \text{remainder when } g_{k-2}(x) \text{ is divided by } g_{k-1}(x). \quad (5.22)$$

Thus $g_{k-2}(x) = Q(x)g_{k-1}(x) - g_k(x)$, for $Q(x)$ the quotient polynomial. The initial conditions are $g_0(x) = f(x)$ and $g_1(x) = f'(x)$. If for some m, $g_m(x)$ is a constant, then we define $g_{m+1}(x) = g_{m+2}(x) = \cdots = g_n(x) = 0$. With the sequence $g_j : j = 0, \ldots, n$ so determined, for any real a we define the quantity $V(a)$ to be the number of sign variations in sequence of numbers $g_0(a), g_1(a), \ldots, g_n(a)$. If $g_m(x)$ is not constant, but $g_{m+j}(x) = 0$ for $j > 1$, then $V(a)$ is defined to be the number of sign variations of

$$\frac{g_0(a)}{g_m(a)}, \quad \frac{g_1(a)}{g_m(a)}, \ldots, \frac{g_{m-1}(a)}{g_m(a)}.$$

In counting sign variations, zeros are ignored.

The significance of these sequences, which are examples of Sturm sequences, is made evident by the following statement:

Theorem 5.3. Let a and b be real numbers such that $f(a)$ and $f(b) \neq 0$. Assume $a < b$, and $V(x)$ is defined as above with respect to some polynomial $f(x)$ having at most simple roots in $[a, b]$. Then the number of real roots of $f(x)$ in the interval $[a, b]$ exactly equals $V(a) - V(b)$.

A proof and commentary is to be found in Chapter 3, Section 4.2, of Isaacson and Keller (1966). Further discussion of this method is to be found in Chapter 3 of Householder (1953) as well as Chapter 4 of Young and Gregory (1972). Both of these works discuss numerical aspects of synthetic division, which must be undertaken to implement (5.22).

Example 5.8. Let $f(x) = g_0(x) = x^2 - x - 2$. Then $g_1(x) = f'(x) = 2x - 1$. Taking the negative of the remainder when g_0 is divided by g_1, we have $g_2(x) = +9/4$. Thus, by way of confirmation, we have

$$g_0(x) = (\tfrac{1}{2}x - \tfrac{1}{4})(2x - 1) - \tfrac{9}{4}.$$

We have that $(g_0(0), g_1(0), g_2(0))$ is $(-2, -1, 9/4)$, and so $V(0) = 1$. $g_0(5) = 18$, $g_1(5) = 9$, $g_2(5) = 9/4$, and thus $V(5) = 0$, since no sign change occurs in this sequence. $g_0(-5) = 28$, $g_1(-5) = -11$, $g_2(-5) = 9/4$, and so $V(-5) = 2$. From this we may conclude that the number of roots in $[-5,5]$ is $V(-5) - V(5) = 2$. Also, the number of roots in $[0,5]$ is $V(0) - V(5) = 1$. The reader may confirm that the roots are actually 2 and -1.

5.2.2. The Lehmer–Schur Method

Basically, the Lehmer–Schur method provides a means of determining whether a given circle in the complex plane contains a root of a given polynomial $f(x)$. The method can be used iteratively to locate roots within smaller and smaller circles, achieving a linear convergence rate thereby, or it can be used to locate a good starting point for methods such as Newton's or Bairstow's (to be described later), which achieve faster convergence.

Given the polynomial

$$f(z) = a_0 z^n + a_1 z^{n-1} + \cdots + a_{n-1} z + a_n,$$

define

$$f^*(z) = \bar{a}_0 + \bar{a}_1 z + \cdots + \bar{a}_{n-1} z^{n-1} + \bar{a}_n z^n,$$

where bars denote complex conjugates. Consider the function

$$T[f(z)] = \bar{a}_n f(z) - a_0 f^*(z).$$

Since the coefficient of z^n on the right-hand side of the above equation is equal to zero, the sequence defined by

$$T^j[f(z)] = T\{T^{j-1}[f(z)]\}, \qquad j = 2, 3, \ldots,$$

is a sequence of polynomials of strictly decreasing degree. Let k denote the smallest positive integer such that $T^k[f(0)] = 0$. The theoretical basis of the Lehmer–Schur method is the following theorem.

Theorem 5.4. Assume that $f(0) \neq 0$. If for some l, $0 < l < k$, $T^l[f(0)] < 0$, then f has at least one zero inside the unit circle. If $T^{k-1}[f(z)]$ is constant and for $0 < l < k$, $T^l[f(0)] > 0$ then no zero of f lies inside the unit circle.

The proof of the theorem uses elementary results from complex variable theory, and it is based on the theorem of Rouché (see Apostol, 1957, p. 538). See Lehmer (1961) for details.

The theorem gives no information in the case in which $T^{k-1}[f(z)]$ is not constant. This incompleteness of the theorem can be eliminated by introducing a new variable ϱz, when ϱ is a positive scalar.

Introducing the function

$$g(z) = f(\varrho z + c)$$

and applying the theorem for polynomial g, we obtain information about the existence of roots of f in the circle $|z - c| < \varrho$.

By using the above theorem and observation, the following algorithm is proposed for determining one root of f.

Lehmer–Schur Algorithm

1. Find the smallest value of M such that f has no root in the circle $|z| < 2^M = R$ but such that f has at least one root in the circle $|z| < 2^{M+1} = 2R$.

The smallest value of M can be determined using the method suggested by the theorem for the circle with center at zero and radii $1, 2, 2^2, 2^3, \ldots$, or, if the unit circle has a root, test circles of radii $2^{-1}, 2^{-2}, \ldots$.

2. Cover the annulus $R < |z| < 2R$ by eight circles with radius $4R/5$ and centers

$$\frac{3R}{2\cos \pi/8} \exp\left(\frac{i2\pi v}{8}\right) \qquad (v = 0, 1, \ldots, 7, i = \sqrt{-1}).$$

Using the method of the theorem we can find a circle containing a root of f.

3. Reduce by half the radius of the circle obtained in the previous step, and by testing this circle, we have two possibilities:

(a) If no root of f lies in the circle, then in the annulus must lie at least one zero of f. Then go to step 2.

(b) If the circle with half-radius has a root of f in its interior, then go to step 3.

Repeating steps 2 and 3, we get a circle with arbitrarily small radius containing a zero of f, and so the center of this circle can be considered as an approximation of a root of polynomial f. The graphical interpretation is shown in Figure 5.6.

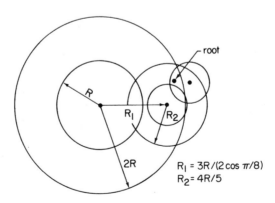

Fig. 5.6. Application of the Lehmer–Schur algorithm.

Example 5.9. In the case of the polynomial of Example 5.1 we have

$$T[f(z)] = 9(z^3 - 3z^2 - z + 9) - 1(9z^3 - z^2 - 3z + 1)$$
$$= -26z^2 - 6z + 80,$$
$$T^2[f(z)] = 80(-26z^2 - 6z + 80) + 26(80z^2 - 6z - 26)$$
$$= -636z + 5724,$$
$$T^3[f(z)] = 5724(-636z + 5724) + 636(5724z + 636)$$
$$= 5724^2 + 636^2 > 0,$$

and so no root of f lies inside the unit circle. Let

$$g(z) = f(2z) = 8z^3 - 12z^2 - 2z + 9,$$

then we have

$$T[g(z)] = 9(8z^3 - 12z^2 - 2z + 9) - 8(9z^3 - 2z^2 - 12z + 8)$$
$$= -90z^2 + 78z + 17,$$
$$T^2[g(z)] = 17(-90z^2 + 78z + 17) + 90(17z^2 + 78z - 90)$$
$$= 8346z - 7811,$$

which implies $T^2[g(0)] < 0$, that is, the circle $|z| < 2$ contains a root of f.

In most instances, one should be tempted to switch from the Lehmer–Schur iterations to one giving faster convergence as soon as possible. For example, one can switch to Newton's method and achieve quadratic convergence (instead of linear) when $d_0 = K\varepsilon_0$, defined in connection with (5.12), is less than 1. Here, ε_0 will be the radius of a circle known to contain a root. In Section 5.2.3, we shall give a prominent technique for finding quadratic factors of f. This method also achieves quadratic convergence, when started from a sufficiently good approximation. An approximation of a complex root immediately provides an approximation of its complex conjugate, which will also be a root if the polynomial has real coefficients.

Problems

5.21. Prove that one iteration of the Lehmer–Schur method requires at most $n^2 + O(n)$ multiplications.

5.22. Prove that the convergence rate of the Lehmer–Schur algorithm is linear.

5.2.3. Bairstow's Method

A real polynomial has either real or complex conjugate roots. Bairstow's method is useful for finding real quadratic factors of the polynomial, all the conjugate roots being among the roots of the factors.

By dividing the polynomial

$$f(z) = a_0 z^n + a_1 z^{n-1} + \cdots + a_{n-1} z + a_n$$

by a polynomial $z^2 + pz + q$, we formally obtain

$$f(z) = (z^2 + pz + q)(b_0 z^{n-2} + b_1 z^{n-3} + \cdots + b_{n-2}) + b_{n-1} z + b_n. \quad (5.23)$$

The coefficients of various powers of z must be the same on the left- and right-hand sides, which implies that all the coefficients $b_0, b_1, \ldots, b_{n-1}, b_n$ can be expressed as functions of p and q. Then polynomial $z^2 + pz + q$ will have roots that are also roots of f if and only if

$$b_{n-1}(p, q) = 0, \qquad b_n(p, q) = 0. \quad (5.24)$$

This is a system of nonlinear equations that can be solved using methods to be discussed in Section 5.3. If the generalized Newton method (Section 5.3) is employed to solve the preceding system, the algorithm is called *Bairstow's method*.

A convenient explicit rule for implementing Bairstow's method is given in Problem 5.24 below. This method is also discussed in Section 5.7 of Henrici (1967).

Example 5.9. In the case of the polynomial of Example 5.1 we have

$$x^3 - 3x^2 - x + 9 = (x^2 + px + q)(b_0 x + b_1) + b_2 x + b_3,$$

which implies, upon matching coefficients of like powers of x, that

$$1 = b_0,$$
$$-3 = b_1 + b_0 p,$$
$$-1 = b_2 + b_1 p + b_0 q,$$
$$9 = b_3 + b_1 q,$$

that is, $b_0 = 1$, $b_1 = -3 - p$, $b_2 = -1 + 3p + p^2 - q$, and $b_3 = 9 + 3q + pq$. Thus the system of nonlinear equations (5.24) has the form

$$-1 + 3p + p^2 - q = 0,$$
$$9 + 3q + pq = 0.$$

In Example 5.10, we shall give a solution to this system and then produce a pair of roots to the polynomial of this example.

Problems

5.23. What is the general form of the system of nonlinear equations obtained by using Bairstow's method for polynomials of degree 3 and 4?

5.24. Show that if $f(z)$ has degree n, then the coefficients b_j in the factorization (5.23) are related to the coefficients a_j of $f(z)$ by the system of equations

$$a_0 = b_0,$$
$$a_1 = b_1 + p,$$
$$a_k = b_k + pb_{k-1} + qb_{k-2}, \qquad 2 \le k \le n-1,$$
$$a_n = b_n + qb_{n-2}.$$

5.2.4. The Effect of Coefficient Errors on the Roots

In this section we prove that the roots of polynomials depend continuosly on the coefficients. This implies that one can tolerate some error in the coefficients of a polynomial whose roots are to be located. But, as can be surmised from the development to follow, for high-degree polynomials the zero locations are very sensitive to change in coefficient value, the bound depending on the nth root of the weighted sum of the magnitudes of coefficient error.

Let the polynomial

$$f(z) = z^n + a_1 z^{n-1} + \cdots + a_{n-1} z + a_n$$

be approximated by the polynomial

$$g(z) = z^n + b_1 z^{n-1} + \cdots + b_{n-1} z + b_n.$$

Let the roots of f be denoted by $x_1, \ldots, x_n$ and let the roots of g be denoted by $y_1, \ldots, y_n$. The numbers x_i, y_i, $1 \le i \le n$, may be complex numbers. Let γ be a positive number such that

$$\max_{1 \le i \le n} \{|x_i|\} \le \gamma.$$

We prove the following theorem:

Theorem 5.5. Let x_k be an arbitrary root of f. Then there exists a root y_i of g such that

$$|x_k - y_i| \le \left\{ \sum_{j=1}^{n} |a_j - b_j| \gamma^{n-j} \right\}^{1/n} \qquad (5.25)$$

Proof. Consider the difference of f and g at x_k:

$$f(x_k) - g(x_k) = \sum_{j=1}^{n} (a_j - b_j)x_k^{n-j}.$$

Since $f(x_k) = 0$,

$$-g(x_k) = -\prod_{j=1}^{n} (x_k - y_j) = \sum_{j=1}^{n} (a_j - b_j)x_k^{n-j},$$

which implies

$$\prod_{j=1}^{n} | x_k - y_j | \leq \sum_{j=1}^{n} | a_j - b_j | \gamma^{n-j}.$$

From this inequality we conclude that at least one factor of the left-hand side must be less than or equal to the nth root of the right-hand side, which was to be proved. ∎

Corollary 5.1. Let $\gamma^* = \max_{1\leq i\leq n} \{| y_i |\}$. Then for an arbitrary root y_i of g there exists a root x_k of f such that

$$| y_i - x_k | \leq \left\{ \sum_{j=1}^{n} | a_j - b_j | \gamma^{*n-j} \right\}^{1/n}.$$

Corollary 5.2. If for $i, j = 1, 2, \ldots, n$, $i \neq j$,

$$| x_i - x_j | > 2\varepsilon,$$

where

$$\varepsilon = \left\{ \sum_{k=1}^{n} | a_k - b_k | \gamma^{n-k} \right\}^{1/n},$$

then the roots of f and g can be ordered so that for $k = 1, 2, \ldots, n$,

$$| x_k - y_k | \leq \varepsilon.$$

By Theorem 5.5 one may prove the following result.

Theorem 5.6. Let $y_1, \ldots, y_k$, $k < n$, be the approximations of some roots of polynomial f of degree n, and assume that the errors of numbers $y_1, \ldots, y_k$ do not exceed a fixed positive number ε. Let

$$g(x) = \prod_{i=1}^{k} (x - y_i)$$

and let

$$h(x) = f(x)/g(x)$$

(the remainder is neglected). If the roots of h are denoted by $y_{k+1}, \ldots, y_n$ and the other roots of f are denoted by $x_{k+1}, \ldots, x_n$, then these roots can be ordered so that for $i = k + 1, \ldots, n$,

$$| x_i - y_i | \leq Q\varepsilon^{1/(n-k)}, \tag{5.26}$$

where Q depends on n, k, and f.

The details of the proof and the calculation of Q are not discussed here (see Szidarovszky, 1971).

In order to illustrate the sensitivity of polynomial roots as a function of coefficients, we note that if γ in (5.25) is 1, and $\varDelta$ is an error bound for $| b_i - a_i |, 1 \leq i \leq n$, then in order to ensure that $| x_k - y_i | < 0.1$, for tenth-degree polynomials, for example, we must insist that $\varDelta \simeq 10^{-10}$.

5.3. Systems of Nonlinear Equations and Nonlinear Programming

The system of nonlinear equations

$$f_i(x_1, \ldots, x_n) = 0, \qquad 1 \leq i \leq n, \tag{5.27}$$

may be written in vector notation as

$$\mathbf{f}(\mathbf{x}) = \mathbf{0},$$

where $\mathbf{f} = (f_1, f_2, \ldots, f_n)^T$, $\mathbf{x} = (x_1, \ldots, x_n)^T$, and $\mathbf{0}$ is the zero vector.

The problems of solving systems of nonlinear equations and optimizing nonlinear multivariate functions are very closely related to each other. Specifically, the problem of solving the system of nonlinear equations can be rewritten as the problem of minimizing the function

$$\phi(\mathbf{x}) = \sum_{i=1}^{n} [f_i(\mathbf{x})]^2. \tag{5.28}$$

Conversely, the problem of finding the local optimum of some multivariate differentiable function ϕ is equivalent to the task of solving the nonlinear system of equations

$$\frac{\partial \phi(\mathbf{x})}{\partial x_i} = 0, \qquad 1 \leq i \leq n. \tag{5.29}$$

Iterative methods for solving nonlinear multivariate optimization problems are called *nonlinear programming* methods, even when they involve first converting the problem to the problem of finding the root of a system of equations, as just described. Constrained optimization problems, under certain assumptions, also can be rephrased as nonlinear equation problems in several variables, because through the use of Lagrange multipliers they can often be transformed into unconstrained optimization problems, to which (5.27) is equivalent, as we have noted.

5.3.1. Iterative Methods for Solution of Systems of Equations

Toward solving the system (5.27) of equations, we shall consider successive approximation rules of the form

$$\mathbf{x}^{(k+1)} = \mathbf{g}(\mathbf{x}^{(k)}), \tag{5.30}$$

where, as in (5.27), $\mathbf{x}$ is an n-tuple, and consequently $\mathbf{g}$ will be an n-tuple-valued function. The bounds in Theorem 5.7 were introduced in Example 4.7. We shall choose our iteration function $\mathbf{g}$ so that $\mathbf{x}$ is a solution of the system (5.27) if and only if

$$\mathbf{x} = \mathbf{g}(\mathbf{x}). \tag{5.31}$$

Theorem 5.7. Assume $\mathbf{g}$ is differentiable and one of the numbers

$$K_\infty = \max_i \sum_{j=1}^{n} \sup_{\mathbf{x} \in M} \left| \frac{\partial g_i(\mathbf{x})}{\partial x_j} \right|,$$

$$K_1 = \max_j \sum_{i=1}^{n} \sup_{\mathbf{x} \in M} \left| \frac{\partial g_i(\mathbf{x})}{\partial x_j} \right|, \tag{5.32}$$

or

$$K_2 = \sum_{i=1}^{n} \left(\sum_{j=1}^{n} \sup_{\mathbf{x} \in M} \left| \frac{\partial g_i(\mathbf{x})}{\partial x_j} \right|^2 \right)^{1/2},$$

is less than one [here M is a nonempty closed convex subset of the domain of $\mathbf{g}$ such that for all $\mathbf{x} \in M$, $\mathbf{g}(\mathbf{x}) \in M$]. Then the sequence defined by (5.30) converges to the unique solution $\mathbf{x}^*$ of (5.31) in M. Furthermore, for the value K chosen and for $j > 1$, we have a priori and a posteriori error bounds for $\mathbf{x}^{(j)}$ given by

$$\varrho(\mathbf{x}^{(j)}, \mathbf{x}^*) \le \frac{K^j}{1 - K} \varrho(\mathbf{x}^{(0)}, \mathbf{x}^{(1)}),$$

$$\varrho(\mathbf{x}^{(j)}, \mathbf{x}^*) \le \frac{K}{1 - K} \varrho(\mathbf{x}^{(j)}, \mathbf{x}^{(j-1)}). \tag{5.33}$$

In the above bounds, the choice of metric ϱ is the metric of Example 4.3 bearing the same subscript as the value of K chosen as being less than 1.

Proof. The theorem is an immediate consequence of Theorem 4.5, the contraction mapping theorem, once it is recognized that

$$\varrho\big(\mathbf{g}(\mathbf{x}_j), \mathbf{g}(\mathbf{x}_{j-1})\big) \leq K\varrho(\mathbf{x}_j, \mathbf{x}_{j-1}).$$

But this inequality is demonstrated in Example 4.7. ∎

The result of Section 4.7 implies that in case $K_\infty < 1$, the Seidel-type iteration

$$x_1^{(k+1)} = g_1(x_1^{(k)}, x_2^{(k)}, \ldots, x_{n-1}^{(k)}, x_n^{(k)}),$$

$$x_2^{(k+1)} = g_2(x_1^{(k+1)}, x_2^{(k)}, \ldots, x_{n-1}^{(k)}, x_n^{(k)}),$$

$$\vdots \tag{5.34}$$

$$x_n^{(k+1)} = g_n(x_1^{(k+1)}, x_2^{(k+1)}, \ldots, x_{n-1}^{(k+1)}, x_n^{(k)}),$$

is convergent and the coefficient of convergence of this algorithm is usually smaller than that of iteration formula (5.30).

With respect to the nonlinear system of equations

$$\mathbf{f}(\mathbf{x}) = \mathbf{0}, \tag{5.35}$$

assume that some given matrix-valued function $\mathbf{H}(\mathbf{x})$ is of order n and is nonsingular on the domain of $\mathbf{x}$. Define the iteration operator $\mathbf{g}$ by the rule

$$\mathbf{g}(\mathbf{x}) = \mathbf{x} + \mathbf{H}(\mathbf{x})\mathbf{f}(\mathbf{x}). \tag{5.36}$$

Then clearly $\mathbf{x} = \mathbf{g}(\mathbf{x})$ if and only if (5.35) is satisfied. From the preceding theorem, the sequence $\{\mathbf{x}_k\}$ of vectors obtained by successive application of $\mathbf{g}$ converges to a solution of (5.31) provided a matrix norm of the partial derivatives of $\mathbf{g}$ is less than 1 on a closed neighborhood of the solution $\mathbf{x}^*$ and $\mathbf{x}_0$ is sufficiently close to $\mathbf{x}^*$. The norm condition is certainly satisfied (assuming the continuity of the partial derivatives) if all of the partial derivatives are zero at $\mathbf{x}^*$; that is, if for all values of i and k,

$$\frac{\partial}{\partial x_k}\left[x_i + \sum_{j=1}^{n} h_{ij}(\mathbf{x})f_j(\mathbf{x})\right]_{\mathbf{x}=\mathbf{x}^*} \tag{5.37}$$

$$= \delta_{ik} + \sum_{=1}^{n}\left[\frac{\partial h_{ij}(\mathbf{x}^*)}{\partial x_k}f_j(\mathbf{x}^*) + h_{ij}(\mathbf{x}^*)\frac{\partial f_j(\mathbf{x}^*)}{\partial x_k}\right] = 0, \tag{5.38}$$

where h_{ij} denote the elements of $\mathbf{H}$ and δ_{ik} is the Kronecker delta; that is,

$$\delta_{ik} = \begin{cases} 0, & \text{if } i \neq k, \\ 1, & \text{if } i = k. \end{cases}$$

Since $f_j(\mathbf{x}^*) = 0$ and the Jacobian matrix $\mathbf{J}(\mathbf{x})$ of $\mathbf{f}(\mathbf{x})$ is defined by

$$(\mathbf{J}(\mathbf{x}^*))_{jk} = \partial f_j(\mathbf{x}^*)/\partial x_k, \tag{5.39}$$

equation (5.38) is equivalent to

$$\mathbf{I} + \mathbf{H}(\mathbf{x}^*)\mathbf{J}(\mathbf{x}^*) = \mathbf{0},$$

where $\mathbf{I}$ is the identity matrix and $\mathbf{0}$ is the zero matrix. Assuming that $\mathbf{J}(\mathbf{x}^*)$ is nonsingular, we have

$$\mathbf{H}(\mathbf{x}^*) = - \mathbf{J}^{-1}(\mathbf{x}^*),$$

and so we can choose

$$\mathbf{H}(\mathbf{x}) = - \mathbf{J}^{-1}(\mathbf{x})$$

for all values of $\mathbf{x}$. The corresponding iteration rule has the form

$$\mathbf{x}^{(k+1)} = \mathbf{g}(\mathbf{x}^{(k)}) = \mathbf{x}^{(k)} - \mathbf{J}^{-1}(\mathbf{x}^{(k)})\mathbf{f}(\mathbf{x}^{(k)}). \tag{5.40}$$

Formula (5.40) is called the *generalized Newton method*.

By construction, if $\mathbf{g}$ is the generalized Newton iteration operator defined in (5.40), then according to (5.37) the first partial derivatives of each coordinate $g_i(\mathbf{x}^*)$ of $\mathbf{g}(\mathbf{x}^*)$ are zero, and consequently its second-degree Taylor's series expansion about $\mathbf{x}^*$ has no linear term and may be written, for each i, $1 \leq i \leq n$, as

$$g_i(\mathbf{x}) - g_i(\mathbf{x}^*) = \frac{1}{2} \sum_{j,k=1}^{n} \frac{\partial^2}{\partial x_j \, \partial x_k} g_i(\boldsymbol{\xi}^{(i)})(x_j - x_j^*)(x_k - x_k^*), \tag{5.41}$$

for $\boldsymbol{\xi}^{(i)}$ on the line joining $\mathbf{x}$ and $\mathbf{x}^*$. Letting M be as in Theorem 5.7, we define

$$C = \sup_{x \in M} \max_{i,j,k} n^2 \left| \frac{\partial^2}{\partial x_j \, \partial x_k} g_i(\mathbf{x}) \right|,$$

and hence in the notation of Theorem 4.7,

$$| g_i(\mathbf{x}) - g_i(\mathbf{x}^*) | \leq \tfrac{1}{2} C (\| \mathbf{x} - \mathbf{x}^* \|_\infty)^2. \tag{5.42}$$

This relationship implies that

$$\| \mathbf{g}(\mathbf{x}) - \mathbf{g}(\mathbf{x}^*) \|_\infty \leq \tfrac{1}{2} C (\| \mathbf{x} - \mathbf{x}^* \|_\infty)^2, \tag{5.43}$$

and one may conclude under the circumstances that if $f(\mathbf{x})$ has continuous second partial derivatives and the Jacobian matrix is nonsingular at the root, and if the iteration process determined by the generalized Newton method converges at all, it converges quadratically, as in the one-dimensional case.

In Chapter 6, it will be noted that the inversion of matrices requires more operations than the solution of systems of linear equations. So instead of applying equation (5.40) at each step we solve the equivalent linear system

$$\mathbf{J}(\mathbf{x}^{(k)})\mathbf{x}^{(k+1)} = \mathbf{J}(\mathbf{x}^{(k)})\mathbf{x}^{(k)} - \mathbf{f}(\mathbf{x}^{(k)}), \qquad (5.44)$$

where the matrix $\mathbf{J}(\mathbf{x}^{(k)})$ and vector $\mathbf{J}(\mathbf{x}^{(k)})\mathbf{x}^{(k)} - \mathbf{f}(\mathbf{x}^{(k)})$ are known.

The *modified generalized Newton method* is obtained if the Jacobian $\mathbf{J}(\mathbf{x}^{(k)})$ is approximated by $\mathbf{J}(\mathbf{x}^{(0)})$.

Example 5.10. Recall that in Section 5.2.3 we concluded that with respect to some given polynomial $f(z)$, if p and q are the roots to the system

$$b_{n-1}(p, q) = 0, \qquad b_n(p, q) = 0,$$

then $x^2 + px + q$ is a factor of the given polynomial $f(x)$. Consequently the roots of $x^2 + px + q$ are also roots of $f(x)$. In Example 5.9, we derived that $b_{n-1}(p, q) = -1 + 3p + p^2 - q$ and $b_n(p, q) = 9 + 3q + pq$ are the functions associated with $f(x) = x^3 - 3x^2 - x + 9$. With respect to the vector-valued function

$$\begin{pmatrix} b_{n-1}(p, q) \\ b_n(p, q) \end{pmatrix},$$

we easily calculate that the Jacobian $\mathbf{J}$ is given by

$$\mathbf{J}(p, q) = \begin{pmatrix} 2p + 3 & -1 \\ q & p + 3 \end{pmatrix}.$$

Letting $p = -4$ and $q = 6$, and applying the generalized Newton method to this example, we obtain Table 5.8.

Table 5.8. Application of the Generalized Newton Method in Bairstow's Method

Iteration	p	q	$b_{n-1}(p, q)$	$b_n(p, q)$
0	-4	6	-3	3
1	-4.545454545	5.727272727	0.2975206611	0.1487603306
2	-4.524910282	5.899660152	0.0004220619	0.0035475739
3	-4.525102279	5.901243759	369×10^{-10}	-304×10^{-10}
4	-4.525102255	5.901243652	—	—

Upon using the binomial formula on $x^2 + px + q$ with p and q as given in the last iteration of Table 5.8, we find the (approximate) roots

$$z_1, z_2 = 2.262551128 \pm i\, 0.8843675975.$$

For these values $f(z_1)$, $f(z_2) = 17 \times 10^{-10} \pm i\, 11 \times 10^{-10}$.

Problems

5.25. Consider equation (5.32), where $\partial g_i(x^*)/\partial x_j = 0$ for $j < i$. Assuming that function g is sufficiently many times differentiable and the partial derivatives are bounded, prove that the corresponding method (5.34) has at least quadratic convergence.

5.26. Apply the generalized Newton method to finding roots of the nonlinear system

$$\sin(x_1 + x_2) - \exp(x_1 - x_2) = 0,$$
$$\cos(x_1 + x_2) - x_1^2 x_2^2 = 0.$$

5.3.2. The Gradient Method and Related Techniques

Let us return to the approach in equation (5.28) of finding the root of the function $\mathbf{f}(\mathbf{x})$ by minimizing another function $\phi(\mathbf{x})$. In this section, we shall discuss a prominent method for function minimization known as the gradient method. The gradient method is one of the first iterative methods for multivariate function minimization. Its convergence properties, some of which are to be explored here, are relatively well understood, and it remains a popular and convenient method. Moreover, some of the more elegant newer nonlinear programming methods are best understood as being modifications of the gradient method. In this section, it will not concern us that the function ϕ is defined in terms of another function $\mathbf{f}$. Our only assumption will be that $\phi(\mathbf{x})$ has first partial derivatives. We define the *gradient* $\nabla\phi(\mathbf{x})$ of ϕ by the following formula:

$$\nabla\phi(\mathbf{x}) = \mathbf{g}(\mathbf{x}) = \left(\frac{\partial\phi(\mathbf{x})}{\partial x_1}, \frac{\partial\phi(\mathbf{x})}{\partial x_2}, \dots, \frac{\partial\phi(\mathbf{x})}{\partial x_n} \right)^{\mathrm{T}}.$$

Note that the usage of $\mathbf{g}$ in this section differs from that of Section 5.3.1. We now describe the *gradient method* (also called the *method of steepest descent*). Let $\mathbf{x}^{(0)}$ be some n-tuple. Assume $\mathbf{g}^{(0)} = \mathbf{g}(\mathbf{x}^{(0)}) \neq \mathbf{0}$. We define the positive number $t^{(0)}$ to be a minimizer over all positive numbers, of the real function

$$h_0(t) = \phi(\mathbf{x}^{(0)} - t\mathbf{g}^{(0)}). \tag{5.45}$$

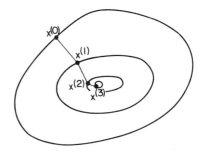

Fig. 5.7. Location of successive points in the gradient method.

We define

$$\mathbf{x}^{(1)} = \mathbf{x}^{(0)} - t^{(0)}\mathbf{g}^{(0)},$$

and more generally, we define recursively

$$\mathbf{x}^{(j+1)} = \mathbf{x}^{(j)} - t^{(j)}\mathbf{g}^{(j)}, \tag{5.46}$$

where $\mathbf{g}^{(j)} = \mathbf{g}(\mathbf{x}^{(j)})$ and $t^{(j)}$ is the positive minimizer of $h_j(t) = \phi(\mathbf{x}^{(j)} - t\mathbf{g}^{(j)})$.

It is clear that if $\mathbf{g}^{(j)} \neq \mathbf{0}$ and the minimizer $t^{(j)}$ of h_j exists, then $\phi(\mathbf{x}^{(j)}) < \phi(\mathbf{x}^{(j-1)})$, for each j. Figure 5.7 illustrates the typical behavior of the gradient method in the neighborhood of a local minimum of a well-behaved function. The closed curves are contour curves, that is, points on which ϕ is constant.

Through two theorems to follow, we investigate the convergence behavior of the gradient method. It is possible to prove that no deterministic procedure can guarantee convergence to a global minimum over such a large class as the set of continuously first differentiable functions. This fact is seen in Yakowitz and Fisher (1973) (where it is proven, on the other hand, that one can achieve convergence in probability to a global minimum through a random search). However, we now see that the gradient method can achieve convergence to *stationary points*; that is, to points at which the gradient $\nabla \phi$ is the zero vector. It is well known (e.g., Luenberger, 1973) that a necessary condition that $\mathbf{x}^*$ minimize ϕ is that $\mathbf{x}^*$ be a stationary point. If ϕ is convex, this is also a sufficient condition.

Theorem 5.8. Let ϕ be a function possessing continuous first partial derivatives. Assume ϕ is bounded from below. If $\{\mathbf{x}^{(m)}\}$ is a sequence of vectors constructed from application of the gradient method, then any limit point $\mathbf{x}^*$ of this sequence is a stationary point of ϕ.

Proof. The proof is by contradiction. Suppose that $\mathbf{x}^*$ is a limit point of $\{\mathbf{x}^{(m)}\}$ and $\mathbf{g}^* = \nabla\phi(\mathbf{x}^*) \neq \mathbf{0}$. Then some subsequence $\{\mathbf{x}^{(n_m)}\}$ converges to $\mathbf{x}^*$ and by the continuity assumption, $\mathbf{g}^{(n_m)} = \mathbf{g}(\mathbf{x}^{(n_m)}) \to \mathbf{g}^*$ as m increases. Since $\mathbf{g}^* \neq \mathbf{0}$, there is some scalar t^* that satisfies.

$$\phi(\mathbf{x}^* - t^*\mathbf{g}^*) < \phi(\mathbf{x}^*). \tag{5.47}$$

It is clear that

$$\phi(\mathbf{x}^{(n_v)}) \to \phi(\mathbf{x}^*) \tag{5.48}$$

for any subsequence $\{\mathbf{x}^{(n_v)}\}$ of $\{\mathbf{x}^{(m)}\}$. But we have

$$\phi(\mathbf{x}^{(n_m+1)}) = \phi(\mathbf{x}^{(n_m)} - t^{n_m}\mathbf{g}^{(n_m)})$$
$$\leq \phi(\mathbf{x}^{(n_m)} - t^*\mathbf{g}^{(n_m)}), \tag{5.49}$$

and thus

$$\lim_{m\to\infty} \phi(\mathbf{x}^{(n_m+1)}) \leq \phi(\mathbf{x}^* - t^*\mathbf{g}^*),$$

in contradiction to (5.48).   ∎

Let $\boldsymbol{\phi}''(\mathbf{x})$ denote the matrix of second partial derivatives of ϕ. That is, $(\boldsymbol{\phi}''(\mathbf{x}))_{ij} = (\partial^2/\partial x_i\,\partial x_j)\phi(\mathbf{x})$. This matrix is known as the *Hessian matrix* of ϕ. A symmetric matrix $\mathbf{M}$ is said to be *positive definite* if the quadratic form $Q(\mathbf{x}) = \mathbf{x}^{\mathrm{T}}\mathbf{M}\mathbf{x}$ is positive for all vectors $\mathbf{x} \neq \mathbf{0}$. (This condition is equivalent to $\mathbf{M}$ having all eigenvalues >0.) If ϕ has continuous second partial derivatives, it is evident by examination of Taylor's representation

$$\phi(\mathbf{x}) - \phi(\mathbf{x}^*) = \nabla\phi(\mathbf{x}^*)^{\mathrm{T}}(\mathbf{x} - \mathbf{x}^*) + \tfrac{1}{2}(\mathbf{x} - \mathbf{x}^*)^{\mathrm{T}}\boldsymbol{\phi}''(\boldsymbol{\xi})(\mathbf{x} - \mathbf{x}^*),$$

that if $\mathbf{x}^*$ is a stationary point, $\mathbf{x}^*$ is a global minimum of ϕ if the Hessian matrix $\boldsymbol{\phi}''(\mathbf{y})$ is positive definite for all arguments $\mathbf{y}$. (This condition is equivalent to the condition that ϕ be strictly convex.) If ϕ is only positive definite at a stationary point $\mathbf{x}^*$, $\mathbf{x}^*$ is a strict *local minimum* of ϕ, that is, there is a sphere S centered at $\mathbf{x}^*$ such that $\phi(\mathbf{y}) > \phi(\mathbf{x}^*)$ for all $\mathbf{y} \in S$, $\mathbf{y} \neq \mathbf{x}^*$.

The convergence property of the gradient method is given in the statement below.

Theorem 5.9. Let ϕ have continuous second partial derivatives. Suppose $\mathbf{x}^*$ is a stationary point at which the Hessian $\boldsymbol{\phi}''(\mathbf{x}^*)$ is positive definite. If $\{\mathbf{x}^{(m)}\}$ is a sequence of points obtained by application of the gradient method that converges to $\mathbf{x}^*$, then the sequence $\{\phi(\mathbf{x}^{(m)})\}$ converges linearly to $\phi(\mathbf{x}^*)$, and the coefficient of convergence is no greater than $[(A - a)/$

$(A + a)]^2$, where a and A are, respectively, the smallest and largest eigenvalues of $\phi''(x^*)$.

We refer the reader to Section 7.6 of Luenberger (1973) for a proof.

Coordinate descent methods require choosing at each iteration m, a coordinate number, say $I(m)$, $1 \le I(m) \le n$. The coordinate $I(m) = i$ having been selected, one defines the next point $x^{(m+1)}$ from the current approximation point $x^{(m)}$ by the rule

$$x^{(m+1)} = (x_1^{(m)}, x_2^{(m)}, \ldots, x_{i-1}^{(m)}, t^*, x_{i+1}^{(m)}, \ldots, x_n^{(m)})^T,$$

where t^* is a minimizer of the function

$$h(t) = \phi((x_1^{(m)}, x_2^{(m)}, \ldots, x_{i-1}^{(m)}, t, x_{i+1}^{(m)}, \ldots, x_n^{(m)})^T).$$

Among the popular plans for choosing coordinates is the *cyclic coordinate descent method*, where $I(m) \triangleq$ [remainder $(m/n)] + 1$. Thus the coordinates chosen are $1, 2, \ldots, n, 1, 2, \ldots, n, 1, \ldots$. If the gradient $\nabla \phi(x^{(m)})$ is readily available, a sensible procedure is to choose $I(m)$ to be the coordinate number of the largest (in magnitude) element of the gradient. This method is known as the *Gauss–Southwell method*. Like the gradient method, it is known (Luenberger, 1973, Section 7.8) that the limit points of coordinate descent methods are stationary points of ϕ. The Gauss–Southwell method converges linearly, but the ratio of convergence of $\phi(x^{(m)})$ to $\phi(x^*)$ is, in the terminology of Theorem 5.9, bounded by $1 - [a/A(n - 1)]$, which is generally larger and never less than the coefficient of convergence of the gradient method.

Another important approach to nonlinear programming is the application of the generalized Newton's method for the problem of finding the stationary point by observing the problem of finding a stationary point that is identical to the problem of finding a root of $\nabla \phi$. Thus the results of the preceding section are immediately applicable to nonlinear programming. However, in the nonlinear programming application, the Jacobian used in the generalized Newton's method turns out to be the Hessian ϕ'' of ϕ. From this fact, we may conclude that the Jacobian is symmetric. Under these circumstances, there are procedures available that allow one to avoid the time-consuming step of directly computing the Jacobian by approximating it sequentially. The most prominent among these methods are the *Davidon–Fletcher–Powell method* and the *rank-one correction procedure*. We refer the reader to Chapter 9 of Luenberger (1973) for development of these techniques. Of course, one pays for the ease of computation by having slower convergence than quadratic.

The conjugate gradient method, which is to be discussed in Section 6.2.3, will be seen to have certain theoretical appeal for quadratic and "nearly quadratic" minimization problems in comparison to the gradient method.

5.4. Supplementary Notes and Discussion

Among the methods studied here, we may distinguish certain trends. The techniques with slower convergence (that is, the bisection and regula falsi methods, which achieve only linear convergence) are the more reliable in the sense that they only require that a root be "straddled." This can usually be done relatively easily, and a test for whether x_0 and x_1 straddle a root is simply to check whether the product of $f(x_0)f(x_1)$ is negative. The secant and Newton's methods, which afford higher-order convergence, generally require a "sufficiently close" approximation to the root, and typically, the condition for being sufficiently close is hard to check. We remark that the requirement of evaluating the first derivative in Newton's method is sometimes expensive and may not be worthwhile. The reader may note that if it is more expensive to evaluate $f'(x)$ than $f(x)$, one can achieve faster convergence with a given amount of computer effort by performing two secant iterations in place of one Newton iteration.

The methods for solution of nonlinear equations frequently do not require great programming effort or computer time. For example, all the calculations reported in this chapter were undertaken on a Texas Instruments Model 58 programmable calculator.

The effect of roundoff error on the calculations has not been discussed. Typically, roundoff error is not so bothersome in iterative methods (such as are used exclusively here) as in direct methods. However, our discussion of the effect of error given in Section 4.6 may be used directly to bound the errors to those methods given in this chapter that are contraction mappings.

The reader may wish to consult Young and Gregory (1972, Chapters 4 and 5), for additional nonlinear equation techniques. In particular, Muller's method described therein is another popular method for approximating roots of polynomials. We note that the trend is to avoid calculating roots of characteristic equations arising in matrix theory by finding eigenvalues directly. Methods for this task will be the topic of Chapter 7.

The book by Ortega and Rheinboldt (1970) is entirely devoted to iterative solution of systems of nonlinear equations. Luenberger (1973) is recommended for further study of nonlinear programming.

6

The Solution of Simultaneous Linear Equations

Simultaneous linear equations are equations of the form

$$\sum_{j=1}^{n} a_{ij}x_j = b_i, \qquad 1 \le i \le m, \tag{6.1}$$

where the coefficients a_{ij} and b_i are real or complex numbers. Equation (6.1) can be written in matrix form as

$$\mathbf{Ax} = \mathbf{b}, \tag{6.2}$$

where $\mathbf{A} = (a_{ij})$ is an $m \times n$ matrix of coefficients $\mathbf{x} = (x_1, \ldots, x_n)^{\mathrm{T}}$ and $\mathbf{b} = (b_1, \ldots, b_m)^{\mathrm{T}}$, the superscript T denoting "transpose."

We have already seen that many of the effective numerical analysis techniques require, as a key step, solving simultaneous linear equations. Specifically, in polynomial approximation, the method of least squares and spline approximation reduced to equations of the form (6.1). In the theory of numerical integration, one method for finding the coefficients for a Newton–Cotes formula involves solving a linear equation, as does the generalized Newton method for nonlinear equations. As we shall see in later chapters, popular methods for solving a great many linear and nonlinear differential equations proceed by approximating the differential equation by linear equations of the form (6.1). These cases give us some inkling of the central role linear equations play in numerical analysis and consequently in applied mathematics and engineering. It is almost impossible to overestimate the importance of linear equations and how much hinges on our being able to find accurate and efficient methods to solve them. This is the subject of the present chapter.

In this study, we shall restrict ourselves to the case in which matrix **A** in equation (6.2) is square and of order n. Some familiar methods for solving "textbook-sized" linear equations with $n = 2$ or 3 by hand calculation are totally inadequate for higher-dimension systems. For example, in order to solve a 20th-order linear system by Cramer's rule, about 5×10^{19} multiplications are required, a task that would require on the order of 10^7 years on a CDC 6400 computer. Yet people routinely solve 100 variable systems and instances have been cited of solution of equations with over 1000 variables.

6.1. Direct Methods

Direct methods, as opposed to iteration methods (which are to be discussed later), are characterized by the property that the order n having been fixed, these methods provide the exact answer, except for error introduced by roundoff, within a predetermined number of steps.

For linear equations of moderate order n (say $n < 100$), direct methods are the methods of choice by virtue of their accuracy and reliability. For larger-order systems, the computational cost of direct methods make iterative methods more competitive, especially for sparse systems typical of certain partial differential equations. The popular direct methods are all motivated by Gaussian elimination, which we now describe.

6.1.1. Gaussian Elimination

Let

$$\mathbf{A}\mathbf{x} = \mathbf{b} \tag{6.3}$$

denote the linear equation to be solved and assume that $m = n$ and the matrix of coefficients is nonsingular. Then the system (6.3) has the form

$$a_{11}x_1 + a_{12}x_2 + \cdots + a_{1n}x_n = a_{1,n+1},$$
$$a_{21}x_1 + a_{22}x_2 + \cdots + a_{2n}x_{2n} = a_{2,n+1},$$
$$\vdots \tag{6.4}$$
$$a_{n1}x_1 + a_{n2}x_2 + \cdots + a_{nn}x_n = a_{n,n+1},$$

where $a_{i,n+1} = b_i$ $1 \le i \le n$. Since matrix $\mathbf{A} = (a_{ij})$ is nonsingular, at least one element in the first column must differ from zero. Interchanging a row containing the nonzero element with the first row (if necessary) we have $a_{11} \ne 0$. By subtracting the multiple a_{i1}/a_{11} of the first equation from the ith equation ($2 \le i \le n$), we obtain the first *derived system*

$$a_{11}x_1 + a_{12}x_2 + \cdots + a_{1n}x_n = a_{1,n+1},$$
$$a_{22}^{(1)}x_2 + \cdots + a_{2n}^{(1)}x_n = a_{2,n+1}^{(1)},$$
$$\vdots \qquad (6.5)$$
$$a_{n2}^{(1)}x_2 + \cdots + a_{nn}^{(1)}x_n = a_{n,n+1}^{(1)},$$

where

$$a_{ij}^{(1)} = a_{ij} - \frac{a_{i1}}{a_{11}}a_{1j}, \qquad 2 \le i \le n, \; 2 \le j \le n + 1.$$

The matrix of coefficients of system (6.5) is nonsingular, because elementary row operations do not cause change in the matrix rank (see Hadley, 1969, p. 144), and so at least one of the numbers $a_{22}^{(1)}$, ..., $a_{n2}^{(1)}$ must differ from zero. If it is necessary (as in the case of the original equations) by interchanging two rows we can have $a_{22}^{(1)} \ne 0$. We subtract $a_{i2}^{(1)}/a_{22}^{(1)}$ times the second equation from the ith equation in (6.5) for $i = 3, 4, \ldots, n$ to obtain the second derived system

$$a_{11}x_1 + a_{12}x_2 + a_{13}x_3 + \cdots + a_{1n}x_n = a_{1,n+1},$$
$$a_{22}^{(1)}x_2 + a_{23}^{(1)}x_3 + \cdots + a_{2n}^{(1)}x_n = a_{2,n+1}^{(1)},$$
$$a_{33}^{(2)}x_3 + \cdots + a_{3n}^{(2)}x_n = a_{3,n+1}^{(2)},$$
$$\vdots \qquad (6.6)$$
$$a_{n3}^{(2)}x_3 + \cdots + a_{nn}^{(2)}x_n = a_{n,n+1}^{(2)}.$$

Repeating this process until the $(n - 1)$th derived system, we compute

$$a_{11}x_1 + a_{12}x_2 + a_{13}x_3 + \cdots + a_{1n}x_n = a_{1,n+1},$$
$$a_{22}^{(1)}x_2 + a_{23}^{(1)}x_3 + \cdots + a_{2n}^{(1)}x_n = a_{2,n+1}^{(1)},$$
$$a_{33}^{(2)}x_3 + \cdots + a_{3n}^{(2)}x_n = a_{3,n+1}^{(2)},$$
$$\vdots \qquad (6.7)$$
$$a_{nn}^{(n-1)}x_n = a_{n,n+1}^{(n-1)},$$

where the recursive relationship for obtaining the coefficients has the form

$$a_{ij}^{(k)} = a_{ij}^{(k-1)} - \frac{a_{ik}^{(k-1)}}{a_{kk}^{(k-1)}} a_{kj}^{(k-1)},$$

$$1 \le k \le n - 1, \ k + 1 \le j \le n + 1, \ k + 1 \le i \le n, \qquad (6.8)$$

where, of course, $a_{ij}^{(0)} = a_{ij}$, $1 \le i \le n$, $1 \le j \le n + 1$.

The value x_n can be obtained from the last equation of (6.7). Knowing the value of x_n, x_{n-1} can be evaluated from the $(n - 1)$th equation, etc. The (backward) recursive formula for obtaining the value of the solution x_i in terms of the already-calculated coordinates x_j, $j > i$, is

$$x_i = \frac{1}{a_{ii}^{(i-1)}} \left[a_{i,n+1}^{(i-1)} - \sum_{j=i+1}^{n} a_{ij}^{(i-1)} x_j \right], \qquad i = n, \ n - 1, \ \ldots, \ 1. \qquad (6.9)$$

Procedure (6.9) is known as *back-substitution*. The entire process (6.8) and (6.9) constitutes *Gaussian elimination*.

Example 6.1. We solve the system of equations

$$2x_1 + x_2 + x_3 = 4$$
$$x_1 + 3x_2 + 2x_3 = 6$$
$$x_1 + 2x_2 + 2x_3 = 5.$$

By using the Gaussian elimination method, we successively obtain the derived systems Table 6.1. From the second derived system and back-substitution, the solution values are seen to be

$$x_1 = x_2 = x_3 = 1.$$

Table 6.1. Derived Systems for Example 6.1

2	1	1	4
1	3	2	6
1	2	2	5
2	1	1	4
0	5/2	3/2	4
0	3/2	3/2	3
2	1	1	4
0	5/2	3/2	4
0	0	3/5	3/5

Now we calculate the total number of arithmetic operations needed for Gaussian elimination. With reference to formula (6.8) for obtaining the kth derived system, for fixed i, j, k, one addition and one multiplication are required. For fixed k, there are $n - k + 1$ values for j and $n - k$ values for i, and so $(n - k)(n - k + 1)$ additions and multiplications are required. The quotient in (6.8) does not depend on j, and so for each k, only $n - k$ divisions are required. Inasmuch as k ranges from 1 to $n - 1$, the number of multiplications and additions needed to evaluate (6.8) for all i, j, k is given by

$$\sum_{k=1}^{n-1} (n - k)(n - k + 1) = \sum_{k=1}^{n-1} (n - k)^2 + \sum_{k=1}^{n-1} (n - k) = \frac{n^3}{3} - \frac{n}{3}.$$

Similarly, the number of divisions equals $\sum_{k=1}^{n-1} (n - k) = n(n - 1)/2$. In the back-substitution formula (6.9), for each i, we require one division and $n - i$ additions and multiplications. Summing over the range of i, we conclude that back-substitution requires $n(n - 1)/2$ additions and multiplications and n divisions. Adding these terms together, we see that the number of additions and multiplications grows at the asymptotic rate of $n^3/3$. Klyuyev and Kokovkin-Shcherbak (1965) have proven that no method constrained to row and column operations can take fewer operations to solve an n-dimensional linear equation. However, Strassen (1969) has provided an arithmetic method in which the number of operations is proportional to

$$n^{\log_2 7} \simeq n^{2.8}.$$

We now discuss computation of the determinant of $\mathbf{A}$ from the $(n - 1)$th derived system, $\mathbf{A}^{(n-1)}$. If $\mathbf{C}$ is any square matrix, we shall denote its determinant by det $\mathbf{C}$. The two basic types of matrix operations involved in proceeding from one derived system to the next are (1) interchange of rows and (2) subtraction of a multiple of one row from a different row. From elementary matrix algebra (Nering, 1963, p. 77) it is known that if $\mathbf{A}$ and $\mathbf{A}_1$ differ by an interchange of two rows, then det $\mathbf{A} = -$ det $\mathbf{A}_1$, and furthermore, operation (2) leaves the determinant unchanged. It is evident that the determinant of an upper-triangular matrix such as $\mathbf{A}^{(n-1)}$ is simply the product of its diagonal elements. In summary, we have the following rule for calculating the determinant of $\mathbf{A}$ from the $(n - 1)$th derived matrix, $\mathbf{A}^{(n-1)} = (a_{ij}^{(i-1)})$:

$$\det \mathbf{A} = (-1)^p \prod_{i=1}^{n} a_{ii}^{(i-1)},$$

p being the number of times the row interchange operation has been used during the entire computation of derived systems.

Example 6.2. The determinant of the matrix of the system given in Example 6.1 is equal to

$$(-1)^0 \times 2 \times \frac{5}{2} \times \frac{3}{5} = 3.$$

We now describe an efficient way to use Gaussian elimination to produce the inverse of a matrix $\mathbf{A}$. Let $\mathbf{A}$ be any nonsingular matrix. Let the columns of the inverse of $\mathbf{A}$ be denoted by $x_1, x_2, \ldots, x_n$, and the columns of the unit matrix be denoted by $e_1, e_2, \ldots, e_n$. By the definition of the inverse of matrices we have

$$\mathbf{A} \cdot \mathbf{A}^{-1} = \mathbf{A} \cdot (x_1, x_2, \ldots, x_n) = (e_1, e_2, \ldots, e_n),$$

which is equivalent to the collection of systems

$$\mathbf{A}x_1 = e_1,$$
$$\mathbf{A}x_2 = e_2,$$
$$\vdots$$
$$\mathbf{A}x_n = e_n.$$

Since these systems have the same matrix of coefficients, Gaussian elimination can be done simultaneously, using n right-hand side vectors, instead of only one as in (6.8).

It is to be emphasized that in calculating the inverse of $\mathbf{A}$, one needs to compute the derived systems (having $2n$ columns) once, but back-substitution must be done n times. From earlier considerations, we see that this inversion procedure requires, asymptotically, $\frac{4}{3}n^3$ additions and multiplications and $\frac{3}{2}n^2$ divisions.

We remark that one seldom needs to actually find the inverse of a matrix. For example, if the term $\mathbf{A}^{-1}\mathbf{b}$ appears in an equation, one should solve the linear equation $\mathbf{A}x = \mathbf{b}$, and substitute x for this term, cutting the number of computations to about $\frac{1}{4}$ of those necessary for inverting $\mathbf{A}$.

Example 6.3. Now we determine the inverse of the matrix of coefficients of the equation given in Example 6.1. The results are shown in Table 6.2.

By using rule (6.9) successively on the three right-hand columns of the final reduced form in Table 6.2, one obtains the following matrix of the associated three solution vectors:

$$\begin{pmatrix} 2/3 & 0 & -1/3 \\ 0 & 1 & -1 \\ -1/3 & -1 & 5/3 \end{pmatrix}.$$

One may directly confirm that this matrix is the inverse of $\mathbf{A}$.

Table 6.2. Derived Systems for Matrix Inversion

2	1	1	1	0	0
1	3	2	0	1	0
1	2	2	0	0	1
2	1	1	1	0	0
0	5/2	3/2	−1/2	1	0
0	3/2	3/2	−1/2	0	1
2	1	1	1	0	0
0	5/2	3/2	−1/2	1	0
0	0	3/5	−1/5	−3/5	1

In applying the method of elimination, at each step we have to divide by $a_{ii}^{(i-1)}$, which is known as a *pivot*. In Table 1.1 we saw that the absolute value of the rounding error of division is proportional to the square of the denominator. This fact implies that the rounding error bound due to division will be smallest if we choose as the ith pivot element the coordinate $a_{j*k*}^{(i-1)}$ that maximizes $|a_{jk}^{(i-1)}|$, $j, k \geq i$. The element $a_{j*k*}^{(i-1)}$ is then put in the pivot position by interchanging rows i and $j*$ and columns i and $k*$. This operation is known as *pivoting for maximum size* or, more simply, *maximal pivoting*. Often, the search involved in maximal pivoting is prohibitively expensive, for it involves finding the maximum of $(n - i)^2$ numbers before computing the $(i - 1)$th derived system. At the expense of some increase in rounding error, a popular procedure is to do partial pivoting. In *partial pivoting*, for computing the pivot of the $(i - 1)$th derived system, one locates the coordinate $\varrho \geq i$ such that $|a_{\varrho i}| \geq |a_{ji}|$, $j \geq i$. Then the ϱth and ith rows are interchanged, and the computation of the $(i - 1)$th derived system proceeds. In summary, partial pivoting allows only row interchanges, whereas maximal pivoting allows row and column interchanges to maximixe the ith pivot.

Let us now consider the general case of systems of linear equations having form (6.2). The elimination of unknowns terminates when there is no nonzero element in the lower rows. Strictly speaking, we may have to interchange column and corresponding coordinates of $\mathbf{x}$ and $\mathbf{b}_{k-1}$ to guarantee this form. That is, the elimination scheme has achieved the form

$$\begin{pmatrix} \mathbf{A}_{k-1} & \mathbf{B} \\ \mathbf{0} & \mathbf{0} \end{pmatrix} \mathbf{x} = \mathbf{b}_{k-1}, \tag{6.10}$$

where $\mathbf{A}_{k-1}$, $k \geq 1$, is an upper triangle matrix of the order k, $\mathbf{b}_{k-1}$ being an m-dimensional vector. Let vectors $\mathbf{b}_{k-1}$ and $\mathbf{x}$ be written in block form as

$$\mathbf{b}_{k-1} = (\mathbf{b}^{(1)}, \mathbf{b}^{(2)})^{\mathrm{T}}, \qquad \mathbf{x} = (\mathbf{x}^{(1)}, \mathbf{x}^{(2)})^{\mathrm{T}},$$

where $\mathbf{b}^{(1)}$, $\mathbf{x}^{(1)}$ are k-dimensional vectors. Then (6.10) can be written as

$$\mathbf{A}_{k-1}\mathbf{x}^{(1)} + \mathbf{B}\mathbf{x}^{(2)} = \mathbf{b}^{(1)}, \qquad \mathbf{0} = \mathbf{b}^{(2)}. \tag{6.11}$$

If $\mathbf{b}^{(2)} \neq \mathbf{0}$, then no solution exists. If $\mathbf{b}^{(2)} = \mathbf{0}$, then system (6.2) is solvable, and the general solution can be obtained as $\mathbf{x} = (\mathbf{x}^{(1)}, \mathbf{x}^{(2)})^{\mathrm{T}}$, where $\mathbf{x}^{(2)}$ is an arbitrary vector, and for each particular value of $\mathbf{x}^{(2)}$ the components of $\mathbf{x}^{(1)}$ can be obtained from the first equation of (6.11) using formula (6.9). By this technique, we have obtained an effective procedure, in the absence of machine error, for checking for the existence of a solution for a linear equation and for finding one, when solutions exist.

Problems

6.1. Solve the system
$$\begin{aligned} 4x_1 - x_2 \quad\quad &= 2, \\ -x_1 + 4x_2 - x_3 &= 6, \\ -x_2 + 4x_3 &= 2, \end{aligned}$$
by Gaussian elimination.

6.2. Compute the determinant of the matrix of the system in Problem 6.1.

6.3. Compute the inverse of the matrix of the system given in Problem 6.1.

6.1.2. Variants of Gaussian Elimination

In this section we assume that the matrix of equation (6.1) is an $n \times n$ nonsingular matrix. We discuss here the properties of several popular modifications of the Gaussian elimination method.

Through Gaussian elimination only the elements under the diagonal are eliminated. The *Gauss–Jordan reduction* can be regarded as a modified version of the Gaussian elimination, where the elements above the diagonal are also eliminated. In this case after the first step in the first derived system, the matrix can be written as (6.5), and after the second step, which involves eliminating terms above the pivot as well as below, the derived system has the form

$$\begin{aligned} a_{11}^{(2)}x_1 + 0 + a_{13}^{(2)}x_3 + \cdots + a_{1n}^{(2)}x_n &= a_{1,n+1}^{(2)}, \\ a_{22}^{(2)}x_2 + a_{23}^{(2)}x_3 + \cdots + a_{2n}^{(2)}x_n &= a_{2,n+1}^{(2)}, \\ a_{33}^{(2)}x + \cdots + a_{3n}^{(2)}x_n &= a_{3,n+1}^{(2)}, \\ &\;\;\vdots \\ a_{n3}^{(2)}x_3 + \cdots + a_{nn}^{(2)}x_n &= a_{n,n+1}^{(2)}. \end{aligned}$$

In Gauss–Jordan reduction therefore, (6.8) is replaced by

$$a_{ij}^{(k)} = \begin{cases} a_{ij}^{(k-1)} - \dfrac{a_{ik}^{(k-1)}}{a_{kk}^{(k-1)}} a_{kj}^{(k-1)}, & 1 \le k \le n-1, \; j \ge k, \; i \ne k, \\ a_{ij}^{(k-1)}, & i = k \quad \text{or} \quad j < k. \end{cases}$$

As before $a_{ij}^{(0)} = a_{ij}$, $1 \le i, j \le n$. At the end of the elimination process we have a system associated with a diagonal matrix of coefficients:

$$\begin{aligned}
a_{11}^{(n)} x_1 && = a_{1,n+1}^{(n)}, \\
a_{22}^{(n)} x_2 && = a_{2,n+1}^{(n)}, \\
a_{33}^{(n)} x_3 && = a_{3,n+1}^{(n)}, \\
&\vdots& \\
a_{nn}^{(n)} x_n &= a_{n,n+1}^{(n)}.
\end{aligned}$$

(6.12)

The values of the unknowns can be obtained from (6.12) by simple division. Of course, this method can be used also for calculating determinants and inverses of matrices.

One may confirm that solution by the Gauss–Jordan elimination requires $n^3/2 + n^2 - n/2$ additions and multiplications, and for this reason, is inferior to Gauss elimination.

Example 6.4. In solving the system of equations given in Example 6.1 by the Gauss–Jordan method, we obtain Table 6.3. Simple division shows that $x_1 = x_2 = x_3 = 1$.

Table 6.3. Gauss–Jordan Derived Systems for Example 6.1

2	1	1	4
1	3	2	6
1	2	2	5
2	1	1	4
0	5/2	3/2	4
0	3/2	3/2	3
2	0	2/5	12/5
0	5/2	3/2	4
0	0	3/5	3/5
2	0	0	2
0	5/2	0	5/2
0	0	3/5	3/5

Let $\mathbf{A}_i$ denote the matrix of coefficients of the ith derived system in the Gaussian elimination procedure. One may confirm that $\mathbf{A}_i = \mathbf{L}_i \mathbf{A}_{i-1}$, where

$$
\mathbf{L}_i =
\begin{bmatrix}
1 & & & & & & \\
 & \ddots & & & & & \\
 & & 1 & & & & \\
 & & -\dfrac{a_{i+1,i}^{(i-1)}}{a_{ii}^{(i-1)}} & 1 & & & \\
 & & \vdots & & \ddots & & \\
 & & -\dfrac{a_{ni}^{(i-1)}}{a_{ii}^{(i-1)}} & & & 1
\end{bmatrix}
$$

(the elements that are not shown here are equal to zero). Using this observation, Gaussian elimination can be written in matrix form as

$$
\mathbf{A}_{n-1} = \mathbf{L}_{n-1}\mathbf{L}_{n-2} \cdots \mathbf{L}_2\mathbf{L}_1\mathbf{A}. \tag{6.13}
$$

Since

$$
\mathbf{L}_i^{-1} =
\begin{bmatrix}
1 & & & & & & \\
 & \ddots & & & & & \\
 & & 1 & & & & \\
 & & \dfrac{a_{i+1,i}^{(i-1)}}{a_{ii}^{(i-1)}} & 1 & & & \\
 & & \vdots & & \ddots & & \\
 & & \dfrac{a_{ni}^{(i-1)}}{a_{ii}^{(i-1)}} & & & 1
\end{bmatrix}
$$

the matrix $\mathbf{L} = (\mathbf{L}_{n-1}\mathbf{L}_{n-2} \cdots \mathbf{L}_2\mathbf{L}_1)^{-1}$ is lower triangular with ones on the diagonal. If the upper-triangular matrix $\mathbf{A}_{n-1}$ is denoted by $\mathbf{U}$, then (6.13) implies

$$
\mathbf{A} = \mathbf{L}\mathbf{U}. \tag{6.14}
$$

Using equation (6.14) we make a direct computation of the triangular matrices $\mathbf{L}$ and $\mathbf{U}$. From the definition of matrix product and the fact that $\mathbf{L}$ and $\mathbf{U}$ are, respectively, lower and upper triangular, we have immediately from (6.14) that

$$a_{ki} = \sum_{v=1}^{\min\{i,k\}} l_{kv} u_{vi}. \tag{6.15}$$

We can successively calculate the elements in the kth row of $\mathbf{U}$ and from the formula

$$u_{ki} = a_{ki} - \sum_{v=1}^{k-1} l_{kv} u_{vi}, \qquad i = k, \ldots, n. \tag{6.16}$$

This formula follows upon rearranging (6.15) and using the constraint that l_{kk} is 1. Having calculated the kth row of $\mathbf{U}$, we find the kth column of $\mathbf{L}$ by observing that for $i > k$, $u_{ik} = 0$, and (6.15) therefore implies that

$$a_{ik} = \sum_{v=1}^{k} l_{iv} u_{vk},$$

whence we may successively calculate l_{ik}, $i > k$, using

$$l_{ik} = \frac{1}{u_{kk}} \left(a_{ik} - \sum_{v=1}^{k-1} l_{iv} u_{vk} \right). \tag{6.17}$$

The computation begins with $k = 1$, and repeats the calculations (6.16) and (6.17) for successively larger k. By this procedure, all values have been computed by the time they appear in the right-hand side of the above equations.

The preceding method for factoring $\mathbf{A}$ into the product of triangular forms is known as *Crout reduction*. Once $\mathbf{L}$ and $\mathbf{U}$ have been found by this procedure, the equation $\mathbf{A}\mathbf{x} = \mathbf{b}$ can be solved in two steps. First solve the equation

$$\mathbf{L}\mathbf{y} = \mathbf{b}.$$

Since matrix $\mathbf{L}$ is lower triangular, an obvious modification of back-substitution (6.9) can be used. Next solve the equation

$$\mathbf{U}\mathbf{x} = \mathbf{y},$$

again using formula (6.9). Then vector $\mathbf{x}$ is the solution of the original equation, because

$$\mathbf{A}\mathbf{x} = \mathbf{L}\mathbf{U}\mathbf{x} = \mathbf{L}(\mathbf{U}\mathbf{x}) = \mathbf{L}\mathbf{y} = \mathbf{b}.$$

Crout reduction is closely related to Gauss elimination. Like Gauss elimination, it requires in the order of $n^3/3$ additions and multiplications to solve an nth-order system. Crout reduction can be completed if and only if for each k, $u_{kk} \neq 0$. Inasmuch as u_{kk} is identical to the pivots $a_{kk}^{(k-1)}$, Crout reduction is effective if and only if Gaussian elimination does not require pivoting. Crout reduction tends to be more subject to the effects of roundoff error for large n than Gaussian elimination with pivoting for maximal size. But this can be compensated for, in part, by accumulating the inner product terms $\sum_v l_{iv} u_{vk}$ in double precision. In Section 6.3, we shall continue this discussion of error.

Example 6.5. Let us consider again the matrix of Example 6.1. Our object is to find the coordinates in the decomposition

$$\begin{pmatrix} 2 & 1 & 1 \\ 1 & 3 & 2 \\ 1 & 2 & 2 \end{pmatrix} = \begin{pmatrix} 1 & 0 & 0 \\ l_{21} & 1 & 0 \\ l_{31} & l_{32} & 1 \end{pmatrix} \begin{pmatrix} u_{11} & u_{12} & u_{13} \\ 0 & u_{22} & u_{23} \\ 0 & 0 & u_{33} \end{pmatrix}.$$

For $k = 1$, (6.16) gives us that

$$u_{11} = a_{11} = 2,$$
$$u_{12} = a_{12} = 1,$$
$$u_{13} = a_{13} = 1.$$

since l_{11} is constrained to be 1. Equation (6.17) gives

$$l_{21} = a_{21}/u_{11} = \tfrac{1}{2}, \qquad l_{31} = a_{31}/u_{11} = \tfrac{1}{2}.$$

For $k = 2$,

$$u_{22} = a_{22} - l_{21}u_{12} = 3 - \tfrac{1}{2} = \tfrac{5}{2},$$
$$u_{23} = a_{23} - l_{21}u_{13} = 2 - \tfrac{1}{2} = \tfrac{3}{2},$$
$$l_{32} = 1/u_{22}(a_{32} - l_{31}u_{12}) = \tfrac{2}{5}(2 - \tfrac{1}{2}) = \tfrac{3}{5}.$$

Finally, for $k = 3$,

$$u_{33} = a_{33} - l_{31}u_{13} - l_{32}u_{23}$$
$$= 2 - \tfrac{1}{2} - \tfrac{3}{5}\tfrac{3}{2} = \tfrac{3}{5}.$$

As a consequence, the representation (6.14) has the following form:

$$\mathbf{A} = \begin{pmatrix} 2 & 1 & 1 \\ 1 & 3 & 2 \\ 1 & 2 & 2 \end{pmatrix} = \mathbf{LU} = \begin{pmatrix} 1 & 0 & 0 \\ \tfrac{1}{2} & 1 & 0 \\ \tfrac{1}{2} & \tfrac{3}{5} & 1 \end{pmatrix} \begin{pmatrix} 2 & 1 & 1 \\ 0 & \tfrac{5}{2} & \tfrac{3}{2} \\ 0 & 0 & \tfrac{3}{5} \end{pmatrix}.$$

In this case equation $\mathbf{Ly} = \mathbf{b}$ can be written as

$$y_1 \qquad\qquad\quad = 4,$$
$$\tfrac{1}{2}y_1 + y_2 \qquad\quad = 6,$$
$$\tfrac{1}{2}y_1 + \tfrac{3}{5}y_2 + y_3 = 5.$$

Consequently $y_1 = 4$, $y_2 = 4$, $y_3 = \frac{3}{5}$. The equation $\mathbf{U}x = y$ has the form

$$2x_1 + x_2 + x_3 = 4,$$
$$\tfrac{3}{2}x_2 + \tfrac{3}{2}x_3 = 4,$$
$$\tfrac{3}{5}x_3 = \tfrac{3}{5},$$

which implies $x_1 = x_2 = x_3 = 1$.

When $\mathbf{A}$ is real and symmetric, the triangular decomposition calculation above can be greatly simplified by using the representation

$$\mathbf{A} = \mathbf{U}^{\mathrm{T}}\mathbf{U}, \tag{6.18}$$

where $\mathbf{U}$ is an upper-triangular matrix. In this case equations (6.15) have the form

$$a_{ij} = u_{1i}u_{1j} + u_{2i}u_{2j} + \cdots + u_{ii}u_{ij}, \qquad i < j,$$
$$a_{ii} = u_{1i}^2 + u_{2i}^2 + \cdots + u_{ii}^2. \tag{6.19}$$

From (6.19), we have

$$u_{ii} = \left(a_{ii} - \sum_{v=1}^{i-1} u_{vi}^2 \right)^{1/2},$$
$$u_{ij} = \frac{1}{u_{ii}}\left(a_{ij} - \sum_{v=1}^{i-1} u_{vi}u_{vj} \right), \qquad j > i, \tag{6.20}$$

from which we can compute successive rows of $\mathbf{U}$.

The use of the decomposition (6.20) is called *Cholesky's method* or the method of square roots. We remark that the elements u_{ij} can be complex numbers even though the matrix $\mathbf{A}$ is real. This fact is an important disadvantage of this method.

The Crout reduction and Cholesky's method may not be applied to certain matrices because it is possible that in either (6.17) or (6.20), the divisor may be 0. If this is the case, Gaussian elimination with pivoting should be used. Crout reduction and Cholesky's method can be carried out whenever the pivots in Gaussian elimination without permutation of rows are all nonzero. A necessary and sufficient condition for nonzero pivots is that the principal leading submatrices (i.e., the matrices consisting of the first k rows and columns of $\mathbf{A}$) all be nonsingular (Stewart, 1973, p. 120). It is well known that $\mathbf{A}$ satisfies this condition if $\mathbf{A}$ is positive or negative definite. The reader may confirm that if $\mathbf{A}$ is symmetric, Gauss elimination can be modified to be as efficient as Cholesky's method.

Example 6.6. Since the matrix of Example 6.1 is symmetric, the decomposition (6.20) can be used. The matrix equation

$$\begin{pmatrix} 2 & 1 & 1 \\ 1 & 3 & 2 \\ 1 & 2 & 2 \end{pmatrix} = \begin{pmatrix} u_{11} & 0 & 0 \\ u_{12} & u_{22} & 0 \\ u_{13} & u_{23} & u_{33} \end{pmatrix} \begin{pmatrix} u_{11} & u_{12} & u_{13} \\ 0 & u_{22} & u_{23} \\ 0 & 0 & u_{33} \end{pmatrix}$$

gives the following relationships:

$$u_{11}^2 = 2, \qquad u_{12}^2 + u_{22}^2 = 3,$$
$$u_{12}u_{11} = 1, \qquad u_{13}u_{12} + u_{23}u_{22} = 2,$$
$$u_{13}u_{11} = 1, \qquad u_{13}^2 + u_{23}^2 + u_{33}^2 = 2,$$

which imply

$$\mathbf{U} = \begin{pmatrix} 2^{1/2} & 2^{1/2}/2 & 2^{1/2}/2 \\ 0 & (10)^{1/2}/2 & 3(10)^{1/2}/2 \\ 0 & 0 & (15)^{1/2}/5 \end{pmatrix}.$$

Problems

6.4. Show that the method of Crout reduction requires $n^3/3 + O(n^2)$ multiplications and divisions.

6.5. Show that Cholesky's method requires the calculation of the square roots of n numbers and $n^3/6 + O(n^2)$ multiplications and divisions.

6.6. Show that the total number of multiplications and divisions in the case of the Gauss–Jordan algorithm is

$$n^3/2 + O(n^2),$$

and as a result for large n requires about 50% more operations than Gaussian elimination.

6.7. Solve Problems 6.1, 6.2, and 6.3 by the Gauss–Jordan method.

6.8. Solve Problem 6.1 by Crout reduction.

6.9. A square matrix **A** is called a *band matrix* of *bandwidth* $2k + 1$ if $|i - j| > k$ implies $a_{ij} = 0$. How many additions, multiplications, and divisions are required for the Gauss elimination solution of a band matrix of bandwidth $2k + 1$? Observe that for k fixed, the required number of arithmetic operations grows only in proportion to n (instead of n^3).

6.1.3. Inversion by Partitioning

Assume that matrix **A** has the form

$$\mathbf{A} = \begin{pmatrix} \mathbf{P} & \mathbf{Q} \\ \mathbf{R} & \mathbf{S} \end{pmatrix}, \qquad (6.21)$$

where $\mathbf{P}$ and $\mathbf{S}$ are square matrices of order k and $n - k$, respectively, and $\mathbf{Q}$ is $k \times (n - k)$ and $\mathbf{R}$ is $(n - k) \times k$, and $\mathbf{P}^{-1}$ is known or easily obtained. In this section we show that the inverse matrix $\mathbf{A}^{-1}$ can be calculated in an efficient way using the inverse of the minor $\mathbf{P}$. Let us represent the matrix $\mathbf{A}^{-1}$ in block form with partitions of the same dimension as (6.21):

$$\mathbf{A}^{-1} = \begin{pmatrix} \mathbf{Y} & \mathbf{Z} \\ \mathbf{U} & \mathbf{V} \end{pmatrix}.$$

The equation

$$\begin{pmatrix} \mathbf{P} & \mathbf{Q} \\ \mathbf{R} & \mathbf{S} \end{pmatrix} \begin{pmatrix} \mathbf{Y} & \mathbf{Z} \\ \mathbf{U} & \mathbf{V} \end{pmatrix} = \begin{pmatrix} \mathbf{I}_k & 0 \\ 0 & \mathbf{I}_{n-k} \end{pmatrix}$$

(where $\mathbf{I}_k$ and $\mathbf{I}_{n-k}$ denote, respectively, the unit matrices of order k and $n - k$) implies

$$\begin{aligned}
\mathbf{PY} + \mathbf{QU} &= \mathbf{I}_k, \\
\mathbf{PZ} + \mathbf{QV} &= 0, \\
\mathbf{RY} + \mathbf{SU} &= 0, \\
\mathbf{RZ} + \mathbf{SV} &= \mathbf{I}_{n-k}.
\end{aligned} \qquad (6.22)$$

From the second equation we see that

$$\mathbf{Z} = -\mathbf{P}^{-1}\mathbf{QV},$$

and combining it with the fourth equation we have

$$\mathbf{V} = (\mathbf{S} - \mathbf{RP}^{-1}\mathbf{Q})^{-1}.$$

The first equation of (6.22) implies

$$\mathbf{Y} = \mathbf{P}^{-1}(\mathbf{I}_k - \mathbf{QU}) = \mathbf{P}^{-1} - (\mathbf{P}^{-1}\mathbf{Q})\mathbf{U},$$

and combining this with the third equation of (6.22) we have

$$\mathbf{U} = -(\mathbf{S} - \mathbf{RP}^{-1}\mathbf{Q})^{-1}\mathbf{RP}^{-1} = -\mathbf{VRP}^{-1}.$$

Thus we have established the following relations

$$\begin{aligned}
\mathbf{V} &= (\mathbf{S} - \mathbf{RP}^{-1}\mathbf{Q})^{-1}, \\
\mathbf{Z} &= -(\mathbf{P}^{-1}\mathbf{Q})\mathbf{V}, \\
\mathbf{U} &= -\mathbf{VRP}^{-1}, \\
\mathbf{Y} &= \mathbf{P}^{-1} - (\mathbf{P}^{-1}\mathbf{Q})\mathbf{U}.
\end{aligned} \qquad (6.23)$$

Observe that $\mathbf{P}^{-1}\mathbf{Q}$ is used in three equations, and so it should be calculated first.

Example 6.7. In the case of the matrix

$$\mathbf{A} = \begin{pmatrix} 1 & 0 & \vdots & 3 & -1 \\ 0 & 0.5 & \vdots & 4 & -2 \\ - & - & - & - & - & - & - & - \\ 5 & -3 & \vdots & -10 & 7 \\ 6 & -4 & \vdots & -14 & 10.5 \end{pmatrix}$$

the leading minor is a diagonal matrix, which can be inverted easily:

$$\mathbf{P}^{-1} = \begin{pmatrix} 1 & 0 \\ 0 & 2 \end{pmatrix}.$$

Equations (6.23) imply

$$\mathbf{P}^{-1}\mathbf{Q} = \begin{pmatrix} 3 & -1 \\ 8 & 4 \end{pmatrix}, \quad \mathbf{S} - \mathbf{R}(\mathbf{P}^{-1}\mathbf{Q}) = \begin{pmatrix} -1 & 0 \\ 0 & 0.5 \end{pmatrix}, \quad \mathbf{V} = \begin{pmatrix} -1 & 0 \\ 0 & 2 \end{pmatrix},$$

$$\mathbf{Z} = -(\mathbf{P}^{-1}\mathbf{Q})\mathbf{V} = \begin{pmatrix} 3 & 2 \\ 8 & 8 \end{pmatrix}, \quad \mathbf{U} = -\mathbf{V}\mathbf{R}\mathbf{P}^{-1} = \begin{pmatrix} 5 & -6 \\ -12 & 16 \end{pmatrix},$$

$$\mathbf{Y} = \mathbf{P}^{-1} - (\mathbf{P}^{-1}\mathbf{Q})\mathbf{U} = \begin{pmatrix} -26 & 34 \\ -88 & 114 \end{pmatrix}.$$

Thus we have

$$\mathbf{A}^{-1} = \begin{pmatrix} -26 & 34 & \vdots & 3 & 2 \\ -88 & 114 & \vdots & 8 & 8 \\ - & - & - & - & - & - & - & - \\ 5 & -6 & \vdots & -1 & 0 \\ -12 & 16 & \vdots & 0 & 2 \end{pmatrix}.$$

We remark that this method can be applied also to the solution of simultaneous linear equations, but the details are not discussed here.

Problems

6.10. Verify, using the definition of matrix multiplication, that the partitions of $\mathbf{A}$ and $\mathbf{A}^{-1}$ imply (6.22).

6.11. Let the order of $\mathbf{P}$ and $\mathbf{S}$ be equal to p and $n - p$, respectively. Show that the method of the inversion by partitioning requires $[5n^3 + 3(n^2p - p^2n)]/6$ multiplications and divisions.

6.12. On the basis of the method of Section 6.1.3, propose a successive method for the inversion of matrices.

6.13. Prove that if the vectors $\mathbf{u}$ and $\mathbf{v}$ satisfy $\mathbf{u}^T\mathbf{B}^{-1}\mathbf{v} = 0$ for some nonsingular matrix $\mathbf{B}$, then for any scalar c,

$$(\mathbf{B} + c\mathbf{u}\mathbf{v}^T)^{-1} = \mathbf{B}^{-1} - c\mathbf{B}^{-1}\mathbf{u}\mathbf{v}^T\mathbf{B}^{-1}.$$

More generally, prove, for any vectors $\mathbf{u}$ and $\mathbf{v}$ and scalar c, that

$$(\mathbf{B} + c\mathbf{u}\mathbf{v}^T)^{-1} = \mathbf{B}^{-1} - \frac{c\mathbf{B}^{-1}\mathbf{u}\mathbf{v}^T\mathbf{B}^{-1}}{1 + \mathbf{v}^T\mathbf{B}^{-1}\mathbf{u}}.$$

Suppose that $\mathbf{A}$ is a nonsingular matrix that has nonzero diagonal elements. Then one may find the inverse of $\mathbf{A}$ by defining $\mathbf{B}^{(0)}$ to be the matrix composed of the diagonal elements of $\mathbf{A}$ and zeros elsewhere and successively "fitting in" the off-diagonal elements by setting $\mathbf{u}$ and $\mathbf{v}$ in the above relation equal to $\mathbf{e}_i$ and $\mathbf{e}_j$, respectively, where $\mathbf{e}_v$ is the vth column of the unit matrix and $i \neq j$. Describe the details of this procedure, which is called the *rank-one corrections* method. How many additions and multiplications are required?

6.2. Iteration Methods

Generally speaking, for linear systems of small or moderate order (say ≤ 100), direct methods are effective and efficient, and are therefore the methods of choice. For larger-order systems, especially those arising in certain numerical differential equation problems, iteration methods are attractive. Unfortunately, some of these methods are not universally applicable, and care must be exercised to see that certain conditions are fulfilled. Iteration methods tend to be stable in the sense that an error due to a faulty computation in an early stage tends to dampen out as computations proceed. One-point iteration methods have the form

$$\mathbf{x}^{(k)} = \mathbf{B}^{(k)}\mathbf{x}^{(k-1)} + \mathbf{c}^{(k)}, \tag{6.24}$$

where matrices $\mathbf{B}^{(k)}$ and vectors $\mathbf{c}^{(k)}$ can change from one iteration to the next. We also require that the solution $\mathbf{x}^*$ of equation (6.2) must satisfy (6.24); that is,

$$\mathbf{x}^* = \mathbf{B}^{(k)}\mathbf{x}^* + \mathbf{c}^{(k)}, \qquad k \geq 1. \tag{6.25}$$

Subtracting (6.25) from (6.24), we have

$$\mathbf{x}^{(k)} - \mathbf{x}^* = \mathbf{B}^{(k)}(\mathbf{x}^{(k-1)} - \mathbf{x}^*).$$

Repeating this process for $k - 1, k - 2, \ldots, 2, 1$, we see that

$$\mathbf{x}^{(k)} - \mathbf{x}^* = \mathbf{B}^{(k)}\mathbf{B}^{(k-1)} \cdots \mathbf{B}^{(2)}\mathbf{B}^{(1)}(\mathbf{x}^{(0)} - \mathbf{x}^*), \tag{6.26}$$

which implies the following theorem:

Theorem 6.1. The iteration process $\{x^{(n)}\}$ given by method (6.24) converges to x^* from every initial approximation $x^{(0)}$, if and only if $B^{(k)}B^{(k-1)} \cdots B^{(2)}B^{(1)} \to 0$ for $k \to \infty$.

The following theorem gives a sufficient condition for the limit vector to be equal to the solution of the equation:

Theorem 6.2. Assume that the sequences of matrices $B^{(k)}$ and vectors $c^{(k)}$ are convergent. Let B and c denote the limits, respectively. Assume that method (6.24) is convergent for an initial approximation $x^{(0)}$. If the matrix $I - B$ is nonsingular, then the limit vector of the iteration process $\{x^{(k)}\}$ is equal to the solution of the linear equation (6.2), with $A = I - B$ and $b = c$.

Proof. Let x denote the limit vector of the sequence $\{x^{(k)}\}$. For $k \to \infty$, equations (6.24) and (6.25) imply

$$x = Bx + c,$$
$$x^* = Bx^* + c,$$

and consequently, after subtraction,

$$(I - B)x = c,$$
$$(I - B)(x^* - x) = 0. \qquad (6.27)$$

Since matrix $I - B$ is nonsingular, equation (6.27) has a unique solution, which implies $x = x^*$. ∎

6.2.1. Stationary Iteration Processes

Following usage in Chapter 4, method (6.24) is called *stationary* if matrix $B^{(k)}$ and vector $c^{(k)}$ are independent of k. Therefore, stationary methods have the general form

$$x^{(k)} = Bx^{(k-1)} + c. \qquad (6.28)$$

The iteration process (6.28) was introduced in Example 4.8.

Theorem 6.1 implies that the necessary and sufficient condition for the convergence of method (6.28) starting from arbitrary initial approximation $x^{(0)}$ is that $B^k \to 0$ for $k \to \infty$. In Example 4.12 we proved that $B^k \to 0$

if and only if the absolute values of the eigenvalues of **B** are less than 1. In most cases the computation of the eigenvalues of **B** is more complicated than the solution of the linear equations themselves, and so convergence criteria based on eigenvalues are not practical, and other tests are sought.

In Example 4.8. we saw that operator

$$A(\mathbf{x}) = \mathbf{B}\mathbf{x} + \mathbf{c}$$

is bounded, and various norms $\| \mathbf{B} \|$ introduced in Section 4.8 provided bounds for this operator. Applying Theorem 4.5 and attendant error bounds, we have the following result:

Theorem 6.3. If for any matrix norm, $\| \mathbf{B} \| < 1$, then the corresponding operator is a contraction and method (6.28) is convergent, and the following error estimations hold:

$$\| \mathbf{x}^{(k)} - \mathbf{x}^* \| \le \frac{\| \mathbf{B} \|^k}{1 - \| \mathbf{B} \|} \, \| \mathbf{x}^{(0)} - \mathbf{x}^{(1)} \|,$$

$$\| \mathbf{x}^{(k)} - \mathbf{x}^* \| \le \frac{\| \mathbf{B} \|}{1 - \| \mathbf{B} \|} \, \| \mathbf{x}^{(k)} - \mathbf{x}^{(k-1)} \|.$$

From the discussion following Theorem 4.6, we may conclude that **B** is a contraction if for any positive integer m, $\| \mathbf{B}^m \| < 1$. From Example 4.12, we may conclude conversely that if **B** is a contraction and $\| \cdot \|$ is a matrix norm, then for some m, $\| \mathbf{B}^m \| < 1$.

In computer computations, the vectors $\mathbf{x}^{(k)}$ cannot be obtained exactly because of the roundoff errors; consequently, instead of sequence (6.28), we compute the sequence

$$\mathbf{z}^{(k)} = \mathbf{B}\mathbf{z}^{(k-1)} + \mathbf{c} + \mathbf{d}^{(k)},$$

where $\mathbf{z}^{(0)} = \mathbf{x}^{(0)}$ and $\mathbf{d}^{(k)}$ is an error vector. Assume that the condition of Theorem 6.3 holds and for $k = 1, 2, \ldots$, $\| \mathbf{d}^{(k)} \| \le \delta$, where $\delta > 0$ is a fixed number. Then error analysis in Section 4.6 is applicable and in particular (4.28) implies

$$\| \mathbf{x}^{(k)} - \mathbf{z}^{(k)} \| \le \frac{\delta}{1 - \| \mathbf{B} \|}, \tag{6.29}$$

which is an upper bound for the propagated errors in the iteration processes.

An application of the contraction theory developed in Section 4.7 implies that, in case that $\| \mathbf{B} \|_\infty < 1$, the *Seidel-type iteration* algorithm

$$x_1^{(k)} = b_{11}x_1^{(k-1)} + b_{12}x_2^{(k-1)} + \cdots + b_{1,n-1}x_{n-1}^{(k-1)} + b_{1n}x_n^{(k-1)} + c_1,$$

$$x_2^{(k)} = b_{21}x_1^{(k)} + b_{22}x_2^{(k-1)} + \cdots + b_{2,n-1}x_{n-1}^{(k-1)} + b_{2n}x_n^{(k-1)} + c_2,$$

$$\vdots \tag{6.30}$$

$$x_n^{(k)} = b_{n1}x_1^{(k)} + b_{n2}x_2^{(k)} + \cdots + b_{n,n-1}x_{n-1}^{(k)} + b_{nn}x_n^{(k-1)} + c_n,$$

is also convergent and usually the speed of convergence of method (6.30) is faster than the speed of convergence of method (6.28). Here b_{ij} and c_i denote the elements of matrix $\mathbf{B}$ and vector $\mathbf{c}$, respectively.

We can easily prove that method (6.30) is equivalent to a specific iteration method of the form (6.28). Let us introduce the matrices

$$\mathbf{L} = \begin{pmatrix} 0 & 0 & 0 & \cdots & 0 & 0 \\ b_{21} & 0 & 0 & \cdots & 0 & 0 \\ b_{31} & b_{32} & 0 & \cdots & 0 & 0 \\ \vdots & & & & & \\ b_{n1} & b_{n2} & b_{n3} & \cdots & b_{n,n-1} & 0 \end{pmatrix}$$

and

$$\mathbf{R} = \mathbf{B} - \mathbf{L}.$$

Then (6.30) can be written as

$$\mathbf{x}^{(k)} = \mathbf{L}\mathbf{x}^{(k)} + \mathbf{R}\mathbf{x}^{(k-1)} + \mathbf{c},$$

which implies

$$(\mathbf{I} - \mathbf{L})\mathbf{x}^{(k)} = \mathbf{R}\mathbf{x}^{(k-1)} + \mathbf{c}. \tag{6.31}$$

Since matrix $\mathbf{I} - \mathbf{L}$ is lower triangular with 1's on the diagonal, $(\mathbf{I} - \mathbf{L})^{-1}$ exists and (6.31) is equivalent to the equation

$$\mathbf{x}^{(k)} = (\mathbf{I} - \mathbf{L})^{-1}\mathbf{R}\mathbf{x}^{(k-1)} + (\mathbf{I} - \mathbf{L})^{-1}\mathbf{c}, \tag{6.32}$$

which has form (6.28), where $\mathbf{B}$ and $\mathbf{c}$ are replaced by $(\mathbf{I} - \mathbf{L})^{-1}\mathbf{R}$ and $(\mathbf{I} - \mathbf{L})^{-1}\mathbf{c}$, respectively.

Systems of linear algebraic equations are assumed given in the form $\mathbf{A}\mathbf{x} = \mathbf{b}$, and different choices of matrix $\mathbf{B}$ and vector $\mathbf{c}$ for (6.28) give different concrete iteration methods:

1. The *classical iteration algorithm* is the choice $\mathbf{B} = \mathbf{I} - \mathbf{A}$, $\mathbf{c} = \mathbf{b}$.
2. For the Jacobi method, let

$$
\mathbf{L} = \begin{pmatrix}
0 & 0 & 0 & \cdots & 0 & 0 \\
a_{21} & 0 & 0 & \cdots & 0 & 0 \\
a_{31} & a_{32} & 0 & \cdots & 0 & 0 \\
\vdots & & & & & \\
a_{n1} & a_{n2} & a_{n3} & \cdots & a_{n,n-1} & 0
\end{pmatrix},
$$

$$
\mathbf{D} = \begin{pmatrix}
a_{11} & 0 & 0 & 0 & 0 \\
0 & a_{22} & 0 & 0 & 0 \\
0 & 0 & a_{33} & 0 & 0 \\
& & & \ddots & \\
0 & 0 & 0 & 0 & a_{nn}
\end{pmatrix},
$$

$$
\mathbf{R} = \begin{pmatrix}
0 & a_{12} & a_{13} & \cdots & a_{1,n-1} & a_{1n} \\
0 & 0 & a_{23} & \cdots & a_{2,n-1} & a_{2n} \\
\vdots & & & & & \\
0 & 0 & 0 & \cdots & 0 & 0
\end{pmatrix}.
$$

Of course $\mathbf{A} = \mathbf{L} + \mathbf{D} + \mathbf{R}$, and with suitable exchanging of the rows of the nonsingular matrix $\mathbf{A}$ we can have a nonsingular matrix $\mathbf{D}$ with nonzero diagonal elements. The equation $\mathbf{A}\mathbf{x} = \mathbf{b}$ can be rewritten as

$$(\mathbf{L} + \mathbf{D} + \mathbf{R})\mathbf{x} = \mathbf{b},$$

which implies

$$\mathbf{x} = -\mathbf{D}^{-1}(\mathbf{L} + \mathbf{R})\mathbf{x} + \mathbf{D}^{-1}\mathbf{b}.$$

The corresponding iteration law

$$\mathbf{x}^{(k)} = -\mathbf{D}^{-1}(\mathbf{L} + \mathbf{R})\mathbf{x}^{(k-1)} + \mathbf{D}^{-1}\mathbf{b} \tag{6.33}$$

is called the *Jacobi method*.

3. The *Gauss–Seidel method* is the Seidel-type modification (as developed in Section 4.7) of algorithm (6.33). That is, in the Gauss–Seidel method, (6.28) takes on the form

$$\mathbf{B} = -(\mathbf{D} + \mathbf{L})^{-1}\mathbf{R}, \qquad \mathbf{c} = (\mathbf{D} + \mathbf{L})^{-1}\mathbf{b}.$$

In the next section we shall prove that the Gauss–Seidel iteration is convergent when matrix $\mathbf{A}$ is positive definite.

Example 6.8. We wish to apply the Jacobi method to the following system:

$$64x_1 - 3x_2 - x_3 = 14,$$
$$x_1 + x_2 + 40x_3 = 20,$$
$$2x_1 - 90x_2 + x_3 = -5.$$

We exchange the second and third equations to obtain large diagonal elements. After dividing by these large diagonal elements we can verify that the numbers b_{ij} will be small enough to satisfy the conditions of Theorem 6.3. After exchanging the second and third rows, we have

$$64x_1 - 3x_2 - x_3 = 14,$$
$$2x_1 - 90x_2 + x_3 = -5,$$
$$x_1 + x_2 + 40x_3 = 20.$$

and the reader will observe that $\| \mathbf{B} \|_\infty = 1/16$.

Using the Jacobi method we obtain the iteration scheme

$$x_1^{(k)} = \qquad\qquad 0.04688 x_2^{(k-1)} + 0.01562 x_3^{(k-1)} + 0.21875,$$
$$x_2^{(k)} = 0.02222 x_1^{(k-1)} \qquad\qquad + 0.01111 x_3^{(k-1)} + 0.05556,$$
$$x_3^{(k)} = -0.02500 x_1^{(k-1)} - 0.02500 x_2^{(k-1)} \qquad\qquad + 0.50000.$$

We take $x_1^{(0)} = x_2^{(0)} = x_3^{(0)} = 0$. The results of calculations are shown in Table 6.4. The results of application of the Gauss–Seidel method is tabulated in Table 6.5. Its superior convergence is evident. Typically, the Jacobi method is inferior to the other methods we discuss.

Since $\| \mathbf{B} \|_\infty = 0.06250$, the error of $\mathbf{x}^{(6)}$ at the Jacobi iteration does not exceed

$$\frac{\| \mathbf{B} \|_\infty^5}{1 - \| \mathbf{B} \|_\infty} \| \mathbf{x}^{(1)} - \mathbf{x}^{(2)} \| \approx \frac{0.06250^5}{0.03750} 0.01042 \approx 3 \times 10^{-7},$$

and thus the error of $\mathbf{x}^{(6)}$ is not greater than

$$3 \times 10^{-7} + \frac{0.5 \times 10^{-5}}{0.9375} \approx 0.56 \cdot 10^{-5}.$$

Table 6.4. Jacobi Method Computations for Example 6.8

Approximation	$k = 1$	$k = 2$	$k = 3$	$k = 4$	$k = 5$	$k = 6$
$x_1^{(k)}$	0.21875	0.22916	0.22959	0.22955	0.22954	0.22954
$x_2^{(k)}$	0.05556	0.06598	0.06616	0.06613	0.06613	0.06613
$x_3^{(k)}$	0.50000	0.49592	0.49262	0.49261	0.49261	0.49261

Table 6.5. Gauss–Seidel Method Computations for Example 6.8

Approximation	$k = 1$	$k = 2$	$k = 3$	$k = 4$	$k = 5$
$x_1^{(k)}$	0.21875	0.22916	0.22955	0.22954	0.22954
$x_2^{(k)}$	0.05556	0.06621	0.06613	0.06613	0.06613
$x_3^{(k)}$	0.50000	0.49262	0.49261	0.49261	0.49261

Observe that the results for the convergence of the methods described in this section were obtained immediately from the general theory of iteration processes developed in Chapter 4. This is further illustration of the importance of general results in metric space theory.

Problems

6.14. Show that each step of iteration method (6.28) requires n^2 multiplications. Discuss the relative efficiency of iteration methods in comparison to Gauss elimination.

6.15. What are the necessary and sufficient conditions for convergence of the Jacobi method and the Gauss–Seidel iteration?

6.16. Prove that the condition

$$| a_{ii} | > \sum_{\substack{j=1 \\ j \neq i}}^{k} | a_{ij} | \qquad (i = 1, 2, \ldots, n)$$

is a sufficient condition for convergence of both the Jacobi method and Gauss–Seidel iteration. If **A** satisfies this condition, **A** is said to be *diagonally dominant*.

6.17. Solve the system

$$
\begin{aligned}
4x_1 - x_2 &= 2, \\
-x_1 + 4x_2 - x_3 &= 6, \\
- x_2 + 4x_3 &= 2,
\end{aligned}
$$

by Jacobi iteration. Do three iterations.

6.18. Solve the previous system by Gauss–Seidel iteration. Do three iterations.

6.2.2. Iteration Processes Based on the Minimization of Quadratic Forms

Let **x** be an approximate solution of the equation (6.2). Then the vector $\mathbf{y} = \mathbf{x} - \mathbf{x}^*$ is the error vector. We can assess the error of the approximate solution $\mathbf{x}$ by the *residual vector*

$$\rho = \mathbf{b} - \mathbf{Ax} = \mathbf{Ax^*} - \mathbf{Ax} = \mathbf{A}(\mathbf{x^*} - \mathbf{x}) = \mathbf{Ay}.$$

The vector $\mathbf{x}$ is a solution of the equation if and only if $\rho = \mathbf{0}$; consequently the problem of minimizing the nonnegative function

$$f(\mathbf{x}) = \| \rho \|_2^2 = \| \mathbf{b} - \mathbf{Ax} \|_2^2 \tag{6.34}$$

is equivalent to solving the equation $\mathbf{Ax} = \mathbf{b}$. Observe that f is a quadratic form in $\mathbf{x}$. Let $\mathbf{C} = \mathbf{A}^\mathrm{T}\mathbf{A}$. If $\mathbf{A}$ is nonsingular, then the matrix $\mathbf{C} = \mathbf{A}^\mathrm{T}\mathbf{A}$ is symmetric and positive definite. Consider the problem of minimizing the function f. It is easy to see that the simultaneous equations

$$\frac{\partial f(\mathbf{x})}{\partial x_i} = 0, \qquad 1 \le i \le n, \tag{6.35}$$

lead to a linear system having the same complexity as the original equations, and therefore the usual calculus method for optimizing functions is not applicable.

The following iteration method can be utilized for minimizing f. Let $\mathbf{x}^{(0)}$ be an approximation of $\mathbf{x^*}$ and let $\{\mathbf{u}^{(i)}\}_{i \ge 0}$ be any sequence of nonzero vectors. The next approximation $\mathbf{x}^{(1)}$ will depend on the minimum point (over the real variable t) of the function $f(\mathbf{x}^{(0)} + t\mathbf{u}^{(0)})$. The minimum of this function can be easily determined, because it is a univariate quadratic function:

$$
\begin{aligned}
f(\mathbf{x}^{(0)} + t\mathbf{u}^{(0)}) &= (\mathbf{x}^{(0)} + t\mathbf{u}^{(0)} - \mathbf{x^*})^\mathrm{T}\mathbf{C}(\mathbf{x}^{(0)} + t\mathbf{u}^{(0)} - \mathbf{x^*}) \\
&= t^2[\mathbf{u}^{(0)\mathrm{T}}\mathbf{Cu}^{(0)}] + 2t[\mathbf{u}^{(0)\mathrm{T}}\mathbf{C}(\mathbf{x}^{(0)} - \mathbf{x^*})] \\
&\quad + [\mathbf{x}^{(0)\mathrm{T}}\mathbf{Cx}^{(0)} + \mathbf{x}^{*\mathrm{T}}\mathbf{Cx^*} - 2\mathbf{x}^{*\mathrm{T}}\mathbf{Cx}].
\end{aligned} \tag{6.36}
$$

In the second expression, we have used the observation that $\mathbf{b} = \mathbf{Ax^*}$. Since the last two terms do not depend on t, the problem of minimizing f can be reduced to the problem of minimizing the function

$$t^2[\mathbf{u}^{(0)\mathrm{T}}\mathbf{Cu}^{(0)}] - 2t[\mathbf{u}^{(0)\mathrm{T}}\mathbf{r}^{(0)}],$$

where $\mathbf{r}^{(0)} = \mathbf{d} - \mathbf{Cx}^{(0)}$, where $\mathbf{d}$ is defined to be $\mathbf{Cx^*} = \mathbf{A}^\mathrm{T}\mathbf{b}$. Note that $\mathbf{r}^{(0)}$ is the residual vector for the problem $\mathbf{Cx} = \mathbf{d}$. Since the preceding displayed equation can be written in the form

$$\mathbf{u}^{(0)\mathrm{T}}\mathbf{Cu}^{(0)} \left[t - \frac{\mathbf{u}^{(0)\mathrm{T}}\mathbf{r}^{(0)}}{\mathbf{u}^{(0)\mathrm{T}}\mathbf{Cu}^{(0)}} \right]^2 - \frac{[\mathbf{u}^{(0)\mathrm{T}}\mathbf{r}^{(0)}]^2}{\mathbf{u}^{(0)\mathrm{T}}\mathbf{Cu}^{(0)}},$$

and $\mathbf{u}^{(0)\mathrm{T}}\mathbf{C}\mathbf{u}^{(0)} > 0$, the minimum of this function is given by

$$t = \frac{\mathbf{u}^{(0)\mathrm{T}}\mathbf{r}^{(0)}}{\mathbf{u}^{(0)\mathrm{T}}\mathbf{C}\mathbf{u}^{(0)}},$$

which implies

$$\mathbf{x}^{(1)} = \mathbf{x}^{(0)} + \frac{\mathbf{u}^{(0)\mathrm{T}}\mathbf{r}^{(0)}}{\mathbf{u}^{(0)\mathrm{T}}\mathbf{C}\mathbf{u}^{(0)}} \mathbf{u}^{(0)}.$$

Repeating the calculation, one obtains the recursive law

$$\mathbf{x}^{(k)} = \mathbf{x}^{(k-1)} + \frac{\mathbf{u}^{(k-1)\mathrm{T}}\mathbf{r}^{(k-1)}}{\mathbf{u}^{(k-1)\mathrm{T}}\mathbf{C}\mathbf{u}^{(k-1)}} \mathbf{u}^{(k-1)}. \tag{6.37}$$

Observe that

$$f(\mathbf{x}^{(0)}) \geq f(\mathbf{x}^{(1)}) \geq f(\mathbf{x}^{(2)}) \geq \cdots \geq 0,$$

which implies the convergence of the sequence $\{f(\mathbf{x}^{(k)})\}$.

Theorem 6.4. If $f(\mathbf{x}^{(k)}) \to 0$ for $k \to \infty$, then the sequence $\{\mathbf{x}^{(k)}\}$ tends to $\mathbf{x}^*$.

Proof. Since matrix $\mathbf{C}$ is positive definite, an elementary property of the eigenvalues of symmetric matrices (see Wilkinson, 1965, pp. 98–99) implies that if $\mathbf{y}^{(k)} = \mathbf{x}^{(k)} - \mathbf{x}^*$,

$$\frac{f(\mathbf{x}^{(k)})}{\| \mathbf{y}^{(k)} \|_2^2} \geq \lambda_1 > 0,$$

where λ_1 is the least eigenvalue of $\mathbf{C}$. This inequality can be rewritten as

$$\| \mathbf{y}^{(k)} \|_2^2 = \| \mathbf{x}^{(k)} - \mathbf{x}^* \|_2^2 \leq \frac{f(\mathbf{x}^{(k)})}{\lambda_1}, \tag{6.38}$$

which implies the assertion. ∎

Different methods can be obtained by different special choices of vectors $\mathbf{u}^{(k)}$. The most prominent techniques based on the iterative minimization of (6.34) are the method of successive relaxation, which will now be discussed, and the gradient and conjugate gradient methods, which are presented in the following subsections.

In *successive relaxation methods*, we choose the successive vectors $\mathbf{u}^{(k)}$ to be the sequence of vectors $\mathbf{e}_1, \ldots, \mathbf{e}_n, \mathbf{e}_1, \ldots, \mathbf{e}_n$, where $\mathbf{e}_j$ is the vector with a 1 in the jth coordinate and 0's elsewhere. Assuming $\mathbf{u}^{(k)} = \mathbf{e}_i$, in this case equation (6.37) can be rewritten as

$$x_i^{(k)} = x_i^{(k-1)} + r_i^{(k-1)}/c_{ii}$$

$$= x_i^{(k-1)} + \frac{1}{c_{ii}}\left[d_i - \sum_{=1}^{i-1} c_{ij}x^{(k)} - \sum_{i+1}^{n} c_{ij}x_j^{(k-1)}\right], \qquad (6.39)$$

$$x_j^{(k)} = x_j^{(k-1)}, \qquad j \neq i,$$

which is identical to performing one row of the Gauss–Seidel iteration (6.30) for the equation $\mathbf{Cx} = \mathbf{d}$.

Simple calculations show that if r_i is the ith coordinate of the residual vector $\mathbf{r}$, then

$$f(\mathbf{x} + t\mathbf{e}_i) - f(\mathbf{x}) = (1/c_{ii})(c_{ii}t - r_i)^2 - r_i^2/c_{ii},$$

which implies that

$$f(\mathbf{x} + t\mathbf{e}_i) < f(\mathbf{x})$$

if and only if

$$|c_{ii}t - r_i| < |r_i|,$$

which is satisfied if

$$t = w_i(r_i/c_{ii}),$$

where $0 < w_i < 2$. Thus we have hereby proved that sequence $\{f(\mathbf{x}^{(k)})\}$ is strictly monotone decreasing in the case of the generalized method of successive relaxation having the form

$$x_i^{(k)} = x_i^{(k-1)} + w_i(r_i^{(k-1)}/c_{ii}), \qquad 0 < w_i < 2. \qquad (6.40)$$

In general, the method (6.40) is not a stationary iteration, because the coefficients w_i can change from one iteration to the next. If the successive relaxation method is stationary, then the following theorem applies:

Theorem 6.5. For $0 < w < 2$ the stationary successive relaxation method converges to the solution of the simultaneous linear equations in the case of a real, positive-definite matrix of coefficients.

Proof. Application of (6.40) for each index i, $1 \leq i \leq n$, can be rewritten in matrix form

$$\mathbf{x}^{(k)} = (\mathbf{D} + w\mathbf{L})^{-1}[(1 - w)\mathbf{D} - w\mathbf{R}]\mathbf{x}^{(k-1)} + (\mathbf{D} + w\mathbf{L})^{-1}w\mathbf{d}, \qquad (6.41)$$

where matrices $\mathbf{L}$, $\mathbf{D}$, $\mathbf{R}$ are defined as in (6.33) with $\mathbf{C}$ replacing $\mathbf{A}$ and $\mathbf{d}$

replacing **b**. Formula (6.41) can be demonstrated by using the argument that was given to justify (6.33). Method (6.41) is a stationary iteration method of the type discussed in Section 6.2.1.

Let

$$\mathbf{B} = (\mathbf{D} + w\mathbf{L})^{-1}[(1 - w)\mathbf{D} - w\mathbf{R}].$$

In order to demonstrate convergence it is sufficient to prove that the absolute values of the eigenvalues of **B** are less than one, because this condition is the necessary and sufficient condition for the convergence of stationary iterative processes (as discussed in Section 6.2.1).

The eigenvalue problem for matrix **B** has the form

$$(\mathbf{D} + w\mathbf{L})^{-1}[(1 - w)\mathbf{D} - w\mathbf{R}]\mathbf{z} = \lambda\mathbf{z}, \tag{6.42}$$

where λ denotes an eigenvalue of **B** and **z** denotes the corresponding eigenvector. Since **C** is positive definite, for $i = 1, 2, \ldots, n$,

$$c_{ii} = \mathbf{e}_i^T\mathbf{C}\mathbf{e}_i > 0.$$

Consequently the matrix $\mathbf{D} + w\mathbf{L}$ is nonsingular. Equation (6.42) implies that

$$\lambda(\mathbf{D} + w\mathbf{L})\mathbf{z} = [(1 - w)\mathbf{D} - w\mathbf{R}]\mathbf{z}. \tag{6.43}$$

Since $\mathbf{C} = \mathbf{L} + \mathbf{D} + \mathbf{R}$, after multiplying (6.43) by 2 we have

$$\lambda[\mathbf{D}(2 - w) + w(\mathbf{C} - \mathbf{R} + \mathbf{L})]\mathbf{z} = [(2 - w)\mathbf{D} - w\mathbf{C} + w(\mathbf{L} - \mathbf{R})]\mathbf{z}. \tag{6.44}$$

Let

$$a = \bar{\mathbf{z}}^T\mathbf{C}\mathbf{z}, \qquad \delta = \bar{\mathbf{z}}^T\mathbf{D}\mathbf{z}, \qquad \varrho = \frac{2}{w} - 1,$$

where $\bar{\mathbf{z}}$ denotes the complex conjugate of **z**. Obviously $a > 0$ and $\delta > 0$. Since $\mathbf{R} = \mathbf{L}^T$, $\mathbf{z}^T(\mathbf{L} - \mathbf{R})\mathbf{z} = 0$, and equation (6.44) implies that

$$\lambda[\varrho\delta + a] = \varrho\delta - a.$$

that is

$$|\lambda| = \left| \frac{\varrho\delta - a}{\varrho\delta + a} \right|. \tag{6.45}$$

Observe that for any positive numbers a, δ, ϱ the right-hand side of (6.45) is less than one. This completes the proof. ∎

For $w = 1$, method (6.41) is identical to the Gauss–Seidel method; consequently, for real, positive-definite matrices of coefficients, the Gauss–Seidel iteration is convergent to the solution of the equation. For w between 1 and 2, which is the most common case, (6.43) is called *successive over-relaxation* (SOR). Young and Gregory (1972, Section 16.6) give considerable attention to the relative performance of the Jacobi, Gauss–Seidel, and successive overrelaxation methods. The optimal choice of relaxation factor depends on the eigenvalues of **C**, which, of course, are not available.

6.2.3. Application of the Gradient Method

If we set $\mathbf{u}^{(k-1)} = \mathbf{r}^{(k-1)}$, the basic rule (6.37) coincides with the gradient method (which was described and analyzed in Section 5.3.1) for minimizing

$$f(\mathbf{x}) = \| \mathbf{A}\mathbf{x} - \mathbf{b} \|^2 = (\mathbf{x} - \mathbf{x}^*)^{\mathrm{T}} \mathbf{C} (\mathbf{x} - \mathbf{x}^*).$$

For one may confirm that in this case, the gradient $\nabla f(\mathbf{x}) = 2\mathbf{C}(\mathbf{x} - \mathbf{x}^*) = -2\mathbf{r}$. Thus the gradient method iteration has the form

$$\mathbf{x}^{(k+1)} = \mathbf{x}^{(k)} - t\nabla f(\mathbf{x}^{(k)}) = \mathbf{x}^{(k)} + 2t\mathbf{r}^{(k)}.$$

The scalar t minimizing $f(\mathbf{x}^{(k)} + 2t\mathbf{r}^{(k)})$ can readily be seen to be

$$t = \tfrac{1}{2}(\mathbf{r}^{(k)})^{\mathrm{T}}\mathbf{r}^{(k)}/(\mathbf{r}^{(k)})^{\mathrm{T}}\mathbf{C}\mathbf{r}^{(k)}. \tag{6.46}$$

Thus we have confirmed that the general iterative formula (6.37), with $\mathbf{u}^{(k)}$ taken to be the residual vector $\mathbf{r}^{(k)}$, coincides with the gradient method for finding the minimum of $f(\mathbf{x})$. Using this observation and Theorem 5.9, we can immediately conclude that the following assertion is true:

Theorem 6.6. Let **A** be nonsingular and $\mathbf{C} = \mathbf{A}^{\mathrm{T}}\mathbf{A}$ and $\mathbf{d} = \mathbf{A}^{\mathrm{T}}\mathbf{b}$. For an arbitrary vector $\mathbf{x}^{(0)}$, let the sequence $\{\mathbf{x}^{(k)}\}$ be determined recursively by the gradient algorithm

$$\mathbf{x}^{(k+1)} = \mathbf{x}^{(k)} + \frac{\mathbf{r}^{(k)\mathrm{T}}\mathbf{r}^{(k)}}{\mathbf{r}^{(k)\mathrm{T}}\mathbf{C}\mathbf{r}^{(k)}}\,\mathbf{r}^{(k)},$$

where $\mathbf{r}^{(k)} = \mathbf{d} - \mathbf{C}\mathbf{x}^{(k)}$. Then $\{\mathbf{x}^{(k)}\}$ converges to the solution $\mathbf{x}^*$ of $\mathbf{A}\mathbf{x} = \mathbf{b}$. Furthermore, letting m and M denote, respectively, the smallest and largest eigenvalues of the positive-definite matrix **C**, we have

$$\| \mathbf{r}^{(k+1)} \|_2^2 \leq \left(\frac{M - m}{M + m} \right)^2 \| \mathbf{r}^{(k)} \|_2^2.$$

Since

$$f(\mathbf{x}^{(k)}) = \| \mathbf{A}(\mathbf{x}^{(k)} - \mathbf{x}^*) \|_2^2$$
$$= (\mathbf{x}^{(k)} - \mathbf{x}^*)^{\mathrm{T}} \mathbf{C}(\mathbf{x}^{(k)} - \mathbf{x}^*)$$
$$\geq m \| \mathbf{x}^{(k)} - \mathbf{x}^* \|_2^2,$$

the bounds in the theorem imply the following result:

$$\| \mathbf{x}^{(k)} - \mathbf{x}^* \| \leq \left[\frac{f(\mathbf{x}^{(k)})}{m} \right]^{1/2} \leq \left(\frac{M - m}{M + m} \right)^k \left[\frac{f(\mathbf{x}^{(0)})}{m} \right]^{1/2}. \qquad (6.47)$$

Notice also that

$$\frac{M - m}{M + m} = 1 - \frac{2m}{M + m} = 1 - \frac{2}{M/m + 1},$$

and so the coefficient of convergence increases with increasing M/m.

6.2.4. The Conjugate Gradient Method

The conjugate gradient method is the most important member of a class of techniques known as conjugate direction methods. A method is a *conjugate direction method* if it is an iterative rule of the form (6.37) and if, in addition, the vectors $\mathbf{u}^{(0)}, \ldots, \mathbf{u}^{(n-1)}$ are conjugate directions with respect to the positive definite matrix $\mathbf{C}$. A set of nonzero vectors V are *conjugate directions* with respect to $\mathbf{C}$ if for any two distinct members $\mathbf{v}$ and $\mathbf{v}'$ of V, $(\mathbf{v}')^{\mathrm{T}} \mathbf{C} \mathbf{v} = 0$. We say that $\mathbf{w}$ is a conjugate direction to a set V if $\mathbf{w}^{\mathrm{T}} \mathbf{C} \mathbf{v} = 0$ for all $\mathbf{v} \in V$. In the absence of roundoff error, conjugate direction methods give the exact solution in at most n iterations, as we now demonstrate.

Theorem 6.7. Let $\mathbf{x}^{(0)}$ be an arbitrary vector and let $\mathbf{x}^{(k)}, k > 0$, be determined recursively by the rule

$$\mathbf{x}^{(k)} = \mathbf{x}^{(k-1)} + \frac{\mathbf{u}^{(k-1)\mathrm{T}} \mathbf{r}^{(k-1)}}{\mathbf{u}^{(k-1)\mathrm{T}} \mathbf{C} \mathbf{u}^{(k-1)}} \mathbf{u}^{(k-1)}, \qquad (6.48)$$

where the vectors $\mathbf{u}^{(0)}, \ldots, \mathbf{u}^{(n-1)}$ are conjugate directions and $\mathbf{r}^{(k-1)} = \mathbf{d} - \mathbf{C}\mathbf{x}^{(k-1)}$. (That is, let $\{\mathbf{x}^{(k)}\}$ be successive vectors determined by a conjugate direction method.) Then if $\mathbf{x}^*$ denotes the solution of the nth-order linear system $\mathbf{C}\mathbf{x} = \mathbf{d}$, we have

(a) $(\mathbf{x}^{(k)} - \mathbf{x}^*)$ is a conjugate direction to $\{\mathbf{u}^{(v)}: v < k\}$,

and

(b) $\mathbf{x}^{(n)} = \mathbf{x}^*$.

Proof. The demonstration of (a) proceeds by induction on k. The assertion holds trivially for $k = 0$. Suppose it holds for $k = j - 1$. Then (6.48) implies that

$$\mathbf{x}^{(j)} - \mathbf{x}^* = (\mathbf{x}^{(j-1)} - \mathbf{x}^*) + \alpha_{j-1}\mathbf{u}^{(j-1)}, \tag{6.49}$$

where $\alpha_{j-1} = (\mathbf{u}^{(j-1)\mathrm{T}}\mathbf{r}^{(j-1)}/\mathbf{u}^{(j)\mathrm{T}}\mathbf{C}\mathbf{u}^{(j-1)})$. By the inductive hypothesis, the first term in parentheses on the right-hand side of (6.49) is a conjugate direction to $\{\mathbf{u}^{(v)}: v < j - 1\}$, and the second term is a conjugate direction to the set of all $\mathbf{u}^{(v)\prime}s$, $v \neq j - 1$. Thus assertion (a) is established if only we can demonstrate that the right-hand side is a conjugate direction to $\mathbf{u}^{(j-1)}$. Toward demonstrating this fact, we observe that $\mathbf{r}^{(j-1)} = \mathbf{C}(\mathbf{x}^* - \mathbf{x}^{(j-1)})$ and that since $(\mathbf{u}^{(j-1)\mathrm{T}}\mathbf{r}^{(j-1)})$ is a real number, it equals its transpose, which may be written $(\mathbf{x}^* - \mathbf{x}^{(j-1)})^\mathrm{T}\mathbf{C}\mathbf{u}^{(j-1)}$. From these relations, we may conclude that if we take the transpose of (6.48) with $k = j$, and right-multiply both sides by $\mathbf{C}\mathbf{u}^{(j-1)}$, the resulting equation is the same as

$$(\mathbf{x}^{(j)} - \mathbf{x}^*)^\mathrm{T}\mathbf{C}\mathbf{u}^{(j-1)}$$
$$= (\mathbf{x}^{(j-1)} - \mathbf{x}^*)^\mathrm{T}\mathbf{C}\mathbf{u}^{(j-1)} + \frac{(\mathbf{x}^* - \mathbf{x}^{(j-1)})^\mathrm{T}\mathbf{C}\mathbf{u}^{(j-1)}}{(\mathbf{u}^{(j-1)\mathrm{T}}\mathbf{C}\mathbf{u}^{(j-1)})}\,\mathbf{u}^{(j-1)\mathrm{T}}\mathbf{C}\mathbf{u}^{(j-1)}.$$

Upon cancelling the denominator, the right-hand side is seen to be zero, implying that $\mathbf{x}^{(j)} - \mathbf{x}^*$ is a conjugate direction to $\mathbf{u}^{(j-1)}$. This completes the proof of (a). Since (a) holds for $k = n$, we have that

$$(\mathbf{x}^{(n)} - \mathbf{x}^*)^\mathrm{T}\mathbf{C}\mathbf{u}^{(j)} = 0, \qquad 0 \leq j \leq n - 1. \tag{6.50}$$

Let $\mathbf{U}$ be a matrix whose columns are the vectors $\mathbf{u}^{(j)}$, $0 \leq j \leq n - 1$. One may verify that a set of conjugate directions is necessarily linearly independent, and consequently $\mathbf{U}$ is nonsingular. Thus simultaneous equations (6.50) imply that

$$(\mathbf{x}^{(n)} - \mathbf{x}^*)^\mathrm{T}\mathbf{C}\mathbf{U} = \mathbf{0}^\mathrm{T},$$

and the nonsingularity of $\mathbf{C}\mathbf{U}$ implies that $\mathbf{x}^{(n)} - \mathbf{x}^* = \mathbf{0}$, thus proving (b). ■

The conjugate gradient method is the most important of the conjugate direction methods, and indeed, is one of the most powerful methods of nonlinear programming.

The conjugate gradient method is the conjugate direction method in which at each stage j, $\mathbf{u}^{(j)}$ is chosen to be the component of the gradient $\nabla f(\mathbf{x}^{(j)})$, which is a conjugate direction to the previous vectors $\{\mathbf{u}^{(v)}, v < j\}$. In taking $\mathbf{u}^{(j)}$ to be the "conjugate" portion of the gradient, the conjugate gradient achieves the rapid descent characteristic of the gradient method, but at the same time it preserves the conjugate direction property that $\mathbf{x}^{(n)}$ is the solution and $\mathbf{x}^{(j)} - \mathbf{x}^*$ is a conjugate direction to $\{\mathbf{u}^{(v)}: v \leq j\}$. Specifically, the *conjugate gradient* method is the following algorithm: Choose $\mathbf{x}^{(0)} \neq \mathbf{0}$ arbitrarily and define $\mathbf{u}^{(0)} = \nabla f(\mathbf{x}^{(0)}) = 2\mathbf{r}^{(0)} = -2\mathbf{C}(\mathbf{x}^* - \mathbf{x}^{(0)})$. For each $j \geq 0$, if $\nabla f(\mathbf{x}^{(j)}) = -2\mathbf{r}^{(j)} = -2(\mathbf{d} - \mathbf{C}\mathbf{x}^{(j)}) = \mathbf{0}$, $\mathbf{x}^{(j)} = \mathbf{x}^*$ (since the solution is easily seen to be the unique point at which the gradient is $\mathbf{0}$). Otherwise define

$$\mathbf{u}^{(j)} = \mathbf{r}^{(j)} - \frac{\mathbf{r}^{(j)\mathrm{T}}\mathbf{C}\mathbf{u}^{(j-1)}}{\mathbf{u}^{(j-1)\mathrm{T}}\mathbf{C}\mathbf{u}^{(j-1)}}\,\mathbf{u}^{(j-1)}, \tag{6.51}$$

and, as in (6.48),

$$\mathbf{x}^{(j+1)} = \mathbf{x}^{(j)} + \frac{\mathbf{r}^{(j)\mathrm{T}}\mathbf{u}^{(j)}}{\mathbf{u}^{(j)\mathrm{T}}\mathbf{C}\mathbf{u}^{(j)}}\,\mathbf{u}^{(j)}. \tag{6.52}$$

Recall that $\mathbf{r}^{(j)} = \mathbf{d} - \mathbf{C}\mathbf{x}^{(j)} = -\frac{1}{2}\nabla f(\mathbf{x}^{(j)})$.

Theorem 6.8. The conjugate gradient method is a conjugate direction method.

Proof. We need only to show that for each j, $\mathbf{u}^{(j)}$ is a conjugate direction to $\{\mathbf{u}^{(k)}: k < j\}$.

We proceed inductively, noting that $\mathbf{u}^{(0)}$ is trivially a conjugate direction to its predecessors. Suppose $\{\mathbf{u}^{(k)}, k < j\}$ is a set of conjugate directions. $\mathbf{u}^{(j)\mathrm{T}}\mathbf{C}\mathbf{u}^{(j-1)}$ is readily seen to be 0 by right-multiplying the transpose of (6.51) by $\mathbf{C}\mathbf{u}^{(j-1)}$ and noting that the terms on the right cancel. For $k < j - 1$, right-multiply the transpose of (6.51) to get an expression of the form

$$\mathbf{u}^{(j)\mathrm{T}}\mathbf{C}\mathbf{u}^{(k)} = (\mathbf{r}^{(j)\mathrm{T}}\mathbf{C}\mathbf{u}^{(k)}) + \alpha\mathbf{u}^{(j-1)\mathrm{T}}\mathbf{C}\mathbf{u}^{(k)},$$

where α is a number. The last term is 0 by the inductive hypothesis. Toward showing that $\mathbf{r}^{(j)\mathrm{T}}\mathbf{C}\mathbf{u}^{(k)} = 0$, we prove that $\mathbf{C}\mathbf{u}^{(k)}$ is a linear combination of the vectors $\mathbf{u}^{(i)}$, $i = k - 1, k, k + 1$, say

$$\mathbf{C}\mathbf{u}^{(k)} = \alpha_1\mathbf{u}^{(k-1)} + \alpha_2\mathbf{u}^{(k)} + \alpha_3\mathbf{u}^{(k+1)}. \tag{6.53}$$

If this is the case, then the theorem follows easily. For

$$\mathbf{r}^{(j)\mathrm{T}}\mathbf{C}\mathbf{u}^{(k)} = (\mathbf{x}^* - \mathbf{x}^{(j)})^{\mathrm{T}}\mathbf{C}\mathbf{C}\mathbf{u}^{(k)} = \sum_{v=1}^{3}(\mathbf{x}^* - \mathbf{x}^{(j)})^{\mathrm{T}}\mathbf{C}\alpha_v\mathbf{u}^{(k+v-2)},$$

and each term in the summation is 0, by the inductive hypothesis and part (a) of Theorem 6.7.

Thus in order to complete the proof, we need only establish the veracity of (6.53). From (6.51), we have

$$\mathbf{u}^{(k+1)} = \mathbf{C}(\mathbf{x}^* - \mathbf{x}^{(k+1)}) + \beta_k \mathbf{u}^{(k)},$$

where here and during the rest of the proof, Greek letters will consistently be used to denote scalars. Multiply both sides of (6.52) by $-\mathbf{C}$ and add $\mathbf{C}\mathbf{x}^*$ to both sides. The resulting expression will have the form

$$\mathbf{C}(\mathbf{x}^* - \mathbf{x}^{(k+1)}) = \mathbf{C}[(\mathbf{x}^* - \mathbf{x}^{(k)}) + \gamma_k \mathbf{u}^{(k)}] = -\mathbf{r}^{(k)} + \gamma_k \mathbf{C}\mathbf{u}^{(k)}.$$

Again using (6.51), one may readily see that

$$\mathbf{r}^{(k)} = (\mathbf{u}^{(k)} + \delta_k \mathbf{u}^{(k-1)}).$$

In summary,

$$\mathbf{u}^{(k+1)} = -(\mathbf{u}^{(k)} + \delta_k \mathbf{u}^{(k-1)}) + \gamma_k \mathbf{C}\mathbf{u}^{(k)} + \beta_k \mathbf{u}^{(k-1)},$$

from which relation (6.53) follows immediately.   ■

The theory behind the conjugate gradient method is as rich and satisfying as that supporting any of the other methods of nonlinear programming. In contrast to the gradient method, the conjugate gradient method gives the exact solution to a linear equation within n iterations, if roundoff error is ignored. But when carried to completion, the number of multiplications and additions is asymptotically $3n^3$, which is clearly worse than the corresponding number for Gauss elimination, which is $n^3/3$. But the conjugate gradient method is attractive as an iterative method where the number of iterations is significantly less than n and the order of the system is so large as to make Gauss elimination unattractive or infeasible. One optimal property of the conjugate gradient method when used for fewer than n iterations is now described. One may readily confirm that if $\mathbf{x}^{(j)}$ is as described in the conjugate gradient algorithm, then

$$\mathbf{x}^{(j)} = \mathbf{x}^{(0)} + P_j(\mathbf{C})\nabla f(\mathbf{x}^{(0)}), \tag{6.54}$$

where P_j is a polynomial in the variable $\mathbf{C}$ having degree at most j. That is, $P_j(\mathbf{C}) = \sum_{i=0}^{j} \alpha_i \mathbf{C}^i$ for some set of scalars $\alpha_0, \alpha_1, \ldots, \alpha_j$. It is known (e.g., Luenberger, 1973, Section 8.4) that among all $x^{(j)}$ having the form (6.54), the one determined by the conjugate gradient method minimizes $f(\cdot)$. Other properties of the conjugate gradient method are also cogent to numerical solution of linear equations, and we refer the interested reader to Luenberger (1973) and Faddeev and Faddeeva (1963). The conjugate gradient method

Table 6.6. Summary of Iterative Methods

Equation to be solved: $\mathbf{Ax} = \mathbf{b}$

I. Methods of form $\mathbf{x}^{(k+1)} = \mathbf{Bx}^{(k)} + \mathbf{c}$

A. Classical iterative method $\qquad \mathbf{B} = \mathbf{I} - \mathbf{A}, \mathbf{c} = \mathbf{b}$

B. Jacobi method $\qquad \mathbf{B} = -\mathbf{D}^{-1}(\mathbf{L} + \mathbf{R}), \mathbf{c} = \mathbf{D}^{-1}\mathbf{b}$

C. Gauss–Seidel method $\qquad \mathbf{B} = -(\mathbf{D} + \mathbf{L})^{-1}\mathbf{R}, \mathbf{c} = (\mathbf{D} + \mathbf{L})^{-1}\mathbf{b}$

D. Successive relaxation method $\qquad \mathbf{B} = (\mathbf{D} + w\mathbf{L})^{-1}[(1 - w)\mathbf{D} - w\mathbf{R}],$

$\qquad \mathbf{c} = (\mathbf{D} + w\mathbf{L})^{-1}w\mathbf{b}, 0 < w < 2$

II. Methods of form $\mathbf{x}^{(k+1)} = \mathbf{x}^{(k)} + t_k\mathbf{u}^{(k)}$ [a]

A. Successive relaxation method $\qquad \mathbf{u}^{(k)}$ determined by repeating sequence of unit vectors $\mathbf{e}_1, \mathbf{e}_2, \ldots, \mathbf{e}_n, \mathbf{e}_1, \ldots,$ and $t_k = w(\mathbf{u}^{(k)T}\mathbf{r}^{(k)})/(\mathbf{u}^{(k)T}\mathbf{C}\mathbf{u}^{(k)})$, for some $0 < w < 2$; for $w = 1$, we have the Gauss–Seidel method

B. Gradient method $\qquad \mathbf{u}^{(k)} = -\tfrac{1}{2}\nabla f(\mathbf{x}^{(k)}) = \mathbf{r}^{(k)}$

$\qquad t_k = \mathbf{r}^{(k)T}\mathbf{r}^{(k)}/(\mathbf{r}^{(k)T}\mathbf{C}\mathbf{r}^{(k)})$

C. Conjugate gradient method $\qquad \mathbf{u}^{(k)} = \dfrac{\mathbf{r}^{(k)} - (\mathbf{r}^{(k)T}\mathbf{C}\mathbf{u}^{(k-1)})}{(\mathbf{u}^{(k-1)T}\mathbf{C}\mathbf{u}^{(k-1)})}\mathbf{u}^{(k-1)}$

$\qquad t_k = \dfrac{-\mathbf{r}^{(k)T}\mathbf{u}^{(k)}}{(\mathbf{u}^{(k)T}\mathbf{C}\mathbf{u}^{(k)})}$

[a] Notation: $\mathbf{C} = \mathbf{A}^{\mathrm{T}}\mathbf{A}$, $\mathbf{d} = \mathbf{A}^{\mathrm{T}}\mathbf{b}$, and $\mathbf{r}^{(k)} = \mathbf{d} - \mathbf{C}\mathbf{x}^{(k)}$.

was initially introduced as a technique for solving linear equations by Hestenes and Stiefel (1952), but has since also proven its value for nonlinear optimization and solution of differential equations.

In Table 6.6, we summarize the formulas of the iterative methods we have discussed.

Problems

6.19. Show that each step of the method of successive relaxation (6.39) requires n^2 multiplications and divisions, and that the modified version (6.40) of the method requires $n^2 + n$ multiplications and divisions at each step.

6.20. Show that the gradient method requires $2n^2 + 3n + 1$ multiplications and divisions at each step.

6.3. Matrix Conditioning and Error Analysis

Consider the equations

$$2x_1 + 6x_2 = 8,$$
$$2x_1 + 6.00001x_2 = 8.00001,$$

and

$$2x_1 + 6x_2 = 8,$$
$$2x_1 + 5.99999x_2 = 8.00002.$$

It is easy to verify that the solution of the first pair of equations is equal to $x^* = (1, 1)^T$ and the solution of the second equations is given by $x^* = (10, -2)^T$.

This simple example shows that the solutions of simultaneous linear equations can be very far from each other even when the coefficients and right sides are very close. We shall prove that the solutions of simultaneous linear equations are continuous functions of the coefficients and the right-hand sides, but it will be observed that the modulus of continuity may be very large. Throughout this section we shall assume that the matrix norm satisfies $\| I \| = 1$, for I the identity matrix.

6.3.1. Bounds for Errors of Perturbed Linear Equations

Theorem 6.9. Let $\| \cdot \|$ designate some fixed matrix norm, and let matrix A be an $n \times n$ nonsingular matrix. Assume that B is any $n \times n$ matrix such that $\| A - B \| < \| A^{-1} \|^{-1}$. Then B is nonsingular and

$$\| A^{-1} - B^{-1} \| \leq \frac{\| A - B \| \, \| A^{-1} \|^2}{1 - \| A - B \| \, \| A^{-1} \|}. \tag{6.55}$$

Proof. The proof consists of several steps.

1. First we prove that any $n \times n$ matrix M with the property that $\| I - M \| < 1$ is nonsingular. It is sufficient to prove that equation $Mx = 0$ has the unique solution $x = 0$. This equation can be written in the form

$$x = (I - M)x.$$

Since $\| I - M \| < 1$, the operator $(I - M)$ is a contraction. The uniqueness of the solution follows immediately from Theorem 4.5, and 0 is the unique solution.

2. Next we prove that if the $n \times n$ matrix $\mathbf{M}$ satisfies the condition $\| \mathbf{M} \| < 1$, then $(\mathbf{I} + \mathbf{M})^{-1}$ exists and

$$\| (\mathbf{I} + \mathbf{M})^{-1} - \mathbf{I} \| \leq \frac{\| \mathbf{M} \|}{1 - \| \mathbf{M} \|}. \tag{6.56}$$

Step 1 implies the nonsingularity of matrix $\mathbf{I} + \mathbf{M}$, because

$$\| \mathbf{I} - (\mathbf{I} + \mathbf{M}) \| = \| \mathbf{M} \| < 1.$$

Since

$$\mathbf{I} = (\mathbf{I} + \mathbf{M})^{-1}(\mathbf{I} + \mathbf{M}) = \mathbf{R} + \mathbf{RM}, \tag{6.57}$$

where $\mathbf{R} = (\mathbf{I} + \mathbf{M})^{-1}$, we have

$$\| \mathbf{R} \| = \| \mathbf{I} - \mathbf{RM} \| \leq \| \mathbf{I} \| + \| \mathbf{R} \| \| \mathbf{M} \| = 1 + \| \mathbf{R} \| \| \mathbf{M} \|,$$

which implies

$$\| \mathbf{R} \| \leq \frac{1}{1 - \| \mathbf{M} \|}. \tag{6.58}$$

Combining this inequality with (6.57) we obtain

$$\| \mathbf{R} - \mathbf{I} \| = \| \mathbf{I} - \mathbf{R} \| = \| \mathbf{RM} \| \leq \| \mathbf{R} \| \| \mathbf{M} \| \leq \frac{\| \mathbf{M} \|}{1 - \| \mathbf{M} \|}.$$

3. Let $\mathbf{C} = \mathbf{B} - \mathbf{A}$. Then

$$\mathbf{A} + \mathbf{C} = \mathbf{A}(\mathbf{I} + \mathbf{A}^{-1}\mathbf{C}).$$

Since $\| \mathbf{A}^{-1}\mathbf{C} \| \leq \| \mathbf{A}^{-1} \| \| \mathbf{C} \|$, which by the theorem hypothesis is less than $\| \mathbf{A}^{-1} \| \| \mathbf{A}^{-1} \|^{-1} = 1$, we have that $\mathbf{I} + \mathbf{A}^{-1}\mathbf{C}$ is nonsingular and

$$(\mathbf{I} + \mathbf{A}^{-1}\mathbf{C})^{-1}\mathbf{A}^{-1} = (\mathbf{A} + \mathbf{C})^{-1},$$

which implies the nonsingularity of $\mathbf{A} + \mathbf{C}$ and the validity of the equation

$$\mathbf{B}^{-1} - \mathbf{A}^{-1} = (\mathbf{A} + \mathbf{C})^{-1} - \mathbf{A}^{-1} = [(\mathbf{I} + \mathbf{A}^{-1}\mathbf{C})^{-1} - \mathbf{I}]\mathbf{A}^{-1}.$$

Using (6.56) for $\mathbf{M} = \mathbf{A}^{-1}\mathbf{C}$, we have

$$\| \mathbf{B}^{-1} - \mathbf{A}^{-1} \| \leq \| (\mathbf{I} + \mathbf{A}^{-1}\mathbf{C})^{-1} - \mathbf{I} \| \| \mathbf{A}^{-1} \|$$

$$\leq \frac{\| \mathbf{A}^{-1}\mathbf{C} \| \| \mathbf{A}^{-1} \|}{1 - \| \mathbf{A}^{-1}\mathbf{C} \|} \leq \frac{\| \mathbf{A}^{-1} \|^2 \| \mathbf{C} \|}{1 - \| \mathbf{A}^{-1} \| \| \mathbf{C} \|},$$

which implies (6.55). ∎

The quantity

$$\gamma(\mathbf{A}) = \|\mathbf{A}\| \|\mathbf{A}^{-1}\|$$

is called the *condition number* of $\mathbf{A}$ with respect to matrix inversion. Obviously $\gamma(\mathbf{A})$ depends on the choice of norm. The condition number will be seen to play a prominent role in majorizing errors. $\gamma(\mathbf{A})$ is always at least 1, and when $\gamma(\mathbf{A})$ is large, one says that the matrix $\mathbf{A}$ is *ill conditioned*. The following corollary is an instance of the prominent appearance of the condition number in an error bound.

Corollary 6.1. Inequality (6.55) implies that

$$\frac{\|\mathbf{A}^{-1} - \mathbf{B}^{-1}\|}{\|\mathbf{A}^{-1}\|} \leq \frac{\gamma(\mathbf{A})\|\mathbf{A} - \mathbf{B}\|/\|\mathbf{A}\|}{1 - \gamma(\mathbf{A})\|\mathbf{A} - \mathbf{B}\|/\|\mathbf{A}\|}. \tag{6.59}$$

If $\|\mathbf{A} - \mathbf{B}\|$ is small, then (6.59) implies that the relative error of $\mathbf{A}^{-1}$ is bounded by the multiple $\gamma(\mathbf{A})$ of the relative error of $\mathbf{A}$.

Theorem 6.10. Assume that the $n \times n$ matrix $\mathbf{A}$ is nonsingular. Let $\mathbf{x}$ denote the solution of the equation $\mathbf{A}\mathbf{x} = \mathbf{b}$. Let $\mathbf{B}$ be any $n \times n$ matrix such that $\|\mathbf{A} - \mathbf{B}\| \leq \|\mathbf{A}^{-1}\|^{-1}$. Then the unique solution of the equation $\mathbf{B}\mathbf{y} = \mathbf{d}$ satisfies the inequality

$$\|\mathbf{x} - \mathbf{y}\| \leq \frac{\gamma(\mathbf{A})(\|\mathbf{A} - \mathbf{B}\| \|\mathbf{x}\| + \|\mathbf{b} - \mathbf{d}\|)}{\|\mathbf{A}\| - \|\mathbf{A} - \mathbf{B}\|\gamma(\mathbf{A})}. \tag{6.60}$$

Proof. First we observe that for an arbitrary $n \times n$ matrix $\mathbf{M}$ such that $(\mathbf{I} - \mathbf{M})^{-1}$ exists, the equation

$$(\mathbf{I} - \mathbf{M})^{-1}(\mathbf{I} - \mathbf{M}) = \mathbf{I}$$

implies

$$(\mathbf{I} - \mathbf{M})^{-1} = \mathbf{I} + (\mathbf{I} - \mathbf{M})^{-1}\mathbf{M}. \tag{6.61}$$

Now

$$\mathbf{y} - \mathbf{x} = \mathbf{B}^{-1}\mathbf{d} - \mathbf{x} = \mathbf{B}^{-1}\mathbf{b} + \mathbf{B}^{-1}(\mathbf{d} - \mathbf{b}) - \mathbf{x}$$

$$= \mathbf{B}^{-1}(\mathbf{d} - \mathbf{b}) - \mathbf{x} + \mathbf{A}^{-1}[\mathbf{I} - (\mathbf{A} - \mathbf{B})\mathbf{A}^{-1}]^{-1}\mathbf{b},$$

and using the identity (6.61) with $\mathbf{M} = -(\mathbf{A} - \mathbf{B})\mathbf{A}^{-1}$, we obtain

$$\mathbf{y} - \mathbf{x} = \mathbf{B}^{-1}(\mathbf{d} - \mathbf{b}) - \mathbf{x} + \mathbf{A}^{-1}\{\mathbf{I} + [\mathbf{I} - (\mathbf{A} - \mathbf{B})\mathbf{A}^{-1}]^{-1}(\mathbf{A} - \mathbf{B})\mathbf{A}^{-1}\}\mathbf{b}$$

$$= \mathbf{B}^{-1}(\mathbf{d} - \mathbf{b}) - \mathbf{x} + \mathbf{A}^{-1}\mathbf{b} + \mathbf{A}^{-1}[\mathbf{I} - (\mathbf{A} - \mathbf{B})\mathbf{A}^{-1}]^{-1}(\mathbf{A} - \mathbf{B})\mathbf{A}^{-1}\mathbf{b}.$$

Since $\mathbf{x} = \mathbf{A}^{-1}\mathbf{b}$, the preceding equation can be written

$$\mathbf{y} - \mathbf{x} = \mathbf{B}^{-1}(\mathbf{d} - \mathbf{b}) + \mathbf{A}^{-1}[\mathbf{I} - (\mathbf{A} - \mathbf{B})\mathbf{A}^{-1}]^{-1}(\mathbf{A} - \mathbf{B})\mathbf{x}. \quad (6.62)$$

Observe that

$$\mathbf{B}^{-1} = (\mathbf{B}^{-1} - \mathbf{A}^{-1}) + \mathbf{A}^{-1}.$$

Then (6.62) and (6.58) imply

$$\| \mathbf{y} - \mathbf{x} \| \leq \{\| \mathbf{B}^{-1} - \mathbf{A}^{-1} \| + \| \mathbf{A}^{-1} \|\} \| \mathbf{d} - \mathbf{b} \|$$

$$+ \| \mathbf{A}^{-1} \| \frac{1}{1 - \|(\mathbf{A} - \mathbf{B})\mathbf{A}^{-1}\|} \| \mathbf{A} - \mathbf{B} \| \| \mathbf{x} \|$$

$$\leq \left\{ \frac{\| \mathbf{A} - \mathbf{B} \| \| \mathbf{A}^{-1} \|^2}{1 - \| \mathbf{A} - \mathbf{B} \| \| \mathbf{A}^{-1} \|} + \| \mathbf{A}^{-1} \| \right\} \| \mathbf{d} - \mathbf{b} \|$$

$$+ \frac{\| \mathbf{A}^{-1} \| \| \mathbf{A} - \mathbf{B} \| \| \mathbf{x} \|}{1 - \| \mathbf{A} - \mathbf{B} \| \| \mathbf{A}^{-1} \|},$$

which is equivalent to inequality (6.60).   ∎

Corollary 6.2. Inequality (6.59) can be rewritten in the form

$$\frac{\| \mathbf{y} - \mathbf{x} \|}{\| \mathbf{x} \|} \leq \frac{\gamma(\mathbf{A})\left(\dfrac{\| \mathbf{A} - \mathbf{B} \|}{\| \mathbf{A} \|} + \dfrac{1}{\| \mathbf{A} \|} \cdot \dfrac{\| \mathbf{b} - \mathbf{d} \|}{\| \mathbf{b} \|} \cdot \dfrac{\| \mathbf{b} \|}{\| \mathbf{x} \|} \right)}{1 - \| \mathbf{A} - \mathbf{B} \| \cdot \| \mathbf{A}^{-1} \|}. \quad (6.63)$$

This corollary explicitly shows the dependence of the relative error of $\| \mathbf{x} \|$ on the relative errors of $\mathbf{A}$ and $\mathbf{b}$, in terms of the condition number.

Another important application of the condition number $\gamma(\mathbf{A})$ is now presented. Let $\mathbf{x}$ be an approximation of the solution of the equation (6.2). The exact solution $\mathbf{x}^*$ being unknown, the error vector $\mathbf{y} = \mathbf{x} - \mathbf{x}^*$ cannot be calculated. The "goodness" of the approximating solution is often characterized by the residual vector $\mathbf{r} = \mathbf{b} - \mathbf{Ax}$. The following theorem can be applied for bounding the unknown error vector $\mathbf{y}$ using the known residual vector $\mathbf{r}$.

Theorem 6.11. The inequality

$$\frac{1}{\gamma(\mathbf{A})} \cdot \frac{\| \mathbf{r} \|}{\| \mathbf{b} \|} \leq \frac{\| \mathbf{x} - \mathbf{x}^* \|}{\| \mathbf{x}^* \|} \leq \gamma(\mathbf{A}) \cdot \frac{\| \mathbf{r} \|}{\| \mathbf{b} \|} \quad (6.64)$$

holds.

Proof. Since

$$\mathbf{A}\mathbf{x}^* = \mathbf{b}, \qquad \mathbf{A}\mathbf{y} = \mathbf{A}\mathbf{x} - \mathbf{b} = -\mathbf{r},$$

we have

$$\mathbf{x}^* = \mathbf{A}^{-1}\mathbf{b}, \qquad \mathbf{y} = -\mathbf{A}^{-1}\mathbf{r},$$

which implies the inequalities

$$\| \mathbf{b} \| \le \| \mathbf{A} \| \, \| \mathbf{x}^* \|, \qquad \| \mathbf{x}^* \| \le \| \mathbf{A}^{-1} \| \, \| \mathbf{b} \|,$$
$$\| \mathbf{r} \| \le \| \mathbf{A} \| \, \| \mathbf{y} \|, \qquad \| \mathbf{y} \| \le \| \mathbf{A}^{-1} \| \, \| \mathbf{r} \|.$$

Thus we have

$$\frac{1}{\gamma(\mathbf{A})} \frac{\| \mathbf{r} \|}{\| \mathbf{b} \|} = \frac{\| \mathbf{r} \|}{\| \mathbf{A} \|} \cdot \frac{1}{\| \mathbf{b} \| \, \| \mathbf{A}^{-1} \|} \le \frac{\| \mathbf{y} \|}{\| \mathbf{x}^* \|}$$

$$\le \| \mathbf{A}^{-1} \| \, \| \mathbf{r} \| \frac{\| \mathbf{A} \|}{\| \mathbf{b} \|} = \gamma(\mathbf{A}) \cdot \frac{\| \mathbf{r} \|}{\| \mathbf{b} \|}. \qquad \blacksquare$$

Problems

6.21. Assuming that the absolute values of the eigenvalues of matrix **B** are less than one, prove that $(\mathbf{I} - \mathbf{B})^{-1}$ exists and can be obtained as the sum of the series

$$\mathbf{I} + \mathbf{B} + \mathbf{B}^2 + \cdots.$$

6.22. Prove that the condition $\| \mathbf{B} \| < 1$ is sufficient for the assertion of the previous problem. Assuming that $\| \mathbf{I} \| = 1$ and using the result of the previous problem prove (6.58).

6.23. Prove that for an arbitrary nonsingular matrix **A**, $\gamma(\mathbf{A}) \ge 1$.

6.3.2. Error Bounds for Rounding in Gaussian Elimination

Using Theorem 6.9, we are in a position to develop bounds for rounding errors in the Gaussian elimination method. Some of the details of the derivation are left out because we feel that these developments are not useful or interesting enough to justify the tedium.

Let $\mathbf{x}^*$ be the solution of the given equation $\mathbf{A}\mathbf{x}^* = \mathbf{b}$. Our approach to bounding the error of the approximation $\mathbf{x}$ computed by the Gaussian elimination method is to exhibit a matrix **B** such that $\mathbf{B}\mathbf{x} = \mathbf{b}$. Then clearly

$$\| \mathbf{x}^* - \mathbf{x} \| \le \| \mathbf{A}^{-1} - \mathbf{B}^{-1} \| \, \| \mathbf{b} \|.$$

We shall be able to bound $\| \mathbf{A} - \mathbf{B} \|$ in terms of the floating-point word length t. From this bound and from Theorem 6.9, we immediately have a bound on $\| x^* - x \|$.

We now construct the matrix $\mathbf{B}$. Define $\mathbf{L}_1$ and $\mathbf{U}_1$ to be the computed approximations of $\mathbf{L}$ and $\mathbf{U}$ of the $\mathbf{LU}$ decomposition described in connection with the Crout reduction method. That is, $\mathbf{U}_1$ is the first n columns of the final computed derived system $\mathbf{A}^{(n-1)}$, and for $i > j$,

$$(\mathbf{L}_1)_{ij} = \frac{\bar{a}_{ij}^{(j-1)}}{\bar{a}_{jj}^{(j-1)}} + \alpha_{ij},$$

where $\bar{a}_{uv}^{(k)}$ denotes the uv coordinate of the computed approximation of the kth derived system and α_{ij} is roundoff in the division operation. $\mathbf{y} = (\bar{a}_{1,n+1}^{(n-1)}, \ldots, \bar{a}_{n,n+1}^{(n-1)})^{\mathrm{T}}$ is the approximation of $\mathbf{L}^{-1}\mathbf{b} = \mathbf{y}^*$.

With $\mathbf{L}_1$ and $\mathbf{U}_1$ so defined, we shall describe the construction of matrices $\delta\mathbf{L}$ and $\delta\mathbf{U}$ such that

$$(\mathbf{L}_1 + \delta\mathbf{L})\mathbf{y} = \mathbf{b}, \qquad (\mathbf{U}_1 + \delta\mathbf{U})\mathbf{x} = \mathbf{y}.$$

From these equations it is evident that

$$(\mathbf{L}_1 + \delta\mathbf{L})(\mathbf{U}_1 + \delta\mathbf{U})\mathbf{x} = \mathbf{b},$$

and so $\mathbf{B} = (\mathbf{L}_1 + \delta\mathbf{L})(\mathbf{U}_1 + \delta\mathbf{U})$ suffices for the matrix $\mathbf{B}$ described above. Let us define

$$\mathbf{E} = \mathbf{LU} - \mathbf{L}_1\mathbf{U}_1.$$

Then $\mathbf{A} - \mathbf{B} = \mathbf{E} - (\delta\mathbf{L})\mathbf{U}_1 - \mathbf{L}_1(\delta\mathbf{U}) - (\delta\mathbf{L})(\delta\mathbf{U})$, and so we have the critical result

$$\| \mathbf{A} - \mathbf{B} \| \leq \| \mathbf{E} \| + \| \delta\mathbf{L} \|\,\| \mathbf{U}_1 \| + \| \mathbf{L}_1 \|\,\| \delta\mathbf{U} \| + \| \delta\mathbf{L} \|\,\| \delta\mathbf{U} \|.$$
$$(6.65)$$

The computation of bounds for the matrices $\mathbf{E}$, $\delta\mathbf{L}$, and $\delta\mathbf{U}$ in terms of the rounding error is procedurally a direct application of the developments in Chapter 1, but the details of these computations are lengthy, and we shall bypass them, referring the interested reader to Appendix 3 of Stewart (1973), Section 4.3 of Blum (1972), or Chapter 9 of Ortega (1972). We summarize results of these calculations in the following statement.

Theorem 6.12. Let $\mathbf{x}$ be the approximate solution of the equation $\mathbf{Ax} = \mathbf{b}$ (where $\mathbf{A}$ is nonsingular and of order n) by Gaussian elimination using floating-point arithmetic with word length t. Assume also that

$n2^{1-t} < 1$. Then there exists a matrix $\mathbf{B}$ satisfying $\mathbf{Bx} = \mathbf{b}$ and such that

$$\max_{ij} |(\mathbf{A} - \mathbf{B})_{ij}| \leq n^2 ag_n 2^{-t+1}/(1 - 2^{-t+1}) + n(n+1)^2 ag_n 2^{-t+1}$$
$$+ \frac{(n+1)^4}{4} ag_n(2^{-t+1})^2,$$

where $a = \max_{ij} |a_{ij}|$ and

$$g_n \leq \begin{cases} 2^{n-1}, & \text{if partial pivoting is used,} \\ \left(n \prod_{j=2}^{n} j^{1/(j-1)}\right)^{1/2}, & \text{if maximal pivoting is used.} \end{cases}$$

The term g_n is called the *growth factor*. For maximal pivoting, the growth factor grows slowly. It is known (e.g., Bunch and Parlett, 1971) that $g_n \leq 2n^{\log(n)/4+1/2}$. The growth factor for partial pivoting grows quickly, and is sharp in the sense that it can be achieved for certain matrices (Ortega, 1972, p. 189). Blum (1972) notes that for partial pivoting, these bounds imply that if one computes, with an eight-decimal-place word, the Gauss elimination solution of a tenth-order system, Theorem 6.12 assures only one decimal place of accuracy in the answer, even if the matrix has the lowest possible condition number (which is 1). Furthermore, under these circumstances, the above theorem implies 45 decimal place words are required to compute the solution of a 100th-order system to seven-place accuracy. People with computing experience will affirm that for well-conditioned matrices, much greater accuracy than this can be anticipated.

It turns out that the growth factor for Crout reduction with partial pivoting is smaller than that for Gauss elimination with partial pivoting, provided accumulation of the inner product terms in Crout reduction is done in double precision. For that reason, many authors recommend Crout reduction over the Gauss elimination algorithm. Gauss elimination with maximal pivoting is the most accurate technique. (Maximal pivoting with Crout reduction is not practicable.) But the search for the maximum pivot can be expensive for large-order matrices. Stewart (1973) asserts that this cost can be reduced by careful programming in machine language.

Problem

6.24. The Hilbert segment matrix of order n was defined in Chapter 2 to be the matrix $\mathbf{A}$ determined by

$$a_{ij} = 1/(i + j - 1), \qquad 1 \leq i, j \leq n.$$

Use Gauss elimination to find an approximation to the solution of

$$\mathbf{A}\mathbf{x} = \mathbf{1},$$

for $n = 10$ and $\mathbf{1} = (1, 1, \ldots, 1)^{\mathrm{T}}$. Compute the residual.

6.4. Supplementary Notes and Discussion

Let us begin this discussion with some rules of thumb about where the various methods given in this chapter are appropriately used. For low- and moderate-order linear systems ($n \leq 100$) or large-banded systems, Gaussian elimination and its variants are to be preferred by virtue of their accuracy and sufficient economy. To some extent, the accuracy of the solution can be inferred by calculating the residual vector $\mathbf{r} = \mathbf{b} - \mathbf{A}\mathbf{x}$. If trouble is indicated by a large residual, one may consider using maximal pivoting and/or double precision, or perhaps *iterative refinement*. Iterative refinement, not discussed in this chapter [but discussed in Stewart (1973, Section 4.5), for example], is the process of having found an approximate solution $\mathbf{x}$ of the linear equation $\mathbf{A}\mathbf{x} = \mathbf{b}$, calculating $\mathbf{r} = \mathbf{b} - \mathbf{A}\mathbf{x}$ (this should be done in double precision), and then solving $\mathbf{A}\mathbf{y} = \mathbf{r}$. Then the vector $\mathbf{x} + \mathbf{y}$ will be a better approximation of the solution of the original equation than $\mathbf{x}$ is, provided that $\mathbf{A}$ is not too ill conditioned. This process may be repeated. Algol codes for the direct methods discussed in this chapter are to be found in Wilkinson and Reinsch (1971).

For certain higher-order systems, iterative procedures are recommended by many authors. Recall that Gaussian elimination requires approximately $n^3/3$ additions and multiplications to solve a linear system, and if this number is excessive, one must look to iterative methods. In speaking about certain problems arising in differential equations, control, and economics, Marchuk (1975, p. 289) writes "direct methods are more or less powerless, although the progress in hardware increases their chances." The Jacobi iterative method is mentioned only by way of introducing the Gauss–Seidel method, to which it is usually inferior. A theorem to this effect by Stein and Rosenburg (Varga, 1962) asserts that if the Jacobi iteration matrix $\mathbf{J} = \mathbf{D}^{-1}(-\mathbf{L} - \mathbf{R})$ has no negative elements and $\varrho(\mathbf{B})$ denotes the maximum magnitude of the eigenvalues of any matrix $\mathbf{B}$, then, provided $0 < \varrho(\mathbf{J}) < 1$ (a necessary condition for convergence), for $\mathbf{G}$, the Gauss–Seidel operator matrix,

$$\varrho(\mathbf{G}) < \varrho(\mathbf{J}).$$

Among the most powerful and effective iterative methods are the Gauss–

Seidel, successive overrelaxation, and conjugate gradient methods. It is known (Luenberger, 1973, p. 186) that the Gauss elimination method is, in fact, the conjugate direction method associated with the conjugate directions obtained by using Gram–Schmidt orthogonalization on the vectors e_j having 1 at the jth coordinate and 0s elsewhere. The books by Varga (1962), Young (1971), Faddeev and Faddeeva (1963), Goult *et al.* (1974), and, with emphasis on applications to differential equations, Marchuk (1975), have supplementary material on iterative methods. We also mention a recent study by Paige and Saunders (1975). The books by Stewart (1973) and Forsythe and Moler (1967) specialize in direct methods and contain many topics not discussed here. One such topic is "equilibration" or scaling of rows and columns of the linear equation to reduce rounding error. From Forsythe (1967), which is a commendable study citing highlights of developments in computational linear algebra, it may be seen that equilibration is effective only insofar as it affects pivoting strategies, and it is not clear how one should proceed with equilibration. In Chapter 9, we shall give some discussion and references to iteration methods for linear equations arising in the numerical solution of partial differential equations.

A weakness of the computational theory of linear equations is that, whereas Gauss elimination is the most powerful method, its error analysis is not wholly satisfying in that in most cases it provides such pessimistic bounds. Some have suggested that a probabilistic theory of error bounds is called for, but we are unaware of any substantial developments in this direction.

Another way in which probability theory impinges upon computational linear algebra is by means of the Monte Carlo method. Following investigations by J. von Neumann, Forsythe, and others, there are many algorithms for solving linear systems by simulating random walks on the set of indices $\{1, 2, \ldots, n\}$. These methods have lost their luster because linear systems can be solved so quickly by other methods. However, as engineers and scientists attempt to solve ever-larger systems, it is possible that these methods may again become competitive. The reader may wish to consult Chapter 3 of Yakowitz (1977) for details of the Monte Carlo approach as well as discussion of its potential power.

7

The Solution of Matrix Eigenvalue Problems

Let **A** be a square matrix of order n. Consider the equation

$$\mathbf{A}\mathbf{x} = \lambda\mathbf{x}, \tag{7.1}$$

where λ is a real or complex number and $\mathbf{x}$ is an n-tuple. For $\mathbf{x} = \mathbf{0}$, equation (7.1) is satisfied by an arbitrary scalar λ. The values of λ for which there exists some nonzero vector $\mathbf{x}$ satisfying (7.1) are called *eigenvalues* and the corresponding vectors $\mathbf{x}$ are called *eigenvectors*. Equation (7.1) can be written as

$$(\mathbf{A} - \lambda\mathbf{I})\mathbf{x} = \mathbf{0}, \tag{7.2}$$

which for fixed λ is a homogeneous system of linear equations. This observation implies that if somehow we can obtain the eigenvalue λ of a given matrix **A**, we can find a corresponding eigenvector by solving (7.2) by methods from Chapter 6. Equation (7.2) has a nontrivial solution if and only if

$$\det(\mathbf{A} - \lambda\mathbf{I}) = \begin{vmatrix} a_{11} - \lambda & a_{12} & \cdots & a_{1n} \\ a_{21} & a_{22} - \lambda & \cdots & a_{2n} \\ \vdots & & & \vdots \\ a_{n1} & a_{n2} & \cdots & a_{nn} - \lambda \end{vmatrix} = 0.$$

After expanding this determinant, we obtain a polynomial of degree n,

$$\phi(\lambda) = (-1)^n(\lambda^n - p_1\lambda^{n-1} - \cdots - p_{n-1}\lambda - p_n) = 0, \tag{7.3}$$

which is called the *characteristic polynomial* of matrix **A**. The entire equation (7.3) is called the *characteristic equation*. From this discussion we conclude that λ is an eigenvalue of **A** if and only if it is a root of the characteristic polynomial. One implication of this observation is that there exist no direct methods for finding an eigenvalue of a given matrix **A**. For one may confirm that the characteristic polynomial of the companion matrix

$$\mathbf{A} = \begin{bmatrix} -a_{n-1} & -a_{n-2} & \cdots & -a_1 & -a_0 \\ 1 & 0 & \cdots & 0 & 0 \\ 0 & 1 & \cdots & 0 & 0 \\ \cdot & \cdot & & \cdot & \cdot \\ \cdot & \cdot & & \cdot & \cdot \\ \cdot & \cdot & & \cdot & \cdot \\ 0 & 0 & \cdots & 1 & 0 \end{bmatrix}$$

is the polynomial $p(\lambda) = \lambda^n + a_{n-1}\lambda^{n-1} + \cdots + a_0$. If we could find an eigenvalue of **A** by a direct method, we would be assured of finding a root of the arbitrary polynomial $p(x)$, in contradiction to the Galois theory result mentioned in the beginning of Chapter 5. Thus in numerical computation of eigenvalues and eigenvectors, we are inexorably constrained to use iteration methods.

7.1. Preliminaries

7.1.1. Some Matrix Algebra Background

Before discussing the numerical methods for determining the eigenvalues and eigenvectors of matrices, some properties of eigenvalues, eigenvectors, and matrix analysis are presented.

With iteration processes, results giving domains that contain the eigenvalues are useful for starting iteration processes. We now cite popular results for eigenvalue localization.

Theorem 7.1. Let $\| \cdot \|$ denote a matrix norm. If λ is an eigenvalue of matrix **A**, then

$$| \lambda | \leq \| \mathbf{A} \|.$$

Proof. In Problem 7.1 below, we give a construct for showing that every matrix norm has a compatible vector norm. Equation (7.1) implies that if we use a vector norm compatible with the matrix norm, then

$$| \lambda | \times \| \mathbf{x} \| = \| \lambda \mathbf{x} \| = \| \mathbf{A} \mathbf{x} \| \leq \| \mathbf{A} \| \times \| \mathbf{x} \|. \tag{7.4}$$

Since $\mathbf{x} \neq \mathbf{0}$, $\| \mathbf{x} \| \neq 0$. After dividing (7.4) by $\| \mathbf{x} \|$ we have the assertion. ∎

Corollary 7.1. Recalling the norms introduced in Section 4.8 we have

$$|\lambda| \leq \begin{cases} \max_i \sum_{j=1}^{n} |a_{ij}| = \| \mathbf{A} \|_\infty, \\ \max_j \sum_{i=1}^{n} |a_{ij}| = \| \mathbf{A} \|_1, \\ \left\{ \sum_{i=1}^{n} \sum_{j=1}^{n} |a_{ij}|^2 \right\}^{1/2} = \| \mathbf{A} \|_2. \end{cases}$$

It is possible to restrict the domains even further by the following result, known as the *Gerschgorin theorem.*

Theorem 7.2. For $i = 1, 2, \ldots, n$, let

$$r_i = \sum_{\substack{j=1 \\ j \neq i}}^{n} |a_{ij}|,$$

and let C_i denote the disk (called a Gerschgorin disk) with center a_{ii} and radius r_i. Then all eigenvalues of $\mathbf{A}$ lie within the domain

$$D = \bigcup_{i=1}^{n} C_i.$$

Proof. Let λ be an eigenvalue of $\mathbf{A}$. Let $\mathbf{x} = (x_1, \ldots, x_n)^T$ be an eigenvector of $\mathbf{A}$ corresponding to λ. Let i be determined by the relation

$$|x_i| = \max_{1 \leq j \leq n} \{|x_j|\}.$$

Then equation (7.1) implies

$$\sum_{j=1}^{n} a_{ij} x_j = \lambda x_i,$$

which can be written as

$$\sum_{\substack{j=1 \\ j \neq i}}^{n} a_{ij} x_j = (\lambda - a_{ii}) x_i,$$

from which we conclude

$$|\lambda - a_{ii}| \leq \sum_{\substack{j=1 \\ j \neq i}}^{n} |a_{ij}| \times \left| \frac{x_j}{x_i} \right| \leq \sum_{\substack{j=1 \\ j \neq i}}^{n} |a_{ij}| = r_i,$$

and this completes the proof. ∎

Corollary 7.2. Let

$$\tilde{r}_j = \sum_{\substack{i=1 \\ i \neq j}}^{n} |a_{ij}|,$$

and let $\tilde{C}_j$ be the disk with center a_{jj} and radius $\tilde{r}_j$. Then all eigenvalues of **A** lie within the domain

$$\tilde{D} = \bigcup_{j=1}^{n} \tilde{C}_j.$$

This result can be obtained by applying the theorem to $\mathbf{A}^{\mathrm{T}}$. Stewart (1973, Section 6.4) cites the following extension of these results.

Theorem 7.3. If k Gerschgorin disks of the matrix **A** are disjoint from the other disks, then exactly k eigenvalues of **A** lie in the union of these k disks.

One implication of this theorem is that any isolated Gerschgorin disk contains exactly one eigenvalue.

If $p(t)$ is a polynomial $\sum_{i=0}^{n} a_i t^i$ with real or complex coefficients a_i, $0 \leq i \leq n$, and if **A** is a square matrix, then $p(\mathbf{A})$ will denote the matrix determined by $\sum_{i=0}^{n} a_i \mathbf{A}^i$.

Theorem 7.4. Let $(\lambda, \mathbf{x})$ be an eigenvalue–eigenvector pair of the square matrix **A**. If $p(t)$ is a polynomial, then $p(\lambda)$ and **x** are the eigenvalue and a corresponding eigenvector of the matrix $p(\mathbf{A})$. Furthermore, if $p(\mathbf{A})$ is nonsingular, then $p(\lambda)^{-1}$ is an eigenvalue and **x** a corresponding eigenvector of $p(\mathbf{A})^{-1}$.

Proof. Repeated use of (7.1) allows us to conclude that

$$\mathbf{A}^2 \mathbf{x} = \mathbf{A}(\lambda \mathbf{x}) = \lambda^2 \mathbf{x},$$

and more generally, that $\mathbf{A}^m \mathbf{x} = \lambda^m \mathbf{x}$, from which we see that

$$p(\mathbf{A})\mathbf{x} = p(\lambda)\mathbf{x}. \tag{7.5}$$

The hypothesis that $p(\mathbf{A})^{-1}$ exists implies that λ is not a root of $p(t)$. Upon multiplying both sides of (7.5) by $p(\lambda)^{-1}$, we have $p(\mathbf{A})p(\lambda)^{-1}\mathbf{x} = \mathbf{x}$. Then multiply both sides by $p(\mathbf{A})^{-1}$ to obtain the desired result that

$$p(\lambda)^{-1}\mathbf{x} = p(\mathbf{A})^{-1}\mathbf{x}. \qquad \blacksquare$$

At the present time, the most important eigenvalue methods depend on the linear transformation of the matrix into a specified form, from which the eigenvalues can be conveniently approximated. We now develop certain basic definitions and concepts regarding these transformations. Let **A** and **T** be square matrices of order n, and suppose **T** is nonsingular. Then

$$\mathbf{B} = \mathbf{TAT}^{-1} \tag{7.6}$$

is a *similarity transformation* of **A**, and a given matrix **B** is said to be *similar* to **A** if there exists a matrix **T** satisfying (7.6).

Theorem 7.5. Let **T** be any nonsingular matrix. (a) The eigenvalues of matrix **A** and the eigenvalues of **B** = $\mathbf{TAT}^{-1}$ are the same. If **x** is an eigenvector of **A** corresponding to λ, then **Tx** is an eigenvector of **B** corresponding to λ. (b) All of the eigenvectors of matrix **B** can be obtained in this way.

Proof. (a) Let λ and **x** be an eigenvalue and the corresponding eigenvector of matrix **A**. Then

$$\mathbf{Ax} = \lambda\mathbf{x},$$

which can be rewritten, after left multiplication by **T**, as

$$\mathbf{TAT}^{-1}\mathbf{Tx} = \lambda\mathbf{Tx}.$$

Let **y** = **Tx**. Then

$$\mathbf{By} = \lambda\mathbf{y},$$

and thus λ is an eigenvalue of **B** with corresponding eigenvector **y** = **Tx**.

(b) Let λ and **y** be a corresponding eigenvalue and eigenvector of matrix **B**; that is,

$$\mathbf{By} = \lambda\mathbf{y}.$$

This equation implies that

$$\mathbf{T}^{-1}\mathbf{BTT}^{-1}\mathbf{y} = \lambda\mathbf{T}^{-1}\mathbf{y}.$$

Since $\mathbf{T}^{-1}\mathbf{BT} = \mathbf{A}$, introducing the vector **x** = $\mathbf{T}^{-1}\mathbf{y}$ we have

$$\mathbf{Ax} = \lambda\mathbf{x}.$$

Consequently λ is an eigenvalue of **A** and **x** is a corresponding eigenvector. ∎

The eigenvalue methods that are currently regarded as most effective involve iteratively performing similarity transformations on the given matrix **A** until the final form is *upper triangular* (that is, $a_{ij} = 0$ for $i > j$) or nearly so, in the sense that the subdiagonal elements are small. The motivation of this approach is the observation that if **B** is upper triangular, then its characteristic polynomial is easily seen to be $\phi(\lambda) = \prod_{i=1}^{n} (b_{ii} - \lambda)$. By virtue of this factorization, the eigenvalues of **B** are diagonal elements b_{ii}, $1 \le i \le n$. If **B** has been constructed so as to be similar to **A**, then by Theorem 7.4, the elements b_{ii}, $1 \le i \le n$, are also eigenvalues of **A**. Furthermore, from the continuity of the coefficients of the characteristic polynomial as a function of the matrix coefficients, if $\mathbf{B}(\varepsilon)$ differs from **B**, for ε a positive number, in a manner such that

$$\max_{ij} | b_{ij} - b(\varepsilon)_{ij} | < \varepsilon,$$

then the eigenvalues of $\mathbf{B}(\varepsilon)$ converge to those of **B**, as $\varepsilon \to 0$. This implies that if $\mathbf{B}(\varepsilon)$ has very small subdiagonal elements, the eigenvalues of **B** will be well approximated by the diagonal elements of $\mathbf{B}(\varepsilon)$.

Unfortunately, many of the methods to be discussed are not applicable to all square matrices, but only those that are not defective. A matrix of order n is said to be *defective* if it has fewer than n linearly independent eigenvectors. To give an example of a defective matrix, we consider the Jordan block

$$\mathbf{J}_1 = \begin{bmatrix} \lambda_1 & & & & & 0 \\ 1 & \lambda_1 & & & & \\ & 1 & \cdot & & & \\ & & \cdot & \cdot & & \\ & & & \cdot & \cdot & \\ 0 & & & & 1 & \lambda_1 \end{bmatrix}$$

introduced in Example 4.12. One may readily see that all eigenvectors of this matrix must be proportional to $\mathbf{e}_n = (0, 0, \ldots, 1)^{\mathrm{T}}$. For if $\mathbf{x} = (x_1, x_2, \ldots, x_n)^{\mathrm{T}}$ is an eigenvector, then $\lambda\mathbf{x} = \mathbf{J}_1\mathbf{x}$ implies that

$$\lambda x_j = \lambda x_j + x_{j-1}, \qquad 1 \le j \le n,$$

which, in turn, implies that $x_k = 0$, for $k < n$. If a matrix is not defective, then it is said to be *diagonalizable*, because then (and only then) it is similar to a diagonal matrix (with the eigenvalues comprising the diagonal elements).

It is well known (Stewart, 1973, p. 276) that if $\mathbf{A}$ has distinct eigenvalues $\lambda_1, \lambda_2, \ldots, \lambda_n$, and if $\mathbf{x}_1, \mathbf{x}_2, \ldots, \mathbf{x}_n$ are eigenvectors corresponding to $\lambda_1, \lambda_2, \ldots, \lambda_n$, respectively, then these eigenvectors are linearly independent. Thus such matrices are diagonalizable. Also, it is known that symmetric matrices are diagonalizable (Bellman, 1970). In summary, the class of diagonalizable matrices form a very important subset of matrices, and although our major results will pertain only to such matrices, they are nevertheless highly useful results.

Orthogonal matrices are of particular importance to eigenvalue computations. A matrix $\mathbf{U}$ having real coefficients is said to be *orthogonal* if $\mathbf{U}^T\mathbf{U} = \mathbf{I}$, $\mathbf{I}$ being the identity matrix. One may verify directly from the definition that if $\mathbf{U}$ is orthogonal, then its columns $\mathbf{u}_i$, $1 \le i \le n$, are orthonormal in the sense that

$$(\mathbf{u}_i, \mathbf{u}_j) = \mathbf{u}_i^T\mathbf{u}_j = \begin{cases} 0, & i \ne j, \\ 1, & i = j. \end{cases} \tag{7.7}$$

If the (not necessarily square) matrix $\mathbf{U}$ has complex coefficients, then we shall let the symbol $\bar{\mathbf{U}}$ denote the matrix whose elements are the complex conjugates of the corresponding elements of $\mathbf{U}$. We define the superscript H to denote the matrix

$$\mathbf{U}^H = \bar{\mathbf{U}}^T. \tag{7.8}$$

If the vector space of n-tuples has complex coefficients, the customary inner product is

$$(\mathbf{x}, \mathbf{y}) = \mathbf{x}^H\mathbf{y}. \tag{7.9}$$

Thus a square matrix $\mathbf{U}$ with complex coefficients has orthonormal columns if

$$\mathbf{U}^H\mathbf{U} = \mathbf{I}. \tag{7.10}$$

Such a matrix is called a *unitary* matrix. Every orthogonal matrix is unitary, if we regard the real numbers as being embedded in the complex numbers. Consequently, our assertions to follow about unitary matrices are also true for orthogonal matrices.

From the definition of the vector norm $\| \cdot \|_2$ given in Theorem 4.7, we see immediately that, for any vector $\mathbf{x}$,

$$\mathbf{x}^H\mathbf{x} = \| \mathbf{x} \|_2^2; \tag{7.11}$$

and, from the definition of the matrix norm $\| \cdot \|_2$ in Theorem 4.8, we see that, for any square matrix $\mathbf{A}$,

$$\mathbf{A}^H \mathbf{A} = \| \mathbf{A} \|_2^2. \tag{7.12}$$

We shall see later that the corollary of the following theorem makes unitary matrices particularly attractive in performing similarity transformations. Observe also that if $\mathbf{a}_1, \ldots, \mathbf{a}_n$ are the columns of $\mathbf{A}$, then $\| \mathbf{A} \|_2^2 = \sum_{i=1}^n \| \mathbf{a}_i \|_2^2$.

Theorem 7.6. Let $\mathbf{U}$ and $\mathbf{A}$ be square matrices and $\mathbf{x}$ a vector, all of order n. Assume that $\mathbf{U}$ is unitary. Then

$$\| \mathbf{U}\mathbf{x} \|_2 = \| \mathbf{x} \|_2, \tag{7.13}$$

$$\| \mathbf{U}\mathbf{A} \|_2 = \| \mathbf{A} \|_2. \tag{7.14}$$

Proof. From relation (7.11), we have that

$$\| \mathbf{U}\mathbf{x} \|_2^2 = (\mathbf{U}\mathbf{x})^H (\mathbf{U}\mathbf{x}) = \mathbf{x}^H \mathbf{U}^H \mathbf{U}\mathbf{x} = \mathbf{x}^H \mathbf{x},$$

which establishes (7.13). Let $\mathbf{a}_1, \mathbf{a}_2, \ldots, \mathbf{a}_n$ denote the successive columns of $\mathbf{A}$.
Then

$$\| \mathbf{U}\mathbf{A} \|_2^2 = \sum_{i=1}^n \| \mathbf{U}\mathbf{a}_i \|_2^2 = \sum_{i=1}^n \| \mathbf{a}_i \|_2^2 = \| \mathbf{A} \|_2^2. \qquad \blacksquare$$

Corollary 7.3. Suppose matrix $\mathbf{B}$ is *unitarily similar* to $\mathbf{A}$. That is, suppose

$$\mathbf{B} = \mathbf{U}\mathbf{A}\mathbf{U}^H,$$

where $\mathbf{U}$ is unitary (and consequently $\mathbf{U}^H = \mathbf{U}^{-1}$). Then

$$\| \mathbf{B} \|_2 = \| \mathbf{A} \|_2.$$

Proof. We shall use the obvious fact that for any matrix $\mathbf{M}$, $\| \mathbf{M}^H \|_2 = \| \mathbf{M} \|_2$. Then from the theorem, and the fact that $\mathbf{U}^H$ is unitary,

$$\| \mathbf{A} \|_2 = \| \mathbf{U}^H \mathbf{B}\mathbf{U} \|_2 = \| \mathbf{B}\mathbf{U} \|_2 = \| \mathbf{U}^H \mathbf{B}^H \|_2 = \| \mathbf{B}^H \|_2 = \| \mathbf{B} \|_2. \qquad \blacksquare$$

The value of this corollary is that it leads to strong error bounds through the backward error analysis technique that was presented in Section 6.3.2 in studying the effect of rounding in Gaussian elimination. Specifically, suppose that in attempting to calculate

$$\mathbf{B} = \mathbf{UAU}^\mathrm{H},$$

due to rounding we actually calculate

$$\mathbf{B} = \mathbf{UAU}^\mathrm{H} + \mathbf{F}, \qquad (7.15)$$

where we can be assured that for some known $\varDelta$, $\| \mathbf{F} \|_2 \leq \varDelta$. We may then set $\mathbf{E} = \mathbf{U}^\mathrm{H}\mathbf{FU}$ and write (7.15) as

$$\mathbf{B} = \mathbf{U}(\mathbf{A} + \mathbf{E})\mathbf{U}^\mathrm{H}, \qquad (7.16)$$

and from the corollary, we shall know that $\| \mathbf{E} \|_2 \leq \varDelta$. Typically, we shall be able to obtain eigenvalues of $\mathbf{B}$, and from Theorem 7.5 we shall know that these are the eigenvalues of $\mathbf{A} + \mathbf{E}$. From standard perturbation results to be discussed in Section 7.4, we shall then be able to bound the difference between our exhibited eigenvalues of $\mathbf{A} + \mathbf{E}$ and the desired eigenvalues of $\mathbf{A}$.

Problems

7.1. For any vector $\mathbf{x}$, let $\mathbf{A}(\mathbf{x})$ denote the square matrix whose first column is $\mathbf{x}$, and whose remaining columns are $\mathbf{0}$ vectors. Assume $\| \cdot \|$ is a given matrix norm. Define the vector norm

$$\| \mathbf{x} \|_v \triangleq \| \mathbf{A}(\mathbf{x}) \|,$$

and prove that $\| \cdot \|_v$ is compatible with the matrix norm $\| \cdot \|$.

7.2. Use Gerschgorin's theorem to find a region containing the eigenvalues of the matrix

$$\mathbf{A} = \begin{bmatrix} -4 & -3 & -7 \\ 2 & 3 & 2 \\ 4 & 2 & 7 \end{bmatrix}.$$

7.3. Let $\mathbf{D}$ be a diagonal matrix with diagonal elements $1, 1, x$. Show that the Gerschgorin circles of the matrix $\mathbf{A}$ in Problem 7.2 can be reduced by using the transformation $\mathbf{D}^{-1}\mathbf{AD}$ with suitable values of x.

7.4. Show that if

$$\mathbf{B} = \mathbf{TAT}^{-1},$$

and if $\mathbf{x}$ is an eigenvector of $\mathbf{B}$, then $\mathbf{T}^{-1}\mathbf{x}$ is an eigenvector of $\mathbf{A}$.

7.5. Prove that if the square matrix $\mathbf{A}$ is diagonalizable, then $\mathbf{A}$ is similar to a diagonal matrix. HINT: Use a matrix $\mathbf{X} = [\mathbf{x}_1, \dots, \mathbf{x}_n]$ composed of linearly independent eigenvalues of $\mathbf{A}$.

7.6. Show that if **U** is orthogonal or unitary, then its columns are orthonormal.

7.7. Show in detail how to set up a linear equation as in (7.2) to find an eigenvector of **A**, once an eigenvalue has been found. HINT: Observe that some further constraint on **x** must be imposed.

7.1.2. The Householder Transformation and Reduction to Hessenberg Form

In most instances, the recommended way to compute eigenvalues of a given diagonalizable matrix **A** is first to reduce **A** to upper Hessenberg form by similarity transformations and then to apply the shifted QR algorithm (Section 7.3). A matrix $\mathbf{B} = \{b_{ij}\}$ is *upper Hessenberg* if $b_{ij} = 0$ for $i > j + 1$. **B** is *tridiagonal* if **B** and $\mathbf{B}^T$ are both upper Hessenberg. By use of Householder transformations, we shall have a direct and fairly efficient way to reduce any square matrix to upper Hessenberg form by unitary similarity transformations.

Definition 7.1. Let **I** be the nth-order identity matrix and **x** a (possibly complex) n-tuple such that $\| \mathbf{x} \|_2 = \mathbf{x}^H \mathbf{x} = 1$. Then the matrix of the form

$$\mathbf{H} = \mathbf{I} - 2\mathbf{x}\mathbf{x}^H$$

is a *Householder transformation*.

We leave the easy verification that any Householder transformation matrix is unitary to the reader (Problem 7.8). The following theorem will be important for us. In the statement below and elsewhere, $\mathbf{e}_j$ will denote the jth column of the identity matrix.

Theorem 7.7. Let **x** be any nonzero vector with real coefficients. Define

(a) $\sigma = \pm \| \mathbf{x} \|_2$ with the sign chosen so that $\mathbf{x} \neq -\sigma\mathbf{e}_1$,

(b) $\mathbf{u} = \mathbf{x} + \sigma\mathbf{e}_1$, and

(c) $\pi = \frac{1}{2} \| \mathbf{u} \|_2^2$.

Then $\mathbf{R} = \mathbf{I} - (1/\pi)\mathbf{u}\mathbf{u}^H$ is a unitary matrix with the property that

$$\mathbf{Rx} = -\sigma\mathbf{e}_1. \qquad (7.17)$$

Proof. The reader will easily confirm that **R** is a Householder transformation and is consequently unitary. We check (7.17) by direct substitution:

$$\begin{aligned}
\mathbf{R}\mathbf{x} &= \mathbf{x} - (1/\pi)(\mathbf{x} + \sigma\mathbf{e}_1)(\mathbf{x} + \sigma\mathbf{e}_1)^{\mathrm{H}}\mathbf{x} \\
&= \mathbf{x} - (1/\pi)(\mathbf{x} + \sigma\mathbf{e}_1)(\sigma^2 + \sigma x_1),
\end{aligned} \qquad (7.18)$$

where x_1 is the first coordinate of $\mathbf{x}$. But

$$\begin{aligned}
\pi &= \tfrac{1}{2}(\mathbf{x} + \sigma\mathbf{e}_1)^{\mathrm{H}}(\mathbf{x} + \sigma\mathbf{e}_1) \\
&= \sigma^2 + \sigma x_1,
\end{aligned}$$

and so upon cancellation of π in (7.18), we have

$$\mathbf{R}\mathbf{x} = -\sigma\mathbf{e}_1. \qquad\blacksquare$$

If $\mathbf{x}$ has complex coefficients, one must modify the Householder transformation rule as follows:

$$\sigma = \|\mathbf{x}\|_2 e^{i\theta}, \qquad \pi = \bar{\sigma}(x_1 + \sigma), \qquad (7.19)$$

where θ is the angle of the polar representation of x_1 and $\bar{\sigma}$ is the complex conjugate of σ. If these modifications are made, (7.17) holds for complex $\mathbf{x}$, as the reader may confirm (Problem 7.11).

We now describe an operation that employs the construct in the preceding theorem to obtain unitary matrices $\mathbf{U}_1, \mathbf{U}_2, \ldots, \mathbf{U}_n$ so that if we set $\mathbf{A}_1 = \mathbf{A}$, then for each $k, 3 \leq k \leq n$, $\mathbf{A}_{k-1} = \mathbf{U}_{k-2} \cdots \mathbf{U}_1 \mathbf{A} \mathbf{U}_1^{\mathrm{H}} \cdots \mathbf{U}_{k-2}$ has the form

$$\mathbf{A}_{k-1} = \begin{bmatrix}
a_{11}^{(k-1)} & a_{12}^{(k-1)} & a_{13}^{(k-1)} & & & & \cdots & & a_{1n}^{(k-1)} \\
a_{21}^{(k-1)} & a_{22}^{(k-1)} & a_{23}^{(k-1)} & a_{24}^{(k-1)} & & & \cdots & & a_{2n}^{(k-1)} \\
0 & a_{32}^{(k-1)} & \ddots & & & & & & \vdots \\
0 & 0 & & \ddots & & & & & \\
 & & & & a_{k-2,k-2}^{(k-1)} & a_{k-2,k-1}^{(k-1)} & a_{k-2,k}^{(k-1)} & \cdots & a_{k-2,n}^{(k-1)} \\
 & & & & a_{k-1,k-2}^{(k-1)} & a_{k-1,k-1}^{(k-1)} & a_{k-1,k}^{(k-1)} & \cdots & a_{k-1,n}^{(k-1)} \\
\vdots & \vdots & & & 0 & a_{k,k-1}^{(k-1)} & a_{kk}^{(k-1)} & \cdots & a_{kn}^{(k-1)} \\
 & & & & & \vdots & \vdots & & \vdots \\
0 & 0 & \cdots & & 0 & a_{n,k-1}^{(k-1)} & a_{nk}^{(k-1)} & \cdots & a_{nn}^{(k-1)}
\end{bmatrix}$$

$$(7.20)$$

which satisfies the Hessenberg condition for $j < k - 1$. Consequently, $\mathbf{A}_{n-1}$ will be the desired Hessenberg matrix.

We shall take $\mathbf{U}_{k-1}$ to be the matrix

$$\mathbf{U}_{k-1} = \begin{bmatrix} \mathbf{I}_{k-1} & \mathbf{0} \\ \mathbf{0} & \mathbf{R}_{k-1} \end{bmatrix}, \tag{7.21}$$

where $\mathbf{I}_{k-1}$ is the $(k-1)$th-order identity matrix and $\mathbf{R}_{k-1}$ is the House-holder transformation as described in Theorem 7.7, which takes the vector $\mathbf{x} = (a_{k,k-1}^{(k-1)}, a_{k+1,k-1}^{(k-1)}, \ldots, a_{n,k-1}^{(k-1)})^{\mathrm{T}}$ into a vector of the form $-(\sigma, 0, \ldots, 0)^{\mathrm{T}}$. Now if we left-multiply matrix $\mathbf{A}_{k-1}$ by $\mathbf{U}_{k-1}$, the first $k-2$ columns will retain zero elements at the positions indicated, and the $(k-1)$th column will have its last $n-k+1$ coordinates determined by the vector $\mathbf{R}_{k-1}\mathbf{x}$. Thus the product matrix $\mathbf{U}_{k-1}\mathbf{A}_{k-1}$ will have zeros in the position needed for $\mathbf{A}_k$. But right multiplication by

$$\mathbf{U}_{k-1}^{\mathrm{H}} = \begin{bmatrix} \mathbf{I}_{k-1} & \mathbf{0} \\ \mathbf{0} & \mathbf{R}_{k-1}^{\mathrm{H}} \end{bmatrix} \tag{7.22}$$

will leave the first $k-1$ columns of $\mathbf{U}_{k-1}\mathbf{A}_{k-1}$ unchanged, and so $\mathbf{A}_k = \mathbf{U}_{k-1}\mathbf{A}_{k-1}\mathbf{U}_{k-1}^{\mathrm{H}}$ is of the requisite form. We leave to the reader the easy verification that $\mathbf{U}_{k-1}$ is unitary.

One may verify that complete reduction of a full nth-order matrix to an upper Hessenberg matrix by the prescription we have given requires about $\frac{5}{3}n^3$ multiplications. If $\mathbf{A}$ is symmetric, the Hessenberg matrix $\mathbf{A}_{n-1}$ will be tridiagonal, and the reduction can be done with about $\frac{2}{3}n^3$ multiplications.

Example 7.1. We illustrate the construction of the Householder transformation conversion technique just described by converting the matrix $\mathbf{A}$ below into upper Hessenberg form. Thus

$$\mathbf{A}_1 = \mathbf{A} = \begin{bmatrix} -4 & -3 & -7 \\ 2 & 3 & 2 \\ 4 & 2 & 7 \end{bmatrix}.$$

We seek a unitary matrix

$$\mathbf{U}_1 = \begin{bmatrix} 1 & 0 & 0 \\ 0 & & \\ 0 & \mathbf{R}_1 & \end{bmatrix}$$

of the form (7.21) in which $\mathbf{R}_1$ is a Householder transformation mapping

$$\mathbf{x} = (a_{2,1}, a_{3,1})^{\mathrm{T}} = (2,4)^{\mathrm{T}} \text{ into } -(20)^{1/2}\mathbf{e}_1, \text{ where } \sigma = (20)^{1/2} = \|\mathbf{x}\|_2.$$

According to the construction described in Theorem 7.7, we set

$$\mathbf{u} = \begin{pmatrix} 2 \\ 4 \end{pmatrix} + \begin{pmatrix} (20)^{1/2} \\ 0 \end{pmatrix},$$

from which we compute that $\pi = \frac{1}{2} \| \mathbf{u} \|_2^2 = 20 + 2(20)^{1/2}$. Thus according to Theorem 7.7, we set

$$\mathbf{R}_1 = I - \pi^{-1}\mathbf{uu}^H$$

$$= \frac{1}{\pi} \begin{bmatrix} \pi - [2 + (20)^{1/2}] & -[8 + 4(20)^{1/2}] \\ -[8 + 4(20)^{1/2}] & -16 + \pi \end{bmatrix}$$

$$= \frac{1}{\pi} \begin{bmatrix} a & a \\ -2a & -2a \end{bmatrix},$$

where $a = 4 + 2(20)^{1/2}$. This in turn implies that if $b = a/\pi = [4 + 2(20)]^{1/2}/[20 + 2(20)^{1/2}]$, then

$$\mathbf{U}_1 = \begin{bmatrix} 1 & 0 & 0 \\ 0 & -b & -2b \\ 0 & -2b & b \end{bmatrix}.$$

We then compute that

$$\mathbf{U}_1\mathbf{A} = \begin{bmatrix} -4 & -3 & -7 \\ -10b & -7b & -16b \\ 0 & -4b & 3b \end{bmatrix},$$

and

$$\mathbf{A}_2 = \mathbf{U}_1\mathbf{A}\mathbf{U}_1^H = \begin{bmatrix} -4 & 17b & -b \\ -10b & 39b^2 & -2b^2 \\ 0 & -2b^2 & 11b^2 \end{bmatrix},$$

is upper Hessenberg.

Problems

7.8. Show that a Householder transformation is a unitary matrix.

7.9. Show that a Householder transformation matrix is involutory; that is, if $\mathbf{H}$ is such a matrix, then $\mathbf{H}^2 = \mathbf{H}$.

7.10. Show that if $\mathbf{A}$ is a Hermitian matrix and $\mathbf{U}$ is unitary, then $\mathbf{B} = \mathbf{U}^H\mathbf{A}\mathbf{U}$ is a Hermitian matrix.

7.11. Show that (7.17) holds for complex x under modifications (7.19).

7.12. Show that the matrix defined by (7.21) is unitary if $\mathbf{R}_{k-1}$ is unitary.

7.13. Let $\mathbf{x} = (1, 2, 3)^T$. Find a unitary matrix $\mathbf{R}$ such that $\mathbf{Rx} = -(14)^{1/2}\mathbf{e}_1$.

7.1.3. Matrix Deflation

Suppose we have somehow computed an eigenvalue–eigenvector pair $(\lambda, \mathbf{x})$ of some given nth-order matrix $\mathbf{A}$. If more eigenvalues of $\mathbf{A}$ are needed, it is then advantageous to compute an $(n-1)$th-order matrix $\mathbf{C}$ whose eigenvalues correspond with the remaining eigenvalues of $\mathbf{A}$. Such a computation is known as matrix *deflation*. We now describe the use of the Householder transformation for performing deflation. Assume the eigenvector $\mathbf{x}$ is scaled so that $\| \mathbf{x} \|_2 = 1$. Let $\mathbf{R}$ be the unitary matrix that, as in Theorem 7.7, provides the mapping $\mathbf{Rx} = -\mathbf{e}_1$. (Since λ is generally complex, we shall usually need to use the complex modification described after the proof of Theorem 7.7.) Since $\mathbf{R}$ is unitary, we have that

$$\mathbf{x} = -\mathbf{R}^H \mathbf{e}_1, \tag{7.23}$$

which implies that $\mathbf{R}^H$ has the block representation $[-\mathbf{x}, \mathbf{U}]$, where $\mathbf{U}$ is a matrix of dimension $n \times (n-1)$. Then

$$
\begin{aligned}
\mathbf{A}\mathbf{R}^H &= [-\mathbf{Ax}, \mathbf{AU}] \\
&= [-\lambda\mathbf{x}, \mathbf{AU}],
\end{aligned} \tag{7.24}
$$

and consequently

$$
\mathbf{R}\mathbf{A}\mathbf{R}^H = \begin{bmatrix} -\mathbf{x}^H \\ \mathbf{U}^H \end{bmatrix} [-\lambda\mathbf{x}, \mathbf{AU}]
$$
$$
= \begin{bmatrix} \lambda & -\mathbf{x}^H\mathbf{AU} \\ -\lambda\mathbf{U}^H\mathbf{x} & \mathbf{U}^H\mathbf{AU} \end{bmatrix} = \begin{bmatrix} \lambda & -\mathbf{x}^H\mathbf{AU} \\ 0 & \mathbf{U}^H\mathbf{AU} \end{bmatrix}. \tag{7.25}
$$

In equation (7.25), we have used the property of unitary matrices that the columns are orthogonal to write that $\mathbf{U}^H\mathbf{x} = 0$. Since $\mathbf{R}\mathbf{A}\mathbf{R}^H$ is similar to $\mathbf{A}$, the $(n-1)$th-order matrix

$$\mathbf{C} = \mathbf{U}^H\mathbf{AU} \tag{7.26}$$

must have the same eigenvalues, other than λ, as $\mathbf{A}$, and so with the construction of this matrix $\mathbf{C}$, we have accomplished the desired deflation.

7.2. Some Basic Eigenvalue Approximation Methods

In this discussion, $\mathbf{A}$ will consistently represent a square matrix whose eigenvalues are sought. Throughout the rest of the chapter, we shall adopt the standing assumption that the matrix $\mathbf{A}$ is diagonalizable, unless a statement is specifically made to the contrary.

The power method, the inverse power method, the method of Rayleigh quotient iterations, and the Jacobi method, which are all discussed in this section, are of interest in themselves, but are of additional interest in that they are all tied to a method to be discussed below in Section 7.3: the QR method with Rayleigh shifts, the method now regarded as the most reliable and efficient technique for most eigenvalue problems. We shall find that this QR method is simultaneously an instance of the inverse power method and the Rayleigh quotient iteration method, and thus shares their convergence properties developed here. Additionally, a technique used in computing the Jacobi method iteration is employed in the QR algorithm.

For both the power method and the method of Rayleigh quotient iterations, it is useful to know the definition and a certain approximation property of Rayleigh quotients. Let $\mathbf{x}$ be a given nonzero vector. The *Rayleigh quotient* of $\mathbf{x}$, with respect to $\mathbf{A}$, is defined to be the number

$$\lambda = R(\mathbf{x}) = \mathbf{x}^H \mathbf{A} \mathbf{x} / (\mathbf{x}^H \mathbf{x}). \tag{7.27}$$

It is evident that if $\mathbf{x}$ is an eigenvector, then $R(\mathbf{x})$ is the corresponding eigenvalue. But, as the following theorem states, if $\mathbf{x}$ is only an approximation of an eigenvector, the Rayleigh quotient provides an estimate of the corresponding eigenvalue. This estimate is, in a certain least-squares sense, optimal.

Theorem 7.8. For a given nonzero vector $\mathbf{x}$ and number λ, define the *residual* $\mathbf{r}$ by

$$\mathbf{r} = \mathbf{A} \mathbf{x} - \lambda \mathbf{x}.$$

Then for fixed $\mathbf{x}$, the Euclidean norm $\| \mathbf{r} \|_2$ of the residual is minimized when $\lambda = R(\mathbf{x})$.

Proof. We have that for any λ,

$$\| \mathbf{r} \|_2^2 = (\lambda \mathbf{x} - \mathbf{A}\mathbf{x})^H (\lambda \mathbf{x} - \mathbf{A}\mathbf{x})$$
$$= \| \mathbf{A}\mathbf{x} \|_2^2 - \lambda^H \mathbf{x}^H \mathbf{A}\mathbf{x} - \lambda \mathbf{x}^H \mathbf{A}\mathbf{x} + | \lambda |^2 \| \mathbf{x} \|_2^2. \tag{7.28}$$

Write λ and $\mathbf{x}^H \mathbf{A} \mathbf{x}$ in terms of their real and imaginary components as $\lambda = a + ib$ and $\mathbf{x}^H \mathbf{A} \mathbf{x} = \alpha + i\beta$. Then we may rewrite our above expression as

$$\| \mathbf{r} \|_2^2 = \| \mathbf{A}\mathbf{x} \|_2^2 - (a - ib)(\alpha + i\beta) - (a + ib)(\alpha - i\beta)$$
$$+ (a^2 + b^2) \| \mathbf{x} \|_2^2.$$

Differentiating this expression with respect to a and b and setting this derivative equal to zero in order to find the extremal point, we obtain

$$0 = \frac{\partial}{\partial a} \parallel \mathbf{r} \parallel_2^2 = -(\alpha + i\beta) - (\alpha - i\beta) + 2a \parallel \mathbf{x} \parallel_2^2,$$

$$0 = \frac{\partial}{\partial b} \parallel \mathbf{r} \parallel_2^2 = i(\alpha + i\beta) - i(\alpha - i\beta) + 2b \parallel \mathbf{x} \parallel_2^2.$$

After multiplying the last equation by i and adding the above equations, we have

$$0 = -2(\alpha + i\beta) + 2\lambda \parallel \mathbf{x} \parallel_2^2,$$

or

$$\mathbf{x}^H \mathbf{A} \mathbf{x} = \lambda \parallel \mathbf{x} \parallel_2^2. \qquad \blacksquare$$

Problems

7.14. Consider the matrix

$$\mathbf{A} = \begin{bmatrix} -4 & -3 & -7 \\ 2 & 3 & 2 \\ 4 & 2 & 7 \end{bmatrix}.$$

Find $R(\mathbf{x})$ for $\mathbf{x} = (1, 1, 1)^T$. Compute the norm of the residual $\parallel \mathbf{r} \parallel_2$ for this values of λ and compare it with the norm of the residual if we use the value $\lambda = 2$.

7.15. Let us allow $\mathbf{x}$ and $\mathbf{A}$ to have complex coordinates, but suppose that λ is constrained to be real. With this constraint, what is the value of λ that minimizes $\parallel \mathbf{A}\mathbf{x} - \lambda\mathbf{x} \parallel_2$?

7.16. Suppose $\mathbf{A}$ is a Hermitian matrix having eigenvalue–eigenvector pair $(\lambda, \mathbf{x})$. Show that

$$R(\mathbf{y}) = \lambda + O(\parallel \mathbf{y} - \mathbf{x} \parallel_2^2).$$

7.2.1. The Power Method

The power method is perhaps the simplest and most intuitive of the eigenvalue methods. It requires, however, that $\mathbf{A}$ have a dominant eigenvalue. Let $\lambda_1, \ldots, \lambda_n$ be the eigenvalues of $\mathbf{A}$. λ_1 is said to be *dominant* if

$$\mid \lambda_1 \mid > \mid \lambda_j \mid, \qquad j \neq 1.$$

We now describe the *power method*. Let $\mathbf{q}^{(0)}$ denote an arbitrarily chosen vector. The power method consists in constructing the iteration process

$\{\mathbf{q}^{(k)}\}$ according to the recursive rule

$$\mathbf{w}^{(k)} = \mathbf{A}\mathbf{q}^{(k-1)},$$
$$\mathbf{q}^{(k)} = (1/\|\mathbf{w}^{(k)}\|_2)\mathbf{w}^{(k)}. \tag{7.29}$$

The principal purpose of the second operation is to avoid overflow, and other scalings are possible. Actually, as the reader will recognize by the arguments to follow, it is sufficient to "renormalize" only after several iterations. A more efficient normalization is to insist that the first coordinate of $\mathbf{q}^{(k)}$ be 1, for example, but this will cause trouble if the eigenvector does, in fact, have 0 as its first component.

A convergence property of the power method is stated in the following theorem.

Theorem 7.9. Assume that the matrix $\mathbf{A}$ has (perhaps repeated) eigenvalues satisfying the ordering $|\lambda_1| > |\lambda_2| \geq |\lambda_3| \geq \cdots |\lambda_n|$, and that $\mathbf{u}_1, \mathbf{u}_2, \ldots, \mathbf{u}_n$ are the corresponding linearly independent eigenvectors. Let $\mathbf{q}^{(0)}$ be a vector having the representation

$$\mathbf{q}^{(0)} = \sum_{j=1}^{n} c_j \mathbf{u}_j, \tag{7.30}$$

in which the scalar c_1 is assumed to be nonzero. Then if $\{\mathbf{q}^{(i)}\}$ is the sequence determined by the power method and $\mathbf{1} = (1, 1, \ldots, 1)^{\mathrm{T}}$, we have

$$\mathbf{q}^{(k)} = (1/\|\mathbf{u}_1\|_2)\mathbf{u}_1 + O(|\lambda_2/\lambda_1|^k)\mathbf{1} \tag{7.31}$$

and

$$R(\mathbf{q}^{(k)}) = \lambda_1 + O(|\lambda_2/\lambda_1|^k). \tag{7.32}$$

Proof. From inspection of (7.29) and (7.30), we may write

$$\mathbf{q}^{(k)} = \alpha_k \mathbf{A}^k \mathbf{q}^{(0)}, \tag{7.33}$$

where α_k is the scalar $(\prod_{j=1}^{k} \|\mathbf{w}^{(j)}\|_2)^{-1}$. From representation (7.30) and Theorem 7.4, we have

$$\mathbf{q}^{(k)} = \alpha_k \sum_{j=1}^{n} c_j \lambda_j{}^k \mathbf{u}_j$$
$$= \alpha_k \lambda_1{}^k \left(c_1 \mathbf{u}_1 + \sum_{j=1}^{n} c_j \frac{\lambda_j{}^k}{\lambda_1{}^k} \mathbf{u}_j \right)$$
$$= \alpha_k \lambda_1{}^k [c_1 \mathbf{u}_1 + \mathbf{1}O(|\lambda_2/\lambda_1|^k)]. \tag{7.34}$$

Equation (7.34) shows that $\mathbf{q}^{(k)}$ is asymptotically collinear with $\mathbf{u}_1$. Since $\mathbf{q}^{(k)}$ is always a unit vector, (7.31) is established.

Equation (7.32) is verified by observing that

$$R(\mathbf{q}^{(k)}) = \mathbf{q}^{(k)\mathrm{H}}\mathbf{A}\mathbf{q}^{(k)} / \| \mathbf{q}^{(k)} \|_2^2.$$

Upon substituting representation (7.34) for $\mathbf{q}^{(k)}$, we have

$$R(\mathbf{q}^{(k)}) = \frac{\alpha_k^2\lambda_1^{2k+1}[c_1^2\mathbf{u}_1^{\mathrm{H}}\mathbf{u}_1 + O(|\lambda_2/\lambda_1|^k)]}{\alpha_k^2\lambda_1^{2k}[c_1^2\mathbf{u}_1^{\mathrm{H}}\mathbf{u}_1 + O(|\lambda_2/\lambda_1|^k)]}$$

$$= \lambda_1 + O(|\lambda_2/\lambda_1|^k). \qquad \blacksquare$$

Example 7.2. Let us consider the matrix $\mathbf{A}$ used in Example 7.1. Thus

$$\mathbf{A} = \begin{bmatrix} -4 & -3 & -7 \\ 2 & 3 & 2 \\ 4 & 2 & 7 \end{bmatrix},$$

and let $\mathbf{q}^{(0)} = (1, 0, 0)^{\mathrm{T}}$. Upon applying the power method, with starting value $\mathbf{q}^{(0)}$ we obtain the values shown in Table 7.1. The eigenvalues of $\mathbf{A}$ are 1, 2, and 3. The exact eigenvector for $\lambda_1 = 3$, to six decimal places, is

$$\mathbf{u} = (-0.707107, 0, 0.707107)^{\mathrm{T}}.$$

Table 7.1. Application of the Power Method

k	$q_1^{(k)}$	$q_2^{(k)}$	$q_3^{(k)}$	$R(\mathbf{q}^{(k)})$
0	1.000000	0.000000	0.000000	−4.000000
1	−0.666667	0.333333	0.666667	4.111111
2	−0.725241	0.241747	0.644658	3.181818
3	−0.732828	0.176890	0.657018	3.006386
4	−0.730242	0.124904	0.671487	2.971960
5	−0.725243	0.087481	0.682913	2.971889
6	−0.720400	0.059796	0.690976	2.978648
7	−0.716492	0.040460	0.696421	2.985124
8	−0.713589	0.027212	0.700036	2.989969
9	−0.711525	0.018237	0.018237	2.993319
10	−0.710094	0.012197	0.012197	2.005567
15	−0.707509	0.001614	0.706702	2.999427
20	−0.707160	0.000213	0.707054	2.999925
25	−0.707114	0.000028	0.707100	2.999990
30	−0.707108	0.000004	0.707106	2.999999
35	−0.707107	0.000000	0.707107	3.000000

We say that the value λ is a *multiple* eigenvalue if it is a multiple root of the characteristic polynomial. Our proof can be easily generalized to show that the power method continues to work when λ_1 is a multiple eigenvalue. However, if $|\lambda_1| = |\lambda_2|$ but $\lambda_1 \neq \lambda_2$, then the power method typically fails to converge. Under the circumstance that eigenvalue λ_1 is dominant, we may allow $\mathbf{A}$ to be deficient, provided the deficiency does not occur for λ_1; that is, provided there are as many independent eigenvectors for λ_1 as its multiplicity in the characteristic polynomial.

It is known that if $\mathbf{A}$ is symmetric, then its eigenvectors can be chosen to be orthogonal (Bellman, 1970, p. 36). By careful examination of the expansion for $R(\mathbf{q}^{(k)})$, one may confirm that in the symmetric case, (7.32) may be replaced by

$$R(\mathbf{q}^{(k)}) = \lambda_1 + O(|\lambda_2/\lambda_1|^{2k}). \tag{7.35}$$

In general, the power method is useful only for finding the largest eigenvalue and corresponding eigenvectors. Through matrix deflation, it is possible to obtain one or two other eigenvalues, but such an approach is fraught with potential trouble due to roundoff error of the deflated matrix.

Problems

7.17. Show in detail why the power method works for the multiple eigenvalue case, even when some of the nondominant eigenvalues are deficient.

7.18. Show how the coefficient of convergence of the power method can be changed by applying the power method to the matrix $\mathbf{A} - \mu\mathbf{I}$, instead of $\mathbf{A}$, for some number μ. What is the lower bound to how small the coefficient of convergence can be made, in terms of the matrix eigenvalues $\lambda_1, \lambda_2, \ldots, \lambda_n$?

7.2.2. The Inverse Power Method

The inverse power method is a variation on the power method. With lucky or skillful choice of the parameter λ, the inverse power method can be effective in computing those eigenvalues of $\mathbf{A}$ having intermediate magnitudes. In the notation of Theorem 7.9 and its proof, if we represent an arbitrary vector $\mathbf{q}^{(0)}$ as

$$\mathbf{q}^{(0)} = \sum_{i=1}^{n} c_i \mathbf{u}_i,$$

then from Theorem 7.4, if λ is some number that is not an eigenvalue of $\mathbf{A}$, we may write

$$(\mathbf{I}\lambda - \mathbf{A})^{-1}\mathbf{q}^{(0)} = \sum_{i=1}^{n} (\mathbf{I}\lambda - \mathbf{A})^{-1}c_i\mathbf{u}_i = \sum_{i=1}^{n} \frac{c_i}{\lambda - \lambda_i}\mathbf{u}_i. \tag{7.36}$$

The *inverse power method* consists of constructing the iteration process $\{q^{(i)}\}$ according to the recursive rule

$$(\mathbf{I}\lambda - \mathbf{A})\mathbf{w}^{(k)} = \mathbf{q}^{(k-1)} \tag{7.37}$$

and

$$\mathbf{q}^{(k)} = \parallel \mathbf{w}^{(k)} \parallel_2^{-1}\mathbf{w}^{(k)}, \tag{7.38}$$

where (7.38) is used solely to avoid overflow. Then with α_k defined to be the scale factor $(\prod_{j=1}^{k} \parallel \mathbf{w}^{(j)} \parallel_2^{-1})$ of terms from step (7.38), we may write that

$$\mathbf{q}^{(k)} = \alpha_k \sum_{i=1}^{n} \frac{c_i}{(\lambda - \lambda_i)^k}\,\mathbf{u}_i.$$

If

$$\mid \lambda - \lambda_i \mid\, <\, \mid \lambda - \lambda_j \mid, \qquad j \neq i, \tag{7.39}$$

then from the technique of the proof of Theorem 7.9, we easily see that if $c_i \neq 0$, then

$$\mathbf{q}^{(k)} = (\parallel \mathbf{u}_i \parallel_2)^{-1}\mathbf{u}_i + O(\gamma^k)\mathbf{1}, \tag{7.40}$$

where

$$\gamma = (\min_{j \neq i} \mid \lambda - \lambda_j \mid)^{-1}\mid \lambda - \lambda_i \mid. \tag{7.41}$$

The satisfaction of (7.39) may require that λ be *complex* (if, for example, the eigenvalues occur in complex conjugate pairs).

Problems

7.19. Apply the inverse power method to the matrix studied in Example 7.1.

7.20. Let $\mathbf{A}$ be a square matrix and $\mathbf{r}$ the residual $\mathbf{r} = \mathbf{Ax} - \lambda\mathbf{x}$. Show that there is a matrix $\mathbf{E}$ such that $\parallel \mathbf{E} \parallel_2 = \parallel r \parallel_2$ and such that $\mathbf{x}$ is an eigenvector of $\mathbf{A} + \mathbf{E}$ corresponding to eigenvalue λ. HINT: Use Theorem 7.7 cleverly.

7.21. Discuss the use of the relative magnitudes of $\parallel \mathbf{x} \parallel_2$ and $\parallel \mathbf{x}^1 \parallel_2$, where $(\lambda\mathbf{I} - \mathbf{A})^{-1}\mathbf{x} = \mathbf{x}^1$, in assessing how close $\mathbf{x}$ is to an eigenvector.

7.2.3. The Rayleigh Quotient Iteration Method

The convergence of the inverse power method, like that of the power method itself, is linear. Unlike the power method, the coefficient of convergence γ of the inverse power method can be arbitrarily small, provided λ is close enough to λ_i. This observation motivates the idea of adaptively changing the parameter λ in (3.37) by making a "best guess" of the eigenvalue at each iteration, as $\{\mathbf{q}^{(k)}\}$ converges to an eigenvector. A natural best guess, in view of Theorem 7.8, is the Rayleigh quotient. Thus the *Rayleigh quotient iteration method* differs from the inverse power method in that at each iteration k, we replace λ in (7.37) by the Rayleigh quotient $R(\mathbf{q}^{(k-1)})$. We shall measure the failure of a vector $\mathbf{x}$ to satisfy the eigenvector condition (7.1) by the Euclidean length of the residual after $\mathbf{x}$ has been normalized. Specifically, our criterion will be

$$\varrho(\mathbf{x}) = \| R(\mathbf{x})\mathbf{x} - \mathbf{A}\mathbf{x} \|_2 / \| \mathbf{x} \|_2. \tag{7.42}$$

The Rayleigh quotient iteration method is remarkable in that it yields quadratic and even cubic (order three) convergence of this error as we now state formally.

Theorem 7.10. Let $\{\mathbf{q}^{(k)}\}$ be constructed by the Rayleigh quotient iteration method and assume that (7.39) is satisfied, for some fixed i, at each iteration. Then if ϱ_k denotes the Euclidean length of the residual, that is, if

$$\varrho_k \triangleq \varrho(\mathbf{q}^{(k)}),$$

then for some constant C, and for $k = 1, 2, \ldots,$

$$\varrho_{k+1} \leq C\varrho_k^2. \tag{7.43}$$

Proof. Let $\mathbf{x}$ denote an estimate of an eigenvector and assume that $\| \mathbf{x} \|_2 = 1$ and that the real part of $x_1 \neq 0$, where x_1 is the first coordinate of $\mathbf{x}$. We shall let $\mathbf{R}$ be the Householder transformation (described in Section 7.1.2) such that

$$\mathbf{R}\mathbf{x} = -\mathbf{e}_1. \tag{7.44}$$

From the discussion in Section 7.1.2, we may partition $\mathbf{R}^H$ as $\mathbf{R}^H = (-\mathbf{x}, \mathbf{U})$. In analogy to equation (7.25), we may conclude that the following block partition is valid:

$$\mathbf{R}\mathbf{A}\mathbf{R}^H = \begin{bmatrix} \mathbf{x}^H \mathbf{A}\mathbf{x} & \mathbf{h}^H \\ \mathbf{g} & \mathbf{C} \end{bmatrix} = \begin{bmatrix} \lambda & \mathbf{h}^H \\ \mathbf{g} & \mathbf{C} \end{bmatrix}, \tag{7.45}$$

where it is evident that $\lambda = \mathbf{x}^H \mathbf{A}\mathbf{x}$ is the Rayleigh quotient of $\mathbf{x}$ with respect to $\mathbf{A}$, and where $\mathbf{g}$ and $\mathbf{h}$ are $(n-1)$-tuples.

Using the property of unitary matrices described in Theorem 7.6, we have that if $\varrho(\mathbf{x})$ is as defined in (7.42), then

$$\varrho(\mathbf{x}) = \frac{\|\lambda\mathbf{x} - \mathbf{A}\mathbf{x}\|_2}{\|\mathbf{x}\|_2} = \|\lambda\mathbf{R}\mathbf{x} - \mathbf{R}\mathbf{A}\mathbf{x}\|_2 = \|\lambda\mathbf{e}_1 + \mathbf{R}\mathbf{A}\mathbf{R}^H\mathbf{R}\mathbf{x}\|_2$$
$$= \|\lambda\mathbf{e}_1 - \mathbf{R}\mathbf{A}\mathbf{R}^H\mathbf{e}_1\|_2, \tag{7.46}$$

which in view of (7.45) gives us the error relation

$$\varrho(\mathbf{x}) = \left\|\begin{pmatrix}\lambda\\0\end{pmatrix} - \begin{pmatrix}\lambda\\\mathbf{g}\end{pmatrix}\right\|_2 = \|\mathbf{g}\|_2. \tag{7.47}$$

Let $\mathbf{x}'$ be the unnormalized successor vector to $\mathbf{x}$ determined by a Rayleigh quotient iteration. That is, let

$$(\lambda\mathbf{I} - \mathbf{A})\mathbf{x}' = \mathbf{x}. \tag{7.48}$$

Again using unitary matrix properties, we find that

$$\varrho(\mathbf{x}') = \frac{\|\lambda'\mathbf{R}\mathbf{x}' - \mathbf{R}\mathbf{A}\mathbf{R}^H\mathbf{R}\mathbf{x}'\|_2}{\|\mathbf{R}\mathbf{x}'\|_2}, \tag{7.49}$$

where $\lambda' = R(\mathbf{x}')$. Upon left multiplication of (7.48) by $\mathbf{R}$, we have

$$(\lambda\mathbf{I} - \mathbf{R}\mathbf{A}\mathbf{R}^H)\mathbf{R}\mathbf{x}' = \mathbf{R}\mathbf{x} = -\mathbf{e}_1; \tag{7.50}$$

and, after comparison with (7.45), we conclude that

$$\begin{bmatrix} 0 & -\mathbf{h}^H \\ -\mathbf{g} & \lambda\mathbf{I} - \mathbf{C} \end{bmatrix}\mathbf{R}\mathbf{x}' = -\mathbf{e}_1. \tag{7.51}$$

Assume that the first coordinate of $\mathbf{R}\mathbf{x}'$ is not 0. This is an assumption that will hold for $\mathbf{x}$ close enough to an eigenvector, for then $\mathbf{x}'$ will be nearly collinear with $\mathbf{x}$, and since $\mathbf{R}\mathbf{x} = -\mathbf{e}_1$ the first coordinate of $\mathbf{R}\mathbf{x}' \neq 0$. If we define the scalar δ and the $(n-1)$-tuple $\mathbf{p}$ so that

$$\mathbf{R}\mathbf{x}' = \delta\begin{bmatrix}1\\\mathbf{p}\end{bmatrix}, \tag{7.52}$$

then the last $n-1$ rows of (7.51) imply that

$$(\lambda\mathbf{I} - \mathbf{C})\mathbf{p} = \mathbf{g}, \tag{7.53}$$

and hence from (7.49), we have

$$\varrho(\mathbf{x}') = \frac{\left\|\lambda'\begin{bmatrix}1\\\mathbf{p}\end{bmatrix} - \begin{bmatrix}\lambda & \mathbf{h}^H\\\mathbf{g} & \mathbf{C}\end{bmatrix}\begin{bmatrix}1\\\mathbf{p}\end{bmatrix}\right\|_2}{(1 + \|\mathbf{p}\|_2^2)^{1/2}}. \tag{7.54}$$

But

$$\lambda' = \frac{\mathbf{x}'^H \mathbf{A} \mathbf{x}'}{\mathbf{x}'^H \mathbf{x}'} = \frac{(\mathbf{R}\mathbf{x}')^H (\mathbf{R}\mathbf{A}\mathbf{R}^H)(\mathbf{R}\mathbf{x}')}{(\mathbf{R}\mathbf{x}')^H (\mathbf{R}\mathbf{x}')} = \frac{\begin{bmatrix} 1 \\ \mathbf{p} \end{bmatrix}^H \begin{bmatrix} \lambda & \mathbf{h}^H \\ \mathbf{g} & \mathbf{C} \end{bmatrix} \begin{bmatrix} 1 \\ \mathbf{p} \end{bmatrix}}{1 + \| \mathbf{p} \|_2^2}$$

$$= \frac{\begin{bmatrix} 1 \\ \mathbf{p} \end{bmatrix}^H \begin{bmatrix} \lambda + \mathbf{h}^H \mathbf{p} \\ \mathbf{g} + \mathbf{C}\mathbf{p} \end{bmatrix}}{1 + \| \mathbf{p} \|_2^2} = \frac{\lambda + \mathbf{h}^H \mathbf{p} + \mathbf{p}^H \mathbf{g} + \mathbf{p}^H \mathbf{C} \mathbf{p}}{1 + \| \mathbf{p} \|_2^2}. \qquad (7.55)$$

Equation (7.55) implies that we may write

$$\lambda' = \lambda(1 - \| \mathbf{p} \|_2^2) + \mathbf{h}^H \mathbf{p} + \mathbf{p}^H \mathbf{g} + \mathbf{p}^H \mathbf{C} \mathbf{p} + O(\| \mathbf{p} \|_2^3). \qquad (7.56)$$

Substituting this approximation into (7.54), and noting from (7.53) that $\| \mathbf{g} \|_2 = O(\| \mathbf{p} \|_2)$, we obtain

$$\varrho(\mathbf{x}') = \frac{\left\| \begin{bmatrix} \lambda' \\ \lambda \mathbf{p} + \mathbf{h}^H \mathbf{p} \mathbf{p} \end{bmatrix} - \begin{bmatrix} \lambda + \mathbf{h}^H \mathbf{p} \\ \mathbf{g} + \mathbf{C}\mathbf{p} \end{bmatrix} \right\|_2}{1 + \frac{1}{2} \| \mathbf{p} \|_2^2} + O(\| \mathbf{p} \|_2^3)$$

$$= \frac{\left\| \begin{bmatrix} -\lambda \| \mathbf{p} \|_2^2 + \mathbf{p}^H \mathbf{g} + \mathbf{p}^H \mathbf{C} \mathbf{p} \\ (\lambda \mathbf{I} - \mathbf{C})\mathbf{p} - \mathbf{g} + \mathbf{h}^H \mathbf{p} \mathbf{p} \end{bmatrix} \right\|_2}{1 + \frac{1}{2} \| \mathbf{p} \|_2^2} + O(\| \mathbf{p} \|_2^3)$$

$$= \frac{\left\| \begin{bmatrix} \mathbf{p}^H [\mathbf{g} - (\lambda \mathbf{I} - \mathbf{C})\mathbf{p}] \\ [(\lambda \mathbf{I} - \mathbf{C})\mathbf{p} - \mathbf{g}] + \mathbf{h}^H \mathbf{p} \mathbf{p} \end{bmatrix} \right\|_2}{1 + \frac{1}{2} \| \mathbf{p} \|_2^2} + O(\| \mathbf{p} \|_2^3)$$

$$= \left\| \begin{bmatrix} 0 \\ \mathbf{h}^H \mathbf{p} \mathbf{p} \end{bmatrix} \right\|_2 + O(\| \mathbf{p} \|_2^3).$$

From this observation, we have

$$\varrho(\mathbf{x}') \leq \| \mathbf{h}^H \|_2 \| \mathbf{p} \|_2^2 + O(\| \mathbf{p} \|_2^3). \qquad (7.57)$$

In this statement, we have generalized our $O(\cdot)$ notation to imply the existence of constants K_1 and K_2 such that if $\| \mathbf{p} \|_2 < K_1$, then $| \varrho(\mathbf{x}') - \| \mathbf{h}^H \|_2 \| \mathbf{p} \|_2 | < K_2 \| \mathbf{p} \|_2^3$. We continue the use of this notation. In view of (7.53), (7.57) implies that

$$\varrho(\mathbf{x}') \leq \| \mathbf{h}^H \|_2 \| (\lambda \mathbf{T} - \mathbf{C})^{-1} \|_2 \| \mathbf{g} \|_2^2 + O(\| \mathbf{p} \|_2^3),$$

which gives the desired inequality that

$$\varrho(\mathbf{x}') \leq K\varrho(\mathbf{x})^2 + O(\varrho(\mathbf{x})^3), \qquad (7.58)$$

where $K = \| \mathbf{h}^H \|_2 \| (\lambda \mathbf{I} - \mathbf{C})^{-1} \|_2^2$ and where we have used result (7.47) that $\varrho(\mathbf{x}) = \| \mathbf{g} \|_2$ and the consequent implication, in view of (7.53), that $\| \mathbf{p} \|_2 \leq \| (\lambda \mathbf{T} - \mathbf{C})^{-1} \|_2 \varrho(\mathbf{x})$. ∎

Corollary 7.4. Under the conditions of the theorem, if $\mathbf{A}$ is Hermitian, then for some constant C, and all positive integers k,

$$\varrho_{k+1} \leq C \varrho_k^3.$$

Proof. If $\mathbf{A}$ is Hermitian, then as the reader may easily verify, the unitarily similar matrix $\mathbf{RAR}^H$ in (7.45) will also be Hermitian. This implies that $\mathbf{h} = \mathbf{g}$. Thus the constant K in (7.58) can be written as $K = \varrho(\mathbf{x}) \| (\lambda \mathbf{I} - \mathbf{C})^{-1} \|_2^2$, which, when substituted into (7.58), gives us that

$$\varrho(\mathbf{x'}) \leq \| (\lambda \mathbf{I} - \mathbf{C})^{-1} \|_2^2 \varrho(\mathbf{x})^3 + O(\varrho(\mathbf{x})^3). \qquad (7.59)$$

∎

Example 7.3. Let us return to the matrix $\mathbf{A}$ given in Example 7.2. We perform Rayleigh quotient iterations with this matrix, taking the initial vector as $\mathbf{q}^{(0)} = (1, 0, 0)^T$; this is also the starting vector for Example 7.2. The successive estimates of an eigenvector and eigenvalue are shown in Table 7.2.

In this computation, after only seven iterations we have obtained an approximation of the eigenvector for $\lambda = 1$ that is accurate to nine decimal places. This is to be compared with the computation in Table 7.1, where about 35 power method iterations were required to achieve six-place accuracy.

Table 7.2. Application of the Rayleigh Quotient Iteration Method

k	$q_1^{(k)}$	$q_2^{(k)}$	$q_3^{(k)}$	$R(\mathbf{q}^{(k)})$
0	1.000000000	0.000000000	0.000000000	−4.000000000
1	−0.934592621	0.179236941	0.307263327	−1.487297164
2	0.873709283	−0.271180103	−0.403848290	0.042331448
3	−0.838069564	0.344291770	0.423205132	0.735276831
4	0.820657422	−0.394309989	−0.413571068	0.963908612
5	−0.816640807	0.407725009	0.408482691	0.999027204
6	0.816496721	−0.408247756	−0.408248544	0.999999184
7	−0.816496581	0.408248290	0.408248290	1.000000000
8	0.816496581	−0.408248290	−0.408248290	1.000000000

Problems

7.22. Use the fact that the eigenvectors of a Hermitian matrix **A** can be chosen to be orthogonal in order to show that for such a matrix $\{\mathbf{q}^{(j)}\}$ converges to an eigenvector cubically in the Euclidean norm, if it converges at all.

7.23. Program the Rayleigh quotient iteration method and reproduce Table 7.2. Try to obtain other eigenvalue–eigenvector pairs by choosing other starting vectors $\mathbf{q}^{(0)}$.

7.24. Discuss heuristic procedures for ensuring convergence by starting with pure inverse power iterations and converting to Rayleigh quotient iterations when the successive vectors begin to converge.

7.25. Discuss how to view the eigenvalue approximation problem as a non-linear programming problem. Discuss the applicability of methods given in Section 5.3.

7.2.4. Jacobi-Type Methods

It is known from the theory of matrices that for an arbitrary symmetric matrix **A** there exists an orthogonal matrix **Q** such that the matrix

$$\mathbf{D} = \mathbf{Q}^{\mathrm{T}}\mathbf{A}\mathbf{Q} \tag{7.60}$$

is diagonal (Bellman, 1970, p. 36). We shall construct a sequence $\mathbf{S}_1, \mathbf{S}_2, \ldots, \mathbf{S}_k, \ldots$ of orthogonal matrices such that

$$\lim_{k \to \infty} \mathbf{S}_1 \mathbf{S}_2 \cdots \mathbf{S}_k = \mathbf{Q}. \tag{7.61}$$

Let $\mathbf{T}_0 = \mathbf{A}$ and $\mathbf{T}_k = \mathbf{S}_k^{\mathrm{T}} \mathbf{T}_{k-1} \mathbf{S}_k$, where $\mathbf{T}_k = (t_{ij}^{(k)})_{ij=1}^n$ and $\mathbf{S}_k = (s_{ij}^{(k)})_{ij=1}^n$ will now be determined. Introduce the notation

$$v_k = \sum_{\substack{i=1 \\ i \neq j}}^n \sum_{j=1}^n [t_{ij}^{(k)}]^2,$$

$$w_k = \sum_{i=1}^n \sum_{j=1}^n [t_{ij}^{(k)}]^2 = v_k + \sum_{i=1}^n [t_{ii}^{(k)}]^2.$$

Assume that an element $t_{pq}^{(k-1)} \neq 0$ for $p \neq q$. In terms of the angle θ_k to be specified later, let the elements of $\mathbf{S}_k$ be defined by

$$\begin{aligned}
s_{pp}^{(k)} &= s_{qq}^{(k)} = \cos \theta_k, \\
s_{pq}^{(k)} &= -s_{pq}^{(k)} = \sin \theta_k, \\
s_{ii}^{(k)} &= 1, \qquad i \neq p,\ i \neq q, \\
s_{ij}^{(k)} &= 0, \qquad \text{otherwise.}
\end{aligned} \tag{7.62}$$

Such a matrix $\mathbf{S}^{(k)}$ is known as a *plane rotation*. An easy calculation shows that

$$t_{pq}^{(k)} = t_{qp}^{(k)} = \tfrac{1}{2}(t_{pp}^{(k-1)} - t_{qq}^{(k-1)}) \sin 2\theta_k + t_{pq}^{(k-1)} \cos 2\theta_k.$$

Let the quantity θ_k be determined by the condition $t_{pq}^{(k)} = 0$; that is,

$$\tan 2\theta_k = \frac{-t_{pq}^{(k-1)}}{\tfrac{1}{2}(t_{pp}^{(k-1)} - t_{qq}^{(k-1)})}. \tag{7.63}$$

Equation (7.63) shows the existence of θ_k. By using well-known identities of trigonometric functions, one can verify that

$$\cos \theta_k = \left(\frac{\nu + |\mu|}{2\nu}\right)^{1/2}, \qquad \sin \theta_k = \frac{\text{sign}(\mu)\lambda}{2\nu \cos \theta_k}, \tag{7.64}$$

where

$$\lambda = -t_{pq}^{(k-1)}, \quad \mu = \tfrac{1}{2}(t_{pp}^{(k-1)} - t_{qq}^{(k-1)}), \qquad \nu = (\lambda^2 + \mu^2)^{1/2}.$$

From (7.64), the relation

$$v_k = v_{k-1} - 2(t_{pq}^{(k-1)})^2 \tag{7.65}$$

can be proved by simple calculation, and from this it is seen that the sequence $v_0, v_1, v_2, \ldots$ is strictly monotone decreasing. Note that an off-diagonal element made zero at one step will generally become non-zero at some later step. The largest decrease is guaranteed in each step if the off-diagonal element of greatest magnitude is annihilated. Since the number of the off-diagonal elements is equal to $n^2 - n$, equation (7.65) implies that

$$v_k \leq \left\{1 - \left[\frac{2}{n^2 - n}\right]\right\} v_{k-1} \leq \cdots \leq \left\{1 - \left[\frac{2}{n^2 - n}\right]\right\}^k v_0. \tag{7.66}$$

For large values of n the coefficient of convergence in the bracket is very close to one and consequently the convergence of the Jacobi method may be very slow. Since finding the off-diagonal element of greatest magnitude needs many operations, in practical applications any element with absolute value not less than a fixed positive bound can be chosen for annihilation. Let this positive bound be denoted by ε. Then (7.65) implies that

$$v_k \leq v_{k-1} - 2\varepsilon^2 \leq \cdots \leq v_0 - 2k\varepsilon^2. \tag{7.67}$$

Since the right-hand side of (7.67) tends to $-\infty$ for $k \to \infty$, after a finite number of steps the method has to terminate at a matrix $\mathbf{T}_k$ with off-diagonal elements less than ε. The number of the necessary steps can be bounded from the inequality

$$v_0 - 2k\varepsilon^2 < 2\varepsilon^2,$$

which is equivalent to the condition

$$k > \frac{v_0 - 2\varepsilon^2}{2\varepsilon^2}. \tag{7.68}$$

Assume that the method terminates at the stage k, and let $\mathbf{R}_k = \mathbf{S}_1 \mathbf{S}_2 \cdots \mathbf{S}_k$, $k \geq 1$. The definition of matrices $\mathbf{T}_k$ and Theorem 7.5 imply that the approximate eigenvectors of matrix $\mathbf{A}$ are given by the columns of the matrix $\mathbf{R}_k$. Here we used the fact that the eigenvectors of diagonal matrices are the unit vectors $(0, \ldots, 0, 1, 0, \ldots, 0)^\mathrm{T}$. The vectors $\mathbf{R}_k$ can be computed using the recursive relation $\mathbf{R}_k = \mathbf{R}_{k-1}\mathbf{S}_k$. Since matrix $\mathbf{S}_k$ contains only four nonzero off-diagonal elements, the computation of this product is very simple.

One can prove that the classical Jacobi method is stable with respect to roundoff error, assuming that the square roots are calculated with appropriate accuracy (see Ralston, 1965 p. 491). One step of the method requires $8n + 7$ multiplications and divisions and two square root computations.

Example 7.4. Let us consider the matrix

$$\mathbf{A} = \begin{bmatrix} 1 & 1 & 0.5 \\ 1 & 1 & 0.25 \\ 0.5 & 0.25 & 2 \end{bmatrix}.$$

The off-diagonal element of largest magnitude is 1.0 in the first row and second column ($p = 1$, $q = 2$). Equation (7.64) implies

$$\mu = 0, \qquad \lambda = -1, \qquad v = 1,$$
$$\sin \theta_1 = -1/2^{1/2}, \qquad \cos \theta_1 = 1/2^{1/2},$$

and consequently

$$\mathbf{T}_1 = \mathbf{S}_1^\mathrm{T} \mathbf{T}_0 \mathbf{S}_1 = \begin{bmatrix} 2 & 0 & \dfrac{3}{4 \cdot 2^{1/2}} \\ 0 & 0 & \dfrac{1}{4 \cdot 2^{1/2}} \\ \dfrac{3}{4 \cdot 2^{1/2}} & \dfrac{1}{4 \cdot 2^{1/2}} & 2 \end{bmatrix}.$$

7.3. The QR Algorithm

7.3.1. Principles and Convergence Rates

The QR algorithm is widely considered to be the eigenproblem method of choice for all matrices except sparse matrices of large order. Efficient implementation of the QR method requires that **A** first be reduced to upper Hessenberg form. In this section we shall assume that this has already been accomplished (by the method given in Section 7.1.2, for example) and that **A** is an upper Hessenberg matrix of order n.

Let $\{\alpha_k\}$, $k \geq 0$, denote some (possibly complex) sequence of numbers. The QR algorithm with origin shift consists of construction of the iterative process of matrices $\{\mathbf{A}_k\}_{k \geq 0}$ according to the following recursive rule:

(1) $\mathbf{A}_0 = \mathbf{A}$.

(2) For $k = 0, 1, \ldots$, compute matrices $\mathbf{Q}_k$ and $\mathbf{R}_k$, with $\mathbf{Q}_k$ unitary and $\mathbf{R}_k$ upper triangular, which satisfy the factorization

$$\mathbf{A}_k - \alpha_k \mathbf{I} = \mathbf{Q}_k \mathbf{R}_k. \tag{7.69}$$

(3) Compute

$$\mathbf{A}_{k+1} = \mathbf{R}_k \mathbf{Q}_k + \alpha_k \mathbf{I}. \tag{7.70}$$

The existence of the $\mathbf{Q}_k \mathbf{R}_k$ factorization will be verified later, when we give an effective construction of the matrices $\mathbf{Q}_k$ and $\mathbf{R}_k$. However, at this point we shall concern ourselves with the convergence properties of the $\{\mathbf{A}_k\}$ process.

First note that for each j, $\mathbf{A}_j$ is similar to **A**, because as we now confirm, $\mathbf{A}_{k+1}$ is similar to $\mathbf{A}_k$. For since $\mathbf{Q}_k^{\mathrm{H}} = \mathbf{Q}_k^{-1}$, the factorization (7.69) implies that

$$\mathbf{R}_k = \mathbf{Q}_k^{\mathrm{H}} \mathbf{A}_k - \alpha_k \mathbf{Q}_k^{\mathrm{H}},$$

and after substituting this relation for $\mathbf{R}_k$ into (7.70) we have

$$\mathbf{A}_{k+1} = (\mathbf{Q}_k^{\mathrm{H}} \mathbf{A}_k - \alpha_k \mathbf{Q}_k^{\mathrm{H}})\mathbf{Q}_k + \alpha_k \mathbf{I} = \mathbf{Q}_k^{\mathrm{H}} \mathbf{A}_k \mathbf{Q}_k. \tag{7.71}$$

If all the origin shifts α_k are identically 0, then process (7.69) and (7.70) is known simply as the QR algorithm. We shall be prescribing a recursive computation of the origin shift sequence $\{\alpha_k\}$ that, when applied through the QR algorithm to a diagonalizable matrix, yields a sequence $\{\mathbf{A}_k\}$ that converges to a matrix $\tilde{\mathbf{A}}$ of the form

$$\tilde{\mathbf{A}} = \begin{bmatrix} \mathbf{C} & \mathbf{h} \\ \mathbf{0}^{\mathrm{T}} & \theta \end{bmatrix} \tag{7.72}$$

where $\mathbf{C}$ is an $(n-1)$th-order matrix, $\mathbf{0}$ the $(n-1)$-dimension zero vector, $\mathbf{h}$ an $(n-1)$-tuple, and θ a (possibly complex) number. Clearly θ is an eigenvalue of $\mathbf{A}$. This can be verified by observing that $\mathbf{e}_n = (0, 0, \ldots, 0, 1)^T$ is an eigenvector corresponding to θ of $\tilde{\mathbf{A}}^T$. But as the reader will easily confirm, the characteristic polynomials of $\tilde{\mathbf{A}}$ and $\tilde{\mathbf{A}}^T$ coincide.

In practice, the QR algorithm process stops at $\mathbf{A}_N$ when the coordinates $(a_{n1}^{(N)}, a_{n2}^{(N)}, \ldots, a_{n,n-1}^{(N)})$ of $\mathbf{A}_N$ are negligible. One then lets $a_{nn}^{(N)}$ serve as an approximation of an eigenvalue. In order to estimate other eigenvalues, the QR algorithm is restarted on the $(n-1)$th-order principal submatrix of $\mathbf{A}_N$, which serves as an approximation of $\mathbf{C}$ in (7.72).

We now connect the shifted QR algorithm with the Rayleigh quotient iteration method. From (7.69) we have that

$$(\mathbf{Q}_k\mathbf{R}_k)^{-1} = \mathbf{R}_k^{-1}\mathbf{Q}_k^{-1} = (\mathbf{A}_k - \alpha_k\mathbf{I})^{-1},$$

so that

$$\mathbf{Q}_k^H = \mathbf{R}_k(\mathbf{A}_k - \alpha_k\mathbf{I})^{-1},$$

or

$$\mathbf{Q}_k = (\mathbf{A}_k - \alpha_k\mathbf{I})^{-H}\mathbf{R}_k^H. \tag{7.73}$$

Since $\mathbf{R}_k^H$ is lower triangular, $\mathbf{R}_k^H\mathbf{e}_n = r_{nn}^{(k)}\mathbf{e}_n$. Let $\mathbf{q}_1^{(k)}, \ldots, \mathbf{q}_n^{(k)}$ denote the successive columns of $\mathbf{Q}_k$. Then from (7.73), we have

$$\mathbf{q}_n^{(k)} = \mathbf{Q}_k\mathbf{e}_n = (\mathbf{A}_k^H - \alpha_k\mathbf{I})^{-1}\mathbf{R}_k^H\mathbf{e}_n$$
$$= r_{nn}^{(k)}(\mathbf{A}_k^H - \alpha_k\mathbf{I})^{-1}\mathbf{e}_n. \tag{7.74}$$

Upon comparing this equation with (7.37), it is evident that if we take the shift α_k to be the Rayleigh quotient of the vector $\mathbf{e}_n$ applied to $\mathbf{A}_k^H$, that is,

$$\alpha_k = \mathbf{R}(\mathbf{e}_n) = \mathbf{e}_n^T\mathbf{A}_k^H\mathbf{e}_n, \tag{7.75}$$

then the vector $\mathbf{q}_n$ is the estimate of the eigenvector that results from one step of the Rayleigh quotient iteration. In order to understand the significance of this observation, for $v = 1, 2, \ldots$, partition $\mathbf{A}_v$ analogously to $\tilde{\mathbf{A}}$ in (7.72), so that

$$\mathbf{A}_v = \begin{bmatrix} \mathbf{C}_v & \mathbf{h}^{(v)} \\ \mathbf{g}^{(v)H} & \theta_v \end{bmatrix}, \tag{7.76}$$

where $\mathbf{C}_v$ is $(n-1) \times (n-1)$, $\mathbf{g}^{(v)}$ and $\mathbf{h}^{(v)}$ are $(n-1) \times 1$, and θ_v is a number. From this representation, we can calculate the residual $\mathbf{r}^{(k)}$ of $\mathbf{e}_n$ as an approximate eigenvector for $\mathbf{A}_k^H$ as

$$\mathbf{r}^{(k)} = (\mathbf{A}_k^H - \theta_k\mathbf{I})\mathbf{e}_n = \mathbf{g}^{(k)}. \tag{7.77}$$

Also we have

$$
\begin{aligned}
\mathbf{g}^{(k+1)} &= (\mathbf{A}_{k+1}^{\mathrm{H}} - \theta_{k+1}\mathbf{I})\mathbf{e}_n \\
&= (\mathbf{Q}_k^{\mathrm{H}}\mathbf{A}_k^{\mathrm{H}}\mathbf{Q}_k - \theta_{k+1}\mathbf{Q}_k^{\mathrm{H}}\mathbf{Q}_k)\mathbf{e}_n \\
&= \mathbf{Q}_k^{\mathrm{H}}(\mathbf{A}_k^{\mathrm{H}} - \theta_{k+1}\mathbf{I})\mathbf{q}_n^{(k)}.
\end{aligned}
\tag{7.78}
$$

Now since $\mathbf{Q}_k^{\mathrm{H}}$ is a unitary matrix, from Theorem 7.6, we have

$$
\| \mathbf{g}^{(k+1)} \|_2 \leq \| (\mathbf{A}_k - \theta_k\mathbf{I}) \|_2 \| \mathbf{q}_n^{(k)} \|_2.
\tag{7.79}
$$

But since from (7.74) $\mathbf{q}_n^{(k)}$ results from the Rayleigh quotient iteration, according to Theorem 7.10, the norm of the residual is reduced quadratically (cubically if $\mathbf{A}$ and hence $\mathbf{A}_k$ is Hermitian) and we may therefore conclude that for some constant C, $\| \mathbf{q}_n^{(k)} \|_2 \leq C \| \mathbf{g}^{(k)} \|_2^2$ and consequently

$$
\| \mathbf{g}^{(k+1)} \|_2 \leq C \| \mathbf{A}_k - \theta_k\mathbf{I} \|_2 \| \mathbf{g}^{(k)} \|_2^2,
\tag{7.80}
$$

which establishes the quadratic convergence afforded by the shifted QR algorithm with Rayleigh quotient shifts.

It is this rapid convergence, the fact that successive $\mathbf{A}_k$ are unitarily similar, and the ease of deflation that makes the shifted QR algorithm especially attractive in comparison to available alternatives. Stewart (1973, Section 7.3) discusses connections between the QR algorithm and the power method, and he shows the implication of this connection is that while, as we have shown, the subdiagonals on the last row of $\mathbf{A}_k$ converge to zero at least quadratically, the other subdiagonal elements are also getting smaller, albeit linearly. Also, in Section 7.4, Stewart discusses modifications of the QR algorithm that allow one to sidestep complex arithmetic when $\mathbf{A}$ is real.

7.3.2. Implementation of the QR Algorithm

Consider first the plane rotation matrix ($1 \leq p \leq n,\ 1 \leq q \leq n$)

$$
\mathbf{S}_{qp} =
\begin{bmatrix}
1 & & & & & & & & & \\
& \ddots & & & & & & & & \\
& & \ddots & & & & & & & \\
& & & 1 & & & & & & \\
& & & & \cos\beta & & & -\sin\beta & & \\
& & & & & 1 & & & & \\
& & & & & & \ddots & & & \\
& & & & & & & 1 & & \\
& & & & \sin\beta & & & \cos\beta & & \\
& & & & & & & & 1 & \\
& & & & & & & & & \ddots \\
& & & & & & & & & & 1
\end{bmatrix}
\tag{7.81}
$$

which was introduced in Section 7.2.4 as a step in the Jacobi method. It is easy to see that $\mathbf{S}_{qp}^{\mathrm{T}} = \mathbf{S}_{qp}^{-1}$, and consequently the matrix $\mathbf{S}_{qp}$ is an orthogonal matrix. Let $\mathbf{A} = (a_{ij})_{ij=1}^{n}$ be any square matrix. The matrix $\mathbf{S}_{qp}\mathbf{A}$ differs from $\mathbf{A}$ only in the pth and qth rows. The jth element of the pth and qth rows can be obtained by $a_{pj} \cos \beta + a_{qj} \sin \beta$, and $-a_{pj} \sin \beta + a_{qj} \cos \beta$, respectively. Consequently the pth element of the qth row of the matrix $\mathbf{S}_{qp}\mathbf{A}$ is equal to zero if

$$-a_{pp} \sin \beta + a_{qp} \cos \beta = 0,$$

that is, if

$$\tan \beta = a_{qp}/a_{pp}.$$

As in the preceding section, let us assume that $\mathbf{A}$ has already been put into upper Hessenberg form. For the moment, assume $\mathbf{A}$ is real. Through performing the above procedure in the order $(2,1)$, $(3,2)$, $\ldots$, $(n, n-1)$ one can prove that the elements made zero at one step will remain zero during the later steps. Thus the matrix

$$\mathbf{R} = [\mathbf{S}_{n-1,n} \cdots \mathbf{S}_{2,3}\mathbf{S}_{1,2}]\mathbf{A} = \mathbf{Q}^{\mathrm{H}}\mathbf{A} \tag{7.82}$$

is upper triangular. Equation (7.82) and the fact that the matrices $\mathbf{S}_{qp}$ are orthogonal matrices implies that $\mathbf{Q}$ is orthogonal, and upon left multiplying by $\mathbf{Q}$, we have

$$\mathbf{A} = \mathbf{QR}. \tag{7.83}$$

In the general case, one will apply this operation to $\mathbf{A}_k - \alpha_k \mathbf{I}$ in (7.69). Since α_k can be complex, we shall need to find QR decompositions of complex matrices $\mathbf{A}$. The only modification necessary is to replace $\cos \beta$ and $\sin \beta$ in (7.81) by a_{pp}/δ and a_{qp}/δ, respectively. Here

$$\delta = (a_{pp}^2 + a_{qp}^2)^{1/2},$$

and either square root will suffice. The reader will be able to confirm that with this generalization, $\mathbf{S}_{qp}$ is unitary.

We now show that if

$$\mathbf{A} = \mathbf{QR}$$

is the QR decomposition just constructed, then $\mathbf{RQ}$ is, like $\mathbf{A}$ itself, upper Hessenberg. In view of (7.70), the consequence of this fact is that the shifted QR algorithm preserves the upper Hessenberg property, a feature that is vital to the QR algorithm, as we have described it. From (7.82), we have that

$$\mathbf{Q} = (\mathbf{S}_{n-1,n}, \mathbf{S}_{n-2,n-1}, \ldots, \mathbf{S}_{1,2})^H = (\mathbf{S}_{1,2}^H, \mathbf{S}_{2,3}^H, \ldots, \mathbf{S}_{n-1,n}^H).$$

Note that $\mathbf{S}_{m-1,m}$ is also a rotation and as a consequence, for any matrix $\mathbf{B}$, $\mathbf{B}\mathbf{S}_{m-1,m}$ differs from $\mathbf{B}$ only in the $(m-1)$th and mth columns, and the new columns in these positions must be linear combinations of the former columns in these locations. Specifically, since $\mathbf{R}$ is upper triangular, $\mathbf{R}\mathbf{S}_{12}^H$ must, for the fifth-order matrix, have the form

$$
\begin{array}{ccccc}
\times & \times & \times & \times & \times \\
\times & \times & \times & \times & \times \\
0 & 0 & \times & \times & \times \\
0 & 0 & 0 & \times & \times \\
0 & 0 & 0 & 0 & \times,
\end{array}
$$

where we have designated the possible locations of nonzero elements by $\times$'s. Continuing this reasoning, one may confirm that each successive right multiplication by a matrix $\mathbf{S}_{m-1,m}^H$ leaves the resulting matrix in upper Hessenberg form.

7.4. Eigenproblem Error Analysis

In the discussion following Theorem 7.6 (Section 7.1.1), we noted that if we have obtained an eigenvalue $\hat{\lambda}$ for

$$\hat{\mathbf{A}} = \mathbf{U}\mathbf{A}\mathbf{U}^H + \mathbf{F}, \qquad (7.84)$$

where $\mathbf{U}$ is unitary, then $\hat{\lambda}$ is exactly an eigenvalue for the matrix $\mathbf{A} + \mathbf{E}$, where $\mathbf{E}$ is some matrix that satisfies the equation

$$\| \mathbf{E} \|_2 = \| \mathbf{F} \|_2. \qquad (7.85)$$

In terms of the implementation of the QR method that we have recommended, the unitary matrix $\mathbf{U}$ will be the resultant of the Householder transformations for obtaining Hessenberg form (as described in Section 7.1.3) in composition with the product of the plane rotations that arise in implementing the QR iterations.

In order to extract benefit from the fact that roundoff error can be accounted for by a perturbation of the same magnitude on the original matrix, we need to obtain bounds for $|\lambda(\mathbf{E}) - \lambda|$ in terms of the norm of the perturbation $\parallel \mathbf{E} \parallel_2$. In the above statement, we intended that $\lambda(\mathbf{E})$ should denote an eigenvalue of the perturbed matrix $\mathbf{A} + \mathbf{E}$, and λ denotes the eigenvalue of the original matrix $\mathbf{A}$ that lies closest to and $\lambda(\mathbf{E})$. Additionally, we shall require bounds for the norm of $\mathbf{F}$ in (7.84). This will, in turn, require error estimates for the Householder transformation and plane rotations. The objective of this section is to present major results pertaining to these error bounding problems.

We begin with bounds for eigenvalue and eigenvector changes as a result of perturbation $\mathbf{E}$ of the matrix $\mathbf{A}$. The most frequently cited (Blum, 1972, p. 232; Isaacson and Keller, 1966, Section 4.1; Stewart, 1973, Section 6.4) result is the following assertion:

Theorem 7.11. Let $\mathbf{A}$ be a square, diagonalizable matrix and let $\mathbf{P}$ be the matrix such that for some scalar ε, $\mathbf{E} = \varepsilon\mathbf{P}$ and $\parallel \mathbf{P} \parallel_2 = 1$. Then for $\hat{\lambda}$ an eigenvalue of $\mathbf{A} + \mathbf{E}$, there exists an eigenvalue λ of $\mathbf{A}$ such that

$$| \lambda - \hat{\lambda} | = O(\varepsilon). \tag{7.86}$$

If λ has multiplicity 1 then

$$\lim_{|\varepsilon| \to 0} \frac{\lambda - \hat{\lambda}}{\varepsilon} = \frac{\mathbf{y}^H \mathbf{P} \mathbf{x}_i}{\mathbf{y}^H \mathbf{x}_i}, \tag{7.87}$$

where $\mathbf{x}_i$ and $\mathbf{y}$ are eigenvectors of $\mathbf{A}$ and $\mathbf{A}^H$, respectively, corresponding to eigenvalue λ.

The proof of this theorem is to be found in any of the above literature citations. Stewart (1973, p. 296) exhibits the ε^2 component of the error term and, additionally, shows that if $\mathbf{x}$ and $\mathbf{x}'$ are unit eigenvectors of $\mathbf{A}$ and $\mathbf{A} + \mathbf{E}$, respectively, corresponding to λ and λ', then

$$\parallel \mathbf{x} - \mathbf{x}' \parallel_2 = O(\varepsilon). \tag{7.88}$$

The denominator of the right-hand side of (7.87) cannot be 0 because, as the reader may easily verify, $\mathbf{y}^H \mathbf{x}_j = 0$ for all eigenvectors $\mathbf{x}_j$ corresponding to eigenvalues λ_j of $\mathbf{A}$ other than λ (which, by hypothesis, has multiplicity 1). If we had also that $\mathbf{y}^H \mathbf{x}_i = 0$, this would imply that $\mathbf{y}$ is the zero vector, in contradiction to our assumption that it is an eigenvector. Let us normalize the vectors $\mathbf{x}_i$ and $\mathbf{y}$ so that $\parallel \mathbf{x}_i \parallel_2 = | \mathbf{y}^H \mathbf{x}_i | = 1$. Then (7.87) implies that

$$| \lambda - \hat{\lambda} | < | \varepsilon | \parallel \mathbf{y} \parallel_2 \parallel \mathbf{P} \parallel_2, \tag{7.89}$$

and for that reason, the quantity $\parallel \mathbf{y} \parallel_2$ (computed after the normalizations just described) is spoken of as the *condition number* for λ of the eigenvalue problem.

The *Bauer-Fike theorem* (Theorem 7.12 below) is usually (e.g., Isaacson and Keller, 1966) derived as a corollary to Theorem 7.11. In our statement of this theorem, we have let $\lambda(\varepsilon)$ represent λ'.

Theorem 7.12. Under the conditions of Theorem 7.11 and the assumption that λ has multiplicity 1, we have

$$| \lambda(\varepsilon) - \lambda | \leq | \varepsilon | \, \| \, \mathbf{P} \, \|_2 \gamma(\mathbf{U}), \tag{7.90}$$

where $\mathbf{U}$ is the matrix of eigenvectors of $\mathbf{A}$ and $\gamma(\mathbf{U}) = \| \, \mathbf{U} \, \|_2 \| \, \mathbf{U}^{-1} \, \|_2$.

Note that $\gamma(\mathbf{U})$ is the condition number for matrix inversion. It arose in Section 6.3.1, in connection with error analysis for linear equation methods.

An important strengthening of Theorem 7.12 can be made in the case that $\mathbf{A}$ is Hermitian. For in this circumstance, the matrix $\mathbf{U}$ of eigenvectors can be chosen to be unitary. By use of Theorem 7.6, we have

$$n^{1/2} = \| \, \mathbf{I} \, \|_2 = \| \, \mathbf{U}^{\mathrm{H}}\mathbf{U} \, \|_2 = \| \, \mathbf{U}^{\mathrm{H}} \, \|_2 = \| \, \mathbf{U} \, \|_2. \tag{7.91}$$

Since $\mathbf{U}^{-1} = \mathbf{U}^{\mathrm{H}}$, we have that $\gamma(\mathbf{U}) = n$, and consequently the following corollary holds.

Corollary 7.5. Let $\mathbf{A}$ be a Hermitian matrix with eigenvalues $\lambda_1 , \ldots ,$ λ_n. Then if λ' is an eigenvalue of $\mathbf{A} + \mathbf{E}$, we have

$$\min_{1 \leq i \leq n} | \lambda' - \lambda_i | < n \, \| \, \mathbf{E} \, \|_2. \tag{7.92}$$

In view of this corollary, it is said that when $\mathbf{A}$ is Hermitian, the eigenvalue problem is well conditioned.

Blum (1972, p. 236) gives an a posteriori error bound for the case that $\mathbf{A}$ is Hermitian. Let $\mathbf{x}$ be an arbitrary vector and $\lambda_1 , \ldots , \lambda_n$ the eigenvalues of $\mathbf{A}$; then

$$\min_{1 \leq i \leq n} | \lambda - \lambda_i | \leq \| \, \mathbf{A}\mathbf{x} - \lambda\mathbf{x} \, \|_2 / \| \, \mathbf{x} \, \|_2. \tag{7.93}$$

Now we turn our attention to bounding the norm of the overall error matrix $\mathbf{F}$ in (7.84). As we have mentioned, $\mathbf{F}$ can be regarded as the composite of the errors arising from the transformation of $\mathbf{A}$ to Hessenberg from the errors arising from the QR iterations. We shall now present bounds for roundoff errors arising from these two operations, and one can then obtain a bound for $\| \mathbf{F} \|_2$ by taking the sum of the bounds for errors arising in the individual transformations.

According to Stewart (1973, p. 333), if the upper Hessenberg matrix $\mathbf{A}_{n-1}$ is computed from the given matrix $\mathbf{A}$ by the Householder transformation technique we described in Section 7.1.2, then $\mathbf{A}_{n-1}$ is unitarily similar to $\mathbf{A} + \mathbf{E}$, where

$$\| \mathbf{E} \|_2 < \gamma n^2 2^{-t} \| \mathbf{A} \|_2, \tag{7.94}$$

where γ is a constant that remains bounded with increasing n, and t is the length of the binary computer word. If all inner product operations (that is, multiplications of the form $\mathbf{a}^H \mathbf{b}$, where $\mathbf{a}$ and $\mathbf{b}$ are vectors) are done in double precision, we have the reduced bound

$$\| \mathbf{E} \|_2 < \gamma n 2^{-t} \| \mathbf{A} \|_2. \tag{7.95}$$

Again according to Stewart (1973, p. 364), if a QR iteration with shift α is applied to a given square matrix $\mathbf{B}$, the QR iteration being performed according to the formula we gave in Section 7.3.2, then the resulting matrix is unitarily similar to the matrix $\mathbf{B} + \mathbf{E}$, where

$$\| \mathbf{E} \|_2 < \gamma(n) 2^{-t} \max \{ \| \mathbf{A} \|_2, | \alpha | \}. \tag{7.96}$$

Here $\gamma(n)$ is a "slowly growing" function of n.

7.5. Supplementary Notes and Discussion

The principal reference for this chapter is clearly the book by Stewart (1973). However the (now somewhat dated) volume by Wilkinson (1965) is valuable as a comprehensive development and a source of cogent supplementary information concerning matrix analysis.

The reader may wish to consult Stewart (1973) for some additional information, such as algorithmic presentations of the fundamental transformations we have used as well as thoughts about their efficient computer implementation. Additionally, one may find in Stewart's book some references to tested programs for eigenvalue calculation, as well as discussion of singular value decomposition and the *generalized eigenvalue problem*; that is, the problem of finding values λ and nonzero vectors $\mathbf{x}$ satisfying the matrix equation

$$\mathbf{A}\mathbf{x} - \lambda \mathbf{B}\mathbf{x} = 0.$$

We alert the reader to the fact that the matrix norm $\| \cdot \|_2$ used by Stewart is a different norm from our norm denoted by that symbol. For the norm $\| \cdot \|_2$ used here is denoted by $\| \cdot \|_F$ in Stewart.

To the best of our knowledge, at this writing, Stewart's book is the only text other than the present volume that describes eigenvalue algorithms that achieve quadratic (cubic in the Hermitian case) convergence. It is thus regrettable that in Stewart's (1973, p. 347) work, there seems to be a serious difficulty with his proof of the key theorem giving quadratic convergence of the Rayleigh quotient iteration method. We are indebted to Professor Warren Ferguson for supplying us with the proof of our Theorem 7.10 on quadratic convergence of the Rayleigh quotient iteration method.

We have purposefully overlooked a large class of methods, known as direct methods, for numerical solution of the eigenvalue problem. Such methods are described in Section 4.2 of Householder (1953), for example. *Direct methods* attack the eigenvalue problem by seeking to compute the characteristic polynomial ϕ (or some factor thereof) directly from the given matrix **A**, and then by applying methods such as described in Chapter 5 for iteratively approximating the roots of the characteristic polynomial, these roots being, of course, the eigenvalues of **A**. Direct methods are now viewed in disfavor, in comparison to the methods discussed in this chapter. This disfavor stems in large part because the roundoff error introduced in computing the coefficients of the characteristic polynomials ϕ tends to introduce very large error in the eigenvalue estimates. This sensitivity of roots to purturbations in the polynomial coefficients was mentioned in Chapter 5, and is reflected in the great sensitivity of the error bound of Theorem 5.5.

8

The Numerical Solution of Ordinary Differential Equations

In this chapter we shall study the following three central problems of the computational theory for ordinary differential equations:

Problem 1. Consider the ordinary differential equation

$$\mathbf{y}'(x) = \mathbf{f}(x, \mathbf{y}(x)), \qquad \mathbf{y}(x_0) = \mathbf{y}_0, \tag{8.1}$$

where x is a scalar variable, and $\mathbf{y}$ and $\mathbf{f}$ are N-tuple-valued functions of the indicated variables. The problem of finding the function $\mathbf{y}(x)$ satisfying (8.1) for a given $\mathbf{f}$ and $\mathbf{y}_0$ is called an *initial-value problem*. The solution of equation (8.1) has to be determined in an interval $[x_0, b]$.

Throughout this chapter, we shall restrict attention to functions $\mathbf{f}(x; \mathbf{y})$ that satisfy a Lipschitz condition in a region D of $(N + 1)$-tuples. That is, there exists a constant L such that if $(x, \mathbf{y}_1)$ and $(x, \mathbf{y}_2) \in D$, then

$$\| \mathbf{f}(x, \mathbf{y}_1) - \mathbf{f}(x, \mathbf{y}_2) \| \le L \| \mathbf{y}_1 - \mathbf{y}_2 \|. \tag{8.2}$$

By generalizing Example 4.11 to functions of N dimensions and using Theorem 4.4, it is easy to prove that Problem 1 has a unique solution defined on $[x_0, b]$, provided $\mathbf{f}$ satisfies a Lipschitz condition on appropriate domain.

Toward simplifying the notation in the following sections, we shall assume that $N = 1$, but the multivariate extension of the methods to follow can be accomplished without difficulty. Furthermore, we shall note that real-valued, higher-order differential equations can be reduced to vector-valued differential equations of first order by an obvious artifice, which we illustrate below for order 2. Let

$$y''(x) = F(x; y(x), y'(x)).$$

Define $z_1(x) = y(x)$, and $z_2(x) = y'(x)$. Then we have the equivalent first-order equation

$$\begin{pmatrix} z_1'(x) \\ z_2'(x) \end{pmatrix} = \begin{pmatrix} z_2(x) \\ F(x; z_1(x), z_2(x)) \end{pmatrix}.$$

Since higher-order differential equations are equivalent to systems of first-order differential equations, their numerical solution can be found by the methods of this chapter.

Problem 2. Consider the nth-order differential equation

$$y^{(n)} = F(x; y(x), y^{(1)}(x), \ldots, y^{(n-1)}(x)), \tag{8.3}$$

where x is a scalar and y is a real function. The solution of (8.3) satisfying the conditions

$$g_i(y(x_1), y^{(1)}(x_1), \ldots, y^{(n-1)}(x_1); \ldots; y(x_k), y^{(1)}(x_k), \ldots, y^{(n-1)}(x_k)) = 0,$$
$$1 \le i \le n, \tag{8.4}$$

has to be computed, for given points $x_1, x_2, \ldots, x_k$, and given functions g_i of the indicated nk variables. This problem is called a *boundary-value problem*.

Problem 3. Linear boundary-value problems very often contain unknown parameters. Consider the specific linear differential equation

$$\sum_{k=1}^{2m} p_k(x) y^{(k)}(x) = \lambda r(x) y(x), \qquad a \le x \le b, \tag{8.5}$$

with the boundary conditions

$$\sum_{k=0}^{2m} [\alpha_{ik} y^{(k)}(a) + \beta_{ik} y^{(k)}(b)] = 0, \qquad 1 \le i \le m, \tag{8.6}$$

where $a, b, \alpha_{ik}, \beta_{ik}$ are given numbers, r and $p_1, \ldots, p_{2m}$, are given functions, λ is an unknown parameter, and y is an unknown real function. Obviously, the function $y \equiv 0$ is a solution of the above problem with an arbitrary value of λ. The functions not identical to zero and the corresponding values of λ are to be determined so as to satisfy equation (8.5) and condition (8.6). The nonzero solution functions are called *eigenfunctions* and the corresponding values of λ are called *eigenvalues*. The problem of finding eigenvalues and eigenfunctions satisfying (8.5) and (8.6) is called, analogously to Chapter 7, an *eigenvalue problem*.

8.1. The Solution of Initial-Value Problems

8.1.1. Picard's Method of Successive Approximation

By integrating both sides of (8.1), Problem 1 can be rewritten in the form

$$y(x) = y_0 + \int_{x_0}^{x} f(t, y(t)) \, dt. \tag{8.7}$$

Under our standing Lipschitz condition assumption, Example 4.11 implies that the operator of the right-hand side satisfies the condition of the fixed-point theorem (Theorem 4.4). For this reason, the fixed-point problem determined by viewing the right-hand side of (8.7) as an operator has exactly one solution under the assumptions made in the introduction of this chapter, and the unique fixed point can be obtained as the limit function of the sequence

$$y_0(x) \equiv y_0,$$

$$y_k(x) = y_0 + \int_{x_0}^{x} f(t, y_{k-1}(t)) \, dt \qquad (k = 1, 2, \ldots). \tag{8.8}$$

This successive approximation rule is known as *Picard's method*. The theory of Chapter 4 implies that the convergence of the method is linear. Since, unlike our convenient example below, the evaluation of the integral in (8.8) usually cannot be made analytically and convergence is slow (relative to quasilinearization, to be described), this method is not popular.

Example 8.1. Consider the equation

$$y' = x^2 + y^2, \qquad y(0) = 0.$$

Using the recursive formula (8.8) we have

$$y_0(x) \equiv 0,$$

$$y_1(x) = \int_0^x t^2 \, dt = \frac{x^3}{3},$$

$$y_2(x) = \int_0^x \left(t^2 + \frac{t^6}{9} \right) dt = \frac{x^3}{3} + \frac{x^7}{63},$$

$$y_3(x) = \int_0^x \left(t^2 + \frac{t^6}{9} + \frac{2t^{10}}{189} + \frac{t^{14}}{3969} \right) dt = \frac{x^3}{3} + \frac{x^7}{63} + \frac{2x^{11}}{2079} + \frac{x^{15}}{59,535},$$

and so on. Letting $y(x_k)$ denote the exact solution, we see in Table 8.1 that y_3 above yields a surprisingly accurate approximation.

Table 8.1. A Computation by Picard Iteration

x_k	Approximation $y_3(x_k)$	Exact $y(x_k)$
0.1	0.000333	0.000333
0.2	0.002667	0.002667
0.3	0.009003	0.009003
0.4	0.021359	0.021359
0.5	0.041791	0.041791
0.6	0.072447	0.072448
0.7	0.115660	0.115660
0.8	0.174079	0.174080
0.9	0.250897	0.250907
1.0	0.350185	0.350232

Problems

8.1. Use Picard iteration for the differential equation system

$$y' = z, \qquad y(0) = 1,$$
$$z' = y, \qquad z(0) = 1.$$

8.2. Use the Seidel-type iteration (introduced in Section 4.7) for the system of the previous example and show that in this case the Seidel-type iteration method converges faster than iteration method (8.8).

8.1.2. The Power Series Method

Assume that the bivariate function f is sufficiently many times differentiable for the terms below to exist. Let $h = (b - x_0)/n$, where n is a positive integer, and consider the points

$$x_k = x_0 + k \cdot h \qquad (k = 0, 1, 2, \ldots, n).$$

Using Taylor's formula for the function y, we obtain the expression

$$y(x_{k+1}) = y(x_k) + hy'(x_k) + \cdots + \frac{h^m}{m!} y^{(m)}(x_k) + H_{m+1}, \qquad (8.9)$$

where

$$H_{m+1} = \frac{h^{m+1}}{(m + 1)!} y^{(m+1)}(\eta_k), \qquad x_k < \eta_k < x_{k+1}.$$

The derivatives of the solution can be obtained recursively by the following equations, which were obtained by use of the chain rule of elementary calculus:

$$y' = f,$$
$$y'' = f_x + f_y \cdot f,$$
$$y''' = f_{xx} + 2f_{xy} \cdot f + f_{yy} \cdot f^2 + f_y f_x + f_y^2 \cdot f,$$
$$\vdots$$

The approximation y_{k+1} of $y(x_{k+1})$ can be computed using the equation

$$y(x_{k+1}) \approx y_{k+1} = y(x_k) + h \cdot y'(x_k) + \cdots + \frac{h^m}{m!} y^{(m)}(x_k), \qquad (8.10)$$

which was obtained from (8.9) by neglecting the error term H_{m+1}.

By starting from x_0 and repeating the use of equation (8.10), we can obtain the approximations of the function values $y(x_0)$, $y(x_1)$, ..., $y(x_n)$.

Since the higher-order derivatives of y usually cannot be determined easily (because in most instances the recursive equations become very complicated), the power series method is applied only in special cases.

Example 8.2. Consider again the equation

$$y' = x^2 + y^2, \qquad y(0) = 0.$$

Since

$$y' = x^2 + y^2,$$
$$y'' = 2x + 2yy',$$
$$y''' = 2 + 2yy'' + 2y'^2,$$
$$y^{(IV)} = 2yy''' + 2y'y'' + 4y'y'',$$

if $h = (x - 0)$, we have the Taylor's series approximation

$$y(x) \approx y(0) + hy'(0) + \frac{h^2}{2!} y''(0) + \frac{h^3}{3!} y'''(0) + \frac{h^4}{4!} y^{(IV)}(0).$$

In Table 8.2, by way of comparison to Example 8.1, we have tabulated the above equation at $x_i = 0.1i$, $1 \le i \le 10$.

We see from Table 8.3 that, for this example, some increase in accuracy can be obtained by use of the iteration formula (8.10) with $h = 0, 1$.

Problem

8.3. Solve the equation given in Example 8.2 by using the following method: Let $y(h) = \sum_{k=0}^{\infty} a_k h^k$ be the power series of y. Then the initial value and the dif-

Table 8.2. A Computation by Taylor's Method

x_i	Approximation y_i	Exact $y(x_i)$
0.1	0.000333	0.000333
0.2	0.002667	0.002667
0.3	0.009000	0.009003
0.4	0.213333	0.021359
0.5	0.041666	0.041791
0.6	0.072000	0.072448
0.7	0.114333	0.115660
0.8	0.170667	0.174080
0.9	0.243000	0.250907
1.0	0.333333	0.350232

Table 8.3. A Computation by Iterative Application of Taylor's Formula

x_i	Approximation y_i	Exact $y(x_i)$
0.1	0.000333	0.000333
0.2	0.002667	0.002667
0.3	0.009001	0.009003
0.4	0.021346	0.021359
0.5	0.041734	0.041791
0.6	0.072268	0.072448
0.7	0.115182	0.115660
0.8	0.172959	0.174080
0.9	0.248501	0.250907
1.0	0.345392	0.350232

ferential equation imply that $a_0 = 0$, and

$$\sum_{k=1}^{\infty} k a_k h^{k-1} = h^2 + \left(\sum_{k=1}^{\infty} a_k h^k \right)^2.$$

Use the fact that the coefficients of the powers of h must be the same on the left- and right-hand sides to solve recursively for the coefficients a_k.

8.1.3. Methods of the Runge–Kutta Type

As in the method of power series, let $h = (b - x_0)/n$, where n is a positive integer, and let

$$x_k = x_0 + kh \qquad (k = 0, 1, 2, \ldots, n).$$

The *Runge–Kutta* methods approximate $y(x_{k+1})$ by the equation

$$y_{k+1} = y_k + h \sum_{i=1}^{t} \alpha_i k_i, \tag{8.11}$$

where t is a suitable integer and the k_i satisfy

$$k_i = f\left(x_k + \mu_i h, \, y_k + h \sum_{j=1}^{t} \lambda_{ij} k_j\right), \qquad 1 \le i \le t. \tag{8.12}$$

In equations (8.11) and (8.12) the constants α_i, μ_i, and λ_{ij}, $1 \le i \le t$, $1 \le j \le t$, are given values. The μ_1 and λ_{1j} terms are constrained to be 0, in order to ensure that the data at the left endpoint are used. The motivation of the Runge–Kutta method is that the term $h \sum_{i=1}^{t} \alpha_i k_i$ can be made to approximate

$$\int_{x_k}^{x_{k+1}} f(x, y(x)) \, dx.$$

The initial value gives y_0, and by using formulas (8.11) and (8.12) for $k = 0, 1, 2, \ldots, n - 1$, the approximations of the functional values $y(x_1), \ldots, y(x_n)$ can be obtained.

The determination of the numbers k_i usually requires the numerical solution of the system (8.12) of nonlinear equations. If we let $A(\mathbf{k})$ denote the operator defined on the right-hand side of (8.12), then in the notation of Theorem 4.7 arbitrary vectors $\mathbf{k} = (k_1, \ldots, k_t)$ and $\tilde{\mathbf{k}} = (\tilde{k}_1, \ldots, \tilde{k}_t)$ satisfy the inequalities

$$\| A(\mathbf{k}) - A(\tilde{\mathbf{k}}) \|_\infty = \max_{1 \le i \le t} \left| f\left(x_k + \mu_i h, \, y_k + h \sum_{j=1}^{t} \lambda_{ij} k_j\right) \right.$$
$$\left. - f\left(x_k + \mu_i h, \, y_k + h \sum_{j=1}^{t} \lambda_{ij} \tilde{k}_j\right) \right|$$
$$\le \max_{1 \le i \le t} Lh \sum_{j=1}^{t} | \lambda_{ij} | \, | k_j - \tilde{k}_j |$$
$$\le \max_{1 \le i \le t} \left(\sum_{j=1}^{t} | \lambda_{ij} |\right) Lh \, \| \mathbf{k} - \tilde{\mathbf{k}} \|_\infty,$$

and therefore for sufficiently small h, operator A is a contraction in t-dimen-

sional Euclidean space. From Theorem 4.5 and from the result of Section
4.7 we conclude that system (8.12) of equations has a unique solution, which
can be computed using the iteration method (5.30) or the Seidel-type iteration algorithm (5.34). Other techniques (most notably the generalized
Newton method) of Section 5.3 are also useful for this task.

If for $j \geq i$, $\lambda_{ij} = 0$, then equations (8.12) recursively give the values k_i
(without solution of nonlinear equations). In this case method (8.11) and
(8.12) is called an *explicit Runge–Kutta-type method*.

Let $x \in [x_0, b - h]$ and replace points x_k by x and y_k by y, where y is
an arbitrary real number, in formula (8.12), and introduce the function

$$g(x, y, h) = \sum_{i=1}^{t} \alpha_i k_i.$$

Using this notation, (8.11) can be rewritten as

$$y_{k+1} = y_k + hg(x_k, y_k, h).$$

The term $hg(x_k, y_k, h)$ is an estimate of

$$\int_{x_k}^{x_{k+1}} f(x, y(x)) \, dx.$$

Definition 8.1. The *local error* of method (8.11) and (8.12) at the
point x is defined by

$$H(x, h) = y(x) + hg(x, y(x), h) - y(x + h),$$

where $y(\cdot)$ denotes the exact solution of Problem 1. Approximation (8.11)
is called a *method of order p* if for $h \to 0$ and for fixed x, y,

$$H(x, h) = O(h^{p+1}), \tag{8.13}$$

and p is the largest positive integer satisfying (8.13).

In the Runge–Kutta method the parameters μ_i, λ_{ij}, α_i are determined
according to the condition that the order of the method must be maximal.
By using Taylor's formula, one can verify that (8.13) is guaranteed by the
equations

$$i \left. \frac{\partial^{i-1} g}{\partial h^{i-1}} \right|_{h=0} - y^{(i)}(x) = 0 \qquad (i = 1, 2, \ldots, p), \tag{8.14}$$

which may be used to obtain a system of relations between the parameters.

To verify the equivalence between (8.13) and (8.14), note that the former equation may be restated as

$$y(x + h) - y(x) = hg(x, y, h) + O(h^{p+1}).$$

Upon taking the Taylor's series expansion in h of both sides, we obtain

$$\sum_{i=1}^{p} y^{(i)}(x) \frac{h^i}{i!} = h \sum_{i=0}^{p-1} \frac{\partial^i}{\partial h^i} g(x, y, h)\Big|_{h=0} \frac{h^i}{i!} + O(h^{p+1})$$

and (8.14) results from equating coefficients of like powers of h.

Particular pth order Runge–Kutta parameters are conveniently constructed by choosing the parameters t, α_i, μ_i, and λ_{ij} so that (8.14) holds for every sufficiently differentiable function f. In discussions to follow, we shall apply such a construction to derivation of some popular Runge–Kutta formulas.

1. First-Order Methods. For $t = 1$ the explicit equations (8.11) and (8.12) have the form

$$k_1 = f(x_k, y_k),$$
$$y_{k+1} = y_k + h\alpha k_1.$$

Thus in this case, we have $g(x_k, y_k, h) = \alpha f(x_k, y_k)$, and equation (8.14) for $p = 1$ can be written as

$$(\alpha - 1)f(x_k, y_k) = 0,$$

which implies $\alpha = 1$. Thus rule (8.11) and (8.12) can be written in the form

$$y_{k+1} = y_k + hf(x_k, y_k), \tag{8.15}$$

and this method is called *Euler's method.* By our construction, it is a method of order 1.

We shall now examine the propagation effects of the local error in conjunction with roundoff, for the case of Euler's method. Error analysis of other Runge–Kutta formulas proceeds in the same manner, but the specific formulas will be more complicated.

In the actual computation, instead of sequence (8.15) the sequence $\{z_k\}$ determined by

$$z_{k+1} = z_k + hf(x_k, z_k) + \Delta_k \tag{8.16}$$

is calculated, where $z_0 = y_0$ and Δ_k is the roundoff error of the computation of the second term in the right-hand side. The propagation of the errors of the method and the roundoff errors are bounded in the following theorem.

Theorem 8.1. Assume that for $x \in [x_0, b]$, $y \in (-\infty, \infty)$, the partial derivatives f_x and f_y exist, and for some fixed constants K, M, and Δ,

$$|f_y(x, y)| \leq K, \qquad |f_x(x, y) + f(x, y)f_y(x, y)| \leq M,$$

and

$$|\Delta_k| \leq \Delta \qquad (k = 1, 2, \ldots, n - 1).$$

Then for $k \geq 0$,

$$\varepsilon_k \equiv |z_k - y(x_k)| \leq \left(\frac{Mh}{2K} + \frac{\Delta}{hK}\right)\{\exp[K(x_k - x_0)] - 1\}. \qquad (8.17)$$

Proof. Using the Taylor formula for the solution of (8.1) we have

$$y(x_{k+1}) = y(x_k) + hy'(x_k) + \frac{y''(\xi)}{2}h^2$$

$$= y(x_k) + hf(x_k, y(x_k)) + \frac{y''(\xi)}{2}h^2.$$

Subtracting this equation from (8.16) we have

$$z_{k+1} - y(x_{k+1}) = z_k - y(x_k) + h[f(x_k, z_k) - f(x_k, y(x_k))]$$

$$+ \Delta_k - \frac{y''(\xi)}{2}h^2,$$

which, upon applying the mean value theorem to the term in brackets, implies

$$\varepsilon_{k+1} = |z_{k+1} - y(x_{k+1})| \leq |z_k - y(x_k)|$$

$$+ h \cdot |f_y(x_k, \eta)| \cdot |z_k - y(x_k)| + |\Delta_k| + \frac{h^2}{2}|y''(\xi)|.$$

Since

$$y''(x) = \frac{d}{dx}[f(x, y(x))] = f_x(x, y(x)) + f_y(x, y(x)) \cdot y'(x)$$

$$= f_x(x, y(x)) + f_y(x, y(x)) \cdot f(x, y(x)),$$

by inserting the hypothesized bounds, we obtain

$$\varepsilon_{k+1} \leq \varepsilon_k(1 + hK) + \left(\Delta + \frac{h^2}{2}M\right).$$

By using induction, it is easy to prove that

$$\varepsilon_k \le \left(\varDelta + \frac{h^2 M}{2}\right) \frac{(1 + hK)^k - 1}{hK}, \qquad (8.18)$$

where we have used the fact that $\varepsilon_0 = 0$.

Since for $x \ge 0$, $1 + x \le e^x$, (8.18) implies that

$$\varepsilon_k \le \left(\varDelta + \frac{h^2 M}{2}\right) \frac{1}{hK} [\exp(Kkh) - 1]$$

$$= \left(\frac{\varDelta}{hK} + \frac{hM}{2K}\right)\{\exp[K(x_k - x_0)] - 1\}. \qquad \blacksquare$$

Corollary 8.1. The maximal value of the right-hand side of (8.17) is equal to

$$\left(\frac{\varDelta}{hK} + \frac{hM}{2K}\right)\{\exp[K(b - x_0)] - 1\}. \qquad (8.19)$$

For large values of h the second term of the first factor is large, and for small values of h the first term can be large. The sensible choice of the step size h is determined by the minimum of this maximal error bound. The second factor is a constant number, and therefore the minimum can be obtained by solving the equation

$$-\frac{\varDelta}{h^2 K} + \frac{M}{2K} = 0,$$

which implies that in terms of the roundoff error $\varDelta$ and derivative bound M, the best step size for Euler's method is

$$h = (2\varDelta/M)^{1/2}.$$

Corollary 8.2. In the special case that $\varDelta = 0$, for $h \to 0$, $\varepsilon_k \to 0$, i.e., the approximating solutions converge to the exact solution for $h \to 0$. In this case the convergence is linear in h.

In Section 3.3.2 of Chapter 3 we saw that Richardson's extrapolation allowed us to increase the order of the error term in various Newton–Cotes formulas. The idea of Richardson's extrapolation is to obtain two estimates of the desired quantity, one being based on a step size h and the other using a step size $2h$. The scaling of these estimates is chosen to eliminate the lowest-order term in the power series expansion of the error term. Richardson's extrapolation can also be applied to reduce the local error of the Runge–Kutta-type methods. Assume that $\varDelta = 0$ and y_k is equal to the exact value of $y(x_k)$. Compute the approximation of $y(x_{k+2})$,

$$\bar{\bar{y}}_{k+2} = y_k + 2hf(x_k, y_k). \tag{8.20}$$

Using the step h, the approximation of $y(x_{k+2})$ can be computed by applying (8.15) two successive times:

$$\bar{y}_{k+2} = y_k + hf(x_k, y_k) + hf(x_{k+1}, \bar{y}_{k+1}), \tag{8.21}$$

where $\bar{y}_{k+1} = y_k + hf(x_k, y_k)$.

Theorem 8.2. Assume that for $x \in [x_0, b]$ and $y \in (-\infty, \infty)$ the function f is twice differentiable and the functions $f, f_x, f_y, f_{xx}, f_{xy}, f_{yy}$ are bounded. Then

$$y(x_{k+2}) = 2\bar{y}_{k+2} - \bar{\bar{y}}_{k+2} + O(h^3).$$

Proof. Let us consider the Taylor formula for y:

$$y(x_{k+2}) = y_k + 2hy'(x_k) + \frac{4h^2}{2}y''(x_k) + \frac{8h^3}{6}y'''(\alpha)$$

$$= y_k + 2hf(x_k, y_k) + 2h^2y''(x_k) + \frac{4h^3}{3}y'''(\alpha), \tag{8.22}$$

for $\alpha \in [x_k, x_{k+2}]$.

Subtracting (8.20) from (8.22) we have

$$y(x_{k+2}) - \bar{\bar{y}}_{k+2} = 2h^2y''(x_k) + \frac{4h^3}{3}y'''(\alpha), \tag{8.23}$$

and subtracting (8.21) from (8.22) and using the bivariate Taylor's formula we obtain

$$y(x_{k+2}) - \bar{y}_{k+2} = h[f(x_k, y_k) - f(x_{k+1}, \bar{y}_{k+1})] + 2h^2y''(x_k) + \frac{4h^3}{3}y'''(\alpha)$$

$$= -h[f_x(x_k, y_k)(x_{k+1} - x_k) + f_y(x_k, y_k)(\bar{y}_{k+1} - y_k)$$

$$+ f_{xx}(\xi, \eta)(x_{k+1} - x_k)^2 + 2f_{xy}(\xi, \eta)(x_{k+1} - x_k)(\bar{y}_{k+1} - y_k)$$

$$+ f_{yy}(\xi, \eta)(\bar{y}_{k+1} - y_k)^2] + 2h^2y''(x_k) + \frac{4h^3}{3}y'''(\alpha),$$

where $\xi \in [x_k, x_{k+1}]$ and η is between y_k and $\bar{y}_{k+1}$. Equation (8.15) and the

Table 8.4. A Computation Using Euler's Method and Its Correction

Euler's method		Corrected Euler's method			Exact
x_k	$\hat{y}_k$	$\bar{y}_k$	$\bar{\bar{y}}_k$	$y_k = 2\bar{y}_k - \bar{\bar{y}}_k$	$y(x_k)$
0.0	0.000000	0.000000	0.000000	0.000000	0.000000
0.1	0.000000				0.000333
0.2	0.001000	0.001000	0.000000	0.002000	0.002667
0.3	0.005000				0.009003
0.4	0.014003	0.015004	0.008001	0.020007	0.021359
0.5	0.030022				0.041791
0.6	0.055112	0.061177	0.052087	0.070267	0.072448
0.7	0.091416				0.115660
0.8	0.141252	0.1569006	0.143255	0.170547	0.174080
0.9	0.207247				0.250907
1.0	0.292542	0.324094	0.304364	0.343824	0.350232

relation $y'' = f_x + f_y f$ allow us to rewrite the above expression as

$$y(x_{k+2}) - \bar{y}_{k+2} = -h^2 y''(x_k) - h^3[f_{xx}(\xi, \eta) + 2f_{xy}''(\xi, \eta)f(x_k, y_k)$$

$$+ f_{yy}(\xi, \eta)f^2(x_k, y_k)] + 2h^2 y''(x_k) + \frac{4h^3}{3} y'''(\alpha). \quad (8.24)$$

From equations (8.23) and (8.24) we conclude that

$$2[y(x_{k+2}) - \bar{y}_{k+2}] = y(x_{k+2}) - \bar{\bar{y}}_{k+2} + O(h^3),$$

which implies the assertion.   ■

Example 8.3. In the case of the equation

$$y' = x^2 + y^2, \qquad y(0) = 0$$

with the choice $h = 0.1$, we have obtained Table 8.1. The values $\hat{y}_k$ are the approximation obtained from Euler's method [equation (8.15)]. At each iteration $\bar{y}_k$ is constructed from y_{k-2} by the rule (8.21) and $\bar{\bar{y}}_k$ is constructed from y_{k-2} by (8.20).

2. *Second-Order Methods.* For $t = 2$, the explicit Runge–Kutta equations (8.11) and (8.12) have the form

$$k_1 = f(x_k, y_k),$$

$$k_2 = f(x_k + \mu_2 h, y_k + \lambda_{21} k_1 h),$$

$$y_{k+1} = y_k + \alpha_1 h k_1 + \alpha_2 h k_2.$$

We have that

$$g(x, y, h) = \alpha_1 f(x, y) + \alpha_2 f(x + \mu_2 h, y + \lambda_{21} h f(x, y)).$$

From equation (8.14) $i = 1$, we obtain

$$g(x, y, 0) = \alpha_1 f(x, y) + \alpha_2 f(x, y) = y'(x) = f(x, y),$$

from which we obtain that $\alpha_1 + \alpha_2 = 1$. Applying (8.14) with $i = 2$ yields

$$2 \frac{d}{dh} g(x, y, h) \mid_{h=0} = y^{(2)}(x).$$

Evaluating the left-hand side and substituting $f_x(x, y) + f_y(x, y)f(x, y)$ for the right-hand side, we have

$$2[\alpha_2 f_x(x, y)\mu_2 + \alpha_2 f_y(x, y)\lambda_{21} f(x, y)] = f_x(x, y) + f_y(x, y)f(x, y).$$

These equations are valid for all differentiable functions f if and only if the following relations are true:

$$\alpha_1 + \alpha_2 = 1, \qquad 2\alpha_2\mu_2 = 1, \qquad 2\alpha_2\lambda_{21} = 1,$$

and the maximal value of p is equal to 2. Thus we have three equations for four unknowns, a system with infinitely many solutions. The *modified* Euler method is defined by $\alpha_1 = \alpha_2 = \frac{1}{2}, \mu_2 = 1, \lambda_{21} = 1$. The *Heun method* is specified by $\alpha_1 = \frac{1}{4}, \alpha_2 = \frac{3}{4}, \mu_2 = \frac{2}{3} = \lambda_{21}$.

3. Third-Order Methods. By methods used in two previous cases, the reader may verify that for $t = 3$ the explicit equations (8.14) are satisfied for an arbitrary function f with maximal value of p if and only if $p = 3$ and the parameters satisfy the equations

$$\mu_2 = \lambda_{21},$$
$$\mu_3 = \lambda_{31} + \lambda_{32},$$
$$\alpha_1 + \alpha_2 + \alpha_3 = 1,$$
$$2\alpha_2\lambda_{21} + 2\alpha_3\lambda_{31} + 2\alpha_3\lambda_{32} = 1,$$
$$3\alpha_2\lambda_{21}^2 + 3\alpha_3(\lambda_{31} + \lambda_{32})^2 = 1,$$
$$6\alpha_3\lambda_{32}\lambda_{21} = 1.$$

Thus we have six equations for eight unknowns, and so again there are infinitely many solutions. The following formulas are most frequently used in applications. *Heun's third-order method* is obtained if $\alpha_1 = \frac{1}{4}, \alpha_2 = 0,$

$\alpha_3 = \frac{3}{4}$, $\mu_2 = \frac{1}{3}$, $\mu_3 = \frac{2}{3}$, $\lambda_{21} = \frac{1}{3}$, $\lambda_{31} = 0$, and $\lambda_{32} = \frac{2}{3}$. The *third-order Runge–Kutta method* corresponds to the parameters $\alpha_1 = \frac{1}{6}$, $\alpha_2 = \frac{2}{3}$, $\alpha_3 = \frac{1}{6}$, $\mu_2 = \frac{1}{2}$, $\mu_3 = 1$, $\lambda_{21} = \frac{1}{2}$, $\lambda_{31} = -1$, and $\lambda_{32} = 2$.

4. Fourth-Order Methods. For $t = 4$ our fundamental system of equations (8.11) and (8.12) gives 16 equations for 18 parameters. The most important solution gives the *classical Runge–Kutta method*:

$$k_1 = f(x_k, y_k),$$

$$k_2 = f\left(x_k + \frac{h}{2}, y_k + \frac{h}{2}k_1\right),$$

$$k_3 = f\left(x_k + \frac{h}{2}, y_k + \frac{h}{2}k_2\right),$$

$$k_4 = f(x_k + h, y_k + hk_3),$$

$$y_{k+1} = y_k + \frac{h}{6}(k_1 + 2k_2 + 2k_3 + k_4).$$

Richardson's extrapolation applied to this formula gives the *corrected Runge–Kutta method*, which is

$$y(x_{k+2}) = \frac{16\bar{y}_{k+2} - \bar{\bar{y}}_{k+2}}{15}.$$

Example 8.4. Let us consider again the equation

$$y' = x^2 + y^2, \qquad y(0) = 0.$$

Our computational results are given for $h = 0.1$ in Table 8.5. We see that the corrected Runge–Kutta method is accurate to six significant places.

Problems

8.4. Solve the equation

$$y' = x, \qquad y(0) = 0,$$

in the interval $[0, Nh]$, where h is a given positive number and N is a positive integer, and estimate the actual error of the method in the case of

    (a) the Euler method,

    (b) the classical Runge–Kutta method.

8.5. Prove the limit relation $\lim_{n \to \infty} [1 + (b/n)]^n = e^b$ using the convergence theorem of the Euler method (Theorem 8.1).

Table 8.5. A Computation by the Classical Runge–Kutta Method and Its Correction

x_k	Runge–Kutta		Exact $y(x_k)$
	y_k	Corrected $(16\bar{y}_k - \bar{\bar{y}}_k)/15$	
0.0	0.000000	0.000000	0.000000
0.1	0.000333		0.000333
0.2	0.002667	0.002667	0.002667
0.3	0.009003		0.009003
0.4	0.021359	0.021359	0.021359
0.5	0.041791		0.041791
0.6	0.072448	0.072448	0.072448
0.7	0.115660		0.115660
0.8	0.174081	0.174080	0.174080
0.9	0.250908		0.250907
1.0	0.350234	0.350232	0.350232

8.6. For $y' = 1 - xy^2$, $y(0) = 0$, approximate $y(0.5)$ by

(a) classical degree three Taylor's series,
(b) iterative degree three Taylor's series,
(c) Heun's third-order method.

In b and c, use $h = 0.5$.

8.7. Verify that the corrected Runge–Kutta method is really Richardson's extrapolation (Section 3.2.2) of the classical Runge–Kutta formula.

8.8. Solve the problem in Exercise 8.6 by the classical Runge–Kutta method and the corrected Runge–Kutta method. Use $h = 0.1, 0.05,$ and 0.01.

8.1.4. Linear Multistep Methods

As in the Runge–Kutta-type methods, consider the points $x_j = x_0 + jh$, $0 \leq j \leq n$, where $n = (b - x_0)/h$. Assume that the approximate function values $y_{k+1-i}, y_{k+2-i}, \ldots, y_{k-1}, y_k$ are known. (In practice, the starting values $y_1, y_2, \ldots, y_{i-1}$, are obtained by Runge–Kutta methods.) The derivatives of the solution can be estimated by the values $f_j = f(x_j, y_j)$, $k + 1 - i \leq j \leq k$. The general formula of the *linear multistep method* is

$$y_{k+1} = -\sum_{j=1}^{i} \alpha_j y_{k+1-j} + h \sum_{j=0}^{i} \beta_j f_{k+1-j}, \qquad (8.25)$$

where the coefficients $\alpha_j, 1 \leq j \leq i, \beta_j, 0 \leq j \leq i$, are fixed numbers. Formula (8.25) can be motivated by viewing the right-hand side as an interpolation formula evaluated so as to approximate $y(x_{k+1})$ by extrapolation. Thus (8.25) is particularly close in spirit to Hermite interpolation (Section 2.1.6 of Chapter 2) inasmuch as derivative information $y_j' = f(x_j, y_j) = f_j$ is used in addition to function values y_j. For $\beta_0 = 0$ the value of y_{k+1} can be obtained immediately, but for $\beta_0 \neq 0$ the value of y_{k+1} appears also in the right-hand side, in the term $f_{k+1} = f(x_{k+1}, y_{k+1})$; therefore the determination of y_{k+1} has to be done in this case by solving the nonlinear equation obtained from (8.25), which we rewrite to emphasize the dependency of the right-hand side on y_{k+1}:

$$y_{k+1} = -\sum_{j=1}^{i} \alpha_j y_{k+1-j} + h \sum_{j=1}^{i} \beta_j f_{k+1-j} + h\beta_0 f(x_{k+1}, y_{k+1}). \qquad (8.26)$$

We shall prove that under the Lipschitz assumptions given in the introduction of this chapter, for sufficiently small values of h, equation (8.26) can be solved by use of our general theory for iteration. Let $g(y_{k+1})$ denote the right-hand side of equation (8.26). Since for arbitrary values of y_{k+1} and $\tilde{y}_{k+1}$,

$$|g(y_{k+1}) - g(\tilde{y}_{k+1})| \leq h\beta_0 |f(x_{k+1}, y_{k+1}) - f(x_{k+1}, \tilde{y}_{k+1})|$$
$$\leq h\beta_0 L |y_{k+1} - \tilde{y}_{k+1}|,$$

for step size h small enough to satisfy $h\beta_0 L < 1$, the operator of the right-hand side is a contraction. Thus the conditions of the contraction theorem (Theorem 4.5) are satisfied. Other nonlinear equation methods in Chapter 5 are also appropriate.

For $\beta_0 = 0$, method (8.25) is called an *explicit formula*, and for $\beta_0 \neq 0$, it is called an *implicit formula*.

Assume that function z is differentiable and introduce the linear difference operator, where we have taken $\alpha_0 = 1$:

$$L_n z = \sum_{j=0}^{i} [\alpha_j z(x_0 + (k + 1 - j)h) - h\beta_j z'(x_0 + (k + 1 - j)h)]. \qquad (8.27)$$

Note that $L_n z = 0$ is equivalent to the absence of truncation error of the multistep formula for the function $z(x)$. By linear translation and scaling of the independent variable we can have $h = 1$, $k = i$, and $x_1 = 0$, and so the operator

$$L^* z = \sum_{j=0}^{i} [\alpha_j z(i - j) - \beta_j z'(i - j)]$$

can be obtained. The order of a multistep method will now be defined to be the highest-order polynomial for which the truncation error is zero.

Definition 8.2. It is said that the *order* of the linear multistep method (8.25) is equal to p, if p is the largest positive integer such that

$$L^*x^l = 0 \qquad (l = 0, 1, 2, \ldots, p). \tag{8.28}$$

If we make the convention that $0^0 = 1$, equation (8.28) can be re-written as

$$\sum_{j=0}^{i} [\alpha_j(i-j)^l - \beta_j l(i-j)^{(l-1)_0}] = 0 \qquad (l = 0, 1, 2, \ldots, p), \tag{8.29}$$

where $(l-1)_0 \triangleq \max\{0, l-1\}$.

Since $\alpha_0 = 1$, we have $p + 1$ equations for the $2i + 1$ unknowns. It therefore is evident that for $p < 2i$ (8.29) has infinitely many solutions, for $p > 2i$ (8.29) cannot be solved, and for $p = 2i$ (8.29) has a unique solution.

Definition 8.3. Let y be the exact solution of equation (8.1). The local *error* of the method (8.25) is defined in terms of (8.27) by

$$H_n = L_n y.$$

Assuming that y is analytic, we can easily calculate the value of H_n. Using Taylor's formula, for any integer ν,

$$y(x_{k+1} + \nu h) = y(x_{k+1}) + \frac{y'(x_{k+1})}{1!}(\nu h) + \cdots + \frac{y^{(p)}(x_{k+1})}{p!}(\nu h)^p + \cdots.$$

For pth-order methods we have

$$L_n y = \frac{y^{(p+1)}(x_{k+1})}{(p+1)!} h^{p+1} L^* t^{p+1} + O(h^{p+2})$$

$$= \frac{y^{(p+1)}(x_{k+1})}{(p+1)!} h^{p+1} \sum_{j=0}^{i} [\alpha_j(i-j)^{p+1} - \beta_j(p+1)(i-j)^p] + O(h^{p+2}).$$

Thus the order of the local error equals the order of the method.

By solving equation (8.29) for different values of p and i, different standard methods can be obtained.

For $p = 2$ and $i = 1$, equation (8.29) becomes

$$1 + \alpha_1 = 0, \qquad \text{for } l = 0,$$
$$1 - \beta_0 - \beta_1 = 0, \qquad \text{for } l = 1,$$
$$1 - 2\beta_0 = 0, \qquad \text{for } l = 2,$$

which implies $\alpha_1 = -1$, $\beta_0 = \frac{1}{2}$, $\beta_1 = \frac{1}{2}$, and the corresponding method, which is known as the *modified Euler formula*, has the form

$$y_{k+1} = y_k + \frac{h}{2} \left[f(x_k, y_k) + f(x_{k+1}, y_{k+1}) \right].$$

It is obviously related to the trapezoidal integration rule given in Chapter 3.

Let $p = i = 2$. Then equation (8.29) has infinitely many solutions, because we obtain three equations for $l = 0$, 1, and 2, respectively:

$$1 + \alpha_1 + \alpha_2 = 0,$$
$$2 + \alpha_1 - \beta_0 - \beta_1 - \beta_2 = 0,$$
$$4 + \alpha_1 - 4\beta_0 - 2\beta_1 = 0,$$

for five unknowns. By fixing $\beta_0 = \alpha_2 = 0$, the solution is restricted to be $\alpha_1 = -1$, $\beta_1 = \frac{3}{2}$, and $\beta_2 = -\frac{1}{2}$, and the corresponding rule is

$$y_{k+1} = y_k + \frac{h}{2} (3f_k - f_{k-1}),$$

which is an extrapolation formula. If $\beta_0 = 0$ and $\alpha_1 = 0$, we have $\alpha_2 = -1$, $\beta_1 = 2$, $\beta_2 = 0$, and the corresponding method

$$y_{k+1} = y_{k-1} + 2hf_k$$

is a generalization of the rectangular formula (3.18) known as the *Nystrom midpoint formula*.

Let $p = 4$, $i = 2$; a derivation analogous to the previous cases yields the formula

$$y_{k+1} = y_{k-1} + \frac{h}{3} (f_{k+1} + 4f_k + f_{k-1}),$$

which is a generalization of Simpson's rule (3.19).

In the repeated application of formula (8.25) the local errors of the multistep formula as well as the roundoff errors are accumulated. It can be proved that the propagation effect of the roundoff errors and the errors of the initial functional values will be small for small values of h and for small roundoff errors if and only if the equation

$$p(z) = \sum_{j=0}^{i} \alpha_j z^{i-j} = 0$$

has no zero outside the unit circle and the zeros on the unit circle must have multiplicity one (see Gear, 1971, pp. 174–180). The methods having this property are called *stable methods*.

The polynomials

$$p(z) = z^i - z^{i-1} \qquad (i = 1, 2, \ldots)$$

obviously satisfy the above property, because the root 0 has multiplicity $i - 1$ and the multiplicity of the root 1 is equal to one. The methods based on these polynomials and for fixed value of i having maximal order are called *Adams-type methods*.

For $\beta_0 = 0$, we have explicit methods, which are known as *Adams–Bashforth* formulas and which can be written as

$$y_{k+1} = \begin{cases} y_k + h f_k & (i = 1), \\[2mm] y_k + \dfrac{h}{2}\,(3f_k - f_{k-1}) & (i = 2), \\[2mm] y_k + \dfrac{h}{12}\,(23f_k - 16f_{k-1} + 5f_{k-2}) & (i = 3), \end{cases}$$

and so on, and for $\beta_0 \neq 0$ we obtain the following implicit formulas, called *Adams–Moulton* formulas:

$$y_{k+1} = \begin{cases} y_k + \dfrac{h}{2}\,(f_{k+1} + f_k) & (i = 1), \\[2mm] y_k + \dfrac{h}{12}\,(5f_{k+1} + 8f_k - f_{k-1}) & (i = 2), \\[2mm] y_k + \dfrac{h}{24}\,(9f_{k+1} + 19f_k - 5f_{k-1} + f_{k-2}) & (i = 3), \end{cases}$$

and so on. It can be proved by easy calculations that the Adams–Moulton formulas have the order $i + 1$ and that the Adams–Bashforth formulas have the order i.

For application of the linear multistep methods (8.25), the initial values $y_0, y_1, \ldots, y_{i-1}$ are needed. They can be computed by using, for instance, a Runge–Kutta-type method. Since the errors of the initial values are accumulated in all further functional values, it is necessary either to choose a relatively accurate formula or to choose a smaller value of h in the computation of the initial values.

A commonly used combination of the explicit and implicit methods can be described as follows. Compute the approximate solution y_{k+1} from the previous values $y_{k+1-i}, \ldots, y_{k-1}, y_k$ using an explicit method, and correct it by applying an implicit method with iteration. The explicit method is called "predictor" and the implicit method is called "corrector." Such algorithms are classified as *predictor–corrector methods*. For example, one may use the (explicit) Nystrom formula

$$\bar{y}_{k+1} = y_{k-1} + 2hf(x_k, y_k)$$

as a predictor of $y(x_{k+1})$, based on y_k, y_{k-1}, and use the (implicit) modified Euler formula

$$\bar{y}'_{k+1} = y_k + (h/2)[f(x_{k+1}, \bar{y}_{k+1}) + f(x_k, y_k)]$$

as corrector. For the first iteration of the corrector, we use the value $\bar{y}_{k+1}$ obtained from the predictor, the values y_k, y_{k-1} having already been calculated. On successive iteration of the corrector, the most recently calculated value $\bar{y}'_{k+1}$ is used in place of $\bar{y}_{k+1}$ in evaluating the corrector. Chapter 4 of Lapidus and Seinfeld (1971) is devoted to analysis of predictor–corrector techniques.

Example 8.5. We solve the equation

$$y' = x^2 + y^2, \qquad y(0) = 0,$$

in the interval $[0, 1]$ applying the explicit Adams–Bashforth method for $i = 3$. Two initial functional values are needed; the values y_0 and y_1 are obtained from the use of the fourth-order Runge–Kutta method (see Example 8.4). The results are shown in Table 8.6.

Problems

8.9. Assume that in the use of the explicit formula (8.25) the roundoff errors of the calculation of f_{k+1-j} ($j = 1, 2, \ldots, i$) are bounded by Δ. Give an upper bound for the roundoff error of the approximation y_{k+1} assuming that the multiplications and summations can be calculated exactly.

8.10. With reference to Definition 8.2, verify that the Adams–Bashforth formula for $i = 2$ does have order 2, and that the Adams–Moulton method for $i = 2$ does have order 3.

8.11. Let $y' = 1 - xy^2, y(0) = 0$. Find $y(0.5)$ by the Adams–Bashforth formulas for $i = 2$ and $i = 3$. Use $h = 0.05, 0.005$, and use the classical Runge–

Table 8.6. A Computation by the Adams–
Bashforth Method

x_k	$(i = 3)$ Y_k	Exact $y(x_k)$
0.0	0.000000	0.000000
0.1	0.000333	0.000333
0.2	0.002667	0.002667
0.3	0.009002	0.009003
0.4	0.021350	0.021359
0.5	0.041760	0.041791
0.6	0.072370	0.072448
0.7	0.115493	0.115660
0.8	0.173758	0.174080
0.9	0.250318	0.250907
1.0	0.349191	0.350232

Kutta formula with $h = 0.001$ to compute starting values. Compute to eight significant figures.

8.12. Repeat the above problem, using instead the Adams–Moulton formulas. Use the iterative method described in this section to solve the implicit equations.

8.1.5. Step Size and Its Adaptive Selection

One may bound the global error arising in the use of an arbitrary Runge–Kutta type formula in terms of local error. We assume that the partial derivative bound $|f_y| \leq K$ holds. Let $l_n \triangleq H(x_n, h)$ denote the local error as in Definition 8.1, with $x = x_k$, $y = y(x_k)$, and $h = x_{k+1} - x_k$. Suppose that

$$l_n \leq \varepsilon(x_{n+1} - x_n), \qquad 0 \leq n < N. \tag{8.30}$$

Then by the proof of Theorem 8.1 (see also Dahlquist and Björck, 1974, p. 351) we have

$$| y_N - y(x_N) | \leq \sum_{n=0}^{N-1} \varepsilon(x_{n+1} - x_n) \exp[K(x_n - x_0)]$$

$$\simeq \frac{\varepsilon}{K} \{\exp[K(x_N - x_0)] - 1\}.$$

Once one has chosen an acceptable value for ε, one can choose a fixed step size to assure that the local error bound (8.30) is not exceeded. For the Runge–Kutta and other popular ordinary differential equation techniques, we find that

$$l_n = c_n h^{k+1} + O(h^{k+2}), \tag{8.31}$$

where k is the order of the method and c_n depends on a derivative (usually the kth) of $y(x)$. In principle, one can hope to find some upper bound C for $|c_n|$, $0 \leq n < N$, and then choose the step size h to satisfy

$$h = (\varepsilon/C)^{1/k}.$$

For this value h, we have

$$l_n \leq \varepsilon h = \varepsilon(x_{n+1} - x_n).$$

There is a serious drawback to using the preceding criterion as a choice for step size, even in the relatively rare circumstances when C can be computed, because the step size necessary to satisfy this condition, since it is a worse case [in terms of the (x_n, y_n) values], tends to be minuscule. This, in turn, entails an enormous number of steps.

By adaptively varying the step size as the computation proceeds, one can maintain a desired global accuracy while requiring vastly fewer steps than in the cited step size case. By way of illustration, we refer the reader to a discussion in Section 4.7 of Stroud (1974), where an example of a solution of the three-body problem is cited that required 106 Runge–Kutta steps when a variable step size was used. Observing that if a fixed step size were used, it would have to be the minimum of the variable step size values, Stroud estimates that to achieve the same accuracy by a fixed step size calculation, 62,000 steps instead of 106 would have been needed.

The plan for variable step size methods is:

(1) Compute an approximation $\hat{c}_n$ of c_n as defined in (8.31).
(2) Adjust $h = h_n$ so that

$$\hat{c}_n h_n^{k+1} = \varepsilon h_n = \varepsilon(x_{n+1} - x_n).$$

That is, take

$$h_n = (\varepsilon/\hat{c}_n)^{1/k}.$$

It is advisable to premultiply the right side of the above equation by a "safety factor" less than 1. Also, some bounds ought to be used to ensure that h_n does not become too large or small.

There are various techniques proffered in the literature for constructing the estimate $\hat{c}_n$. Dahlquist and Björck (1974) mention approximating the requisite derivatives of y by differences. Stroud (1974) describes a procedure due to Zonneveld, which approximates the needed kth derivative of y by an appropriate linear combination of the k_j terms computed in the course of the Runge–Kutta calculation (8.11). Gear (1971, Section 5.4) gives a related technique due to Merson. For an adaptive mesh-size algorithm for multistep methods, see Section 8.12 of Young and Gregory (1972).

The technique that we shall now outline is known as the *step-doubling* technique. It should become apparent to the reader that this method is closely allied to Richardson's extrapolation (Section 3.2.2). We have that for a kth-order method, the local error associated with a step size h at x_n is

$$c_n h^{k+1} + O(h^{k+2}).$$

It may be seen (e.g., Dahlquist and Björck, 1974, p. 351 and Section 7.2.2) that if we use step size $h/2$ twice with the same Runge–Kutta formula, the associated local error will be

$$2c_n(h/2)^{k+1} + O(h^{k+2}).$$

Letting y_{n+1} and y_{n+1}^* denote, respectively, the estimate based on step size h and step size $h/2$ (two steps), from the above equations we have

$$y_{n+1} - y_{n+1}^* = c_n h^{k+1}(1 - 2^{-k}) + O(h^{k+2}),$$

whence we obtain the estimate

$$\hat{c}_n = \frac{y_{n+1} - y_{n+1}^*}{h^{k+1}(1 - 2^{-k})}.$$

8.1.6. The Method of Quasilinearization

The method of quasilinearization, as described and developed by Bellman and Kalaba (1965), is an iterative scheme that has proven to be an interesting and valuable method for numerical solution of certain classes of ordinary and partial differential equations. The method of quasilinearization can properly be viewed as an application of Newton's method to function spaces, and as one might then hope, under suitable differentiability assumptions, it achieves a quadratic convergence rate, provided convergence occurs at all. As Bellman and Kalaba (1965) acknowledge, these developments were known to the Soviet mathematician L. V. Kanto-rovich. An important contribution due to Bellman and Kalaba (1965) is the exploitation of a monotonicity property that the iterative solutions obtained by quasilinearization sometimes exhibit. In this section, we shall describe the quasilinearization method and give sufficient conditions for quadratic convergence and monotonic convergence to occur.

Assume that the right-hand side of equation (8.1) is twice differentiable in y and that the partial derivatives f_y and f_{yy} are bounded. The *quasilinearization method* is the following iteration scheme: Let $y_0(x)$ be the initial approximation and for $k = 0, 1, 2, \ldots$ let $y_{k+1}(x)$ be the solution of the linear differential equation

$$y'_{k+1}(x) = f(x, y_k(x)) + [y_{k+1}(x) - y_k(x)]f_y(x, y_k(x)),$$
$$y_{k+1}(x_0) = y_0. \tag{8.32}$$

First we prove that for a sufficiently good initial approximation $y_0(x)$ and for a sufficiently small interval $[x_0, b]$, the sequence $\{y_k(x)\}$ obtained from quasilinearization converges to $y(x)$, and that the convergence is quadratic.

For a fixed value of x, Taylor's formula applied to $f(x, y(x))$ as a function of y implies that for some number η between $y(x)$ and $y_k(x)$,

$$y'(x) = f(x, y(x)) = f(x, y_k(x)) + [y(x) - y_k(x)]f_y'(x, y_k(x))$$
$$+ \frac{[y(x) - y_k(x)]^2}{2!}f_{yy}(x, \eta), \tag{8.33}$$

and combining the above with (8.32) we have

$$y'(x) - y'_{k+1}(x) = [y(x) - y_{k+1}(x)]f_y(x, y_k(x))$$
$$+ \frac{[y(x) - y_k(x)]^2}{2}f_{yy}(x, \eta). \tag{8.34}$$

Upon integrating both sides, we obtain

$$y(x) - y_{k+1}(x) = \int_{x_0}^{b} \left\{ [y(x) - y_{k+1}(x)]f_y(x, y_k(x)) + \frac{[y(x) - y_k(x)]^2}{2}f_{yy}(x, \eta) \right\} dx. \tag{8.35}$$

Assume that

$$|f_y| \le A, \qquad |f_{yy}| \le B,$$

and introduce the notation

$$\varepsilon_k = \max_{x \in [x_0, b]} |y(x) - y_k(x)| \qquad (k = 0, 1, 2, \ldots).$$

Equation (8.35) implies that

$$\varepsilon_{k+1} \leq (b - x_0)\varepsilon_{k+1}A + \varepsilon_k{}^2 B(b - x_0)/2.$$

Assume that $(b - x_0)A < 1$. Then

$$\varepsilon_{k+1} \leq \frac{B(b - x_0)}{2[1 - (b - x_0)A]}\, \varepsilon_k{}^2 = K \cdot \varepsilon_k{}^2. \qquad (8.36)$$

Let the initial approximation $y_0(x)$ satisfy the condition $K\varepsilon_0 < 1$. Then from (8.36) we conclude that

$$\delta_{k+1} \leq \delta_k{}^2 \leq \cdots \leq \delta_0{}^{2^{k+1}},$$

where $\delta_l = K \cdot \varepsilon_l$, $l \geq 0$. This proves the quadratic convergence of the quasilinearization method.

Roberts and Shipman (1972, Section 6.11) give some sufficient conditions for convergence, regardless of the length of the interval $[0, b]$. A proof of convergence, regardless of interval length, provided f is convex in y, will be given in the discussion to follow.

We shall now demonstrate that the solution sequence $\{y_k(x)\}$ obtained from the quasilinearization method is monotonic under certain important circumstances. Before proceeding to these developments, it is first necessary to show that the first-order linear differential equation

$$z'(x) = p(x)z(x) + q(x), \qquad z(x_0) = 0,$$

has the following *positivity* property: If $q(x) \geq 0$ for all $x \in [x_0, b]$, then $z(x) \geq 0$, $x \in [x_0, b]$. But positivity is apparent from inspection of the general solution formula (Kaplan, 1959, p. 447):

$$z(x) = e^{s(x)} \int_{x_0}^{x} e^{-s(v)} q(v)\, dv,$$

where $s(x) = \int_{x_0}^{x} p(v)\, dv$.

For the case that $f_{yy} \geq 0$ (i.e., f is convex in y) we prove that the sequence $\{y_k(x)\}$ determined by quasilinearization is monotone. First we demonstrate that equations (8.35) and (8.36) and the positivity property of the solutions of linear differential equations imply for $x \in [x_0, b]$, that

$$y(x) \geq y_{k+1}(x) \qquad (k = 0, 1, 2 \ldots). \qquad (8.37)$$

If we identify $y(x) - y_{k+1}(x)$ in (8.34) with $z(x)$, $f_y(x, y_k(x))$ with $p(x)$, and the nonnegative term $[y(x) - y_k(x)]^2 f_{yy}(x, \eta)$ with $q(x)$, then the positivity property implies the desired inequality (8.37).

Consider the defining equation for y_{k+1} and y_{k+2}. Since $f_{yy}(x, \eta)$ is nonnegative the Taylor expansion with second-order remainder implies the following inequality:

$$y'_{k+1}(x) = f(x, y_k(x)) + [y_{k+1}(x) - y_k(x)]f_y(x, y_k(x))$$
$$\leq f(x, y_{k+1}(x))$$
$$= f(x, y_{k+1}(x)) + [y_{k+1}(x) - y_{k+1}(x)]f_y(x, y_{k+1}(x)).$$

Thus, y_{k+1} is the solution of the equation

$$y'_{k+1}(x) = f(x, y_{k+1}(x)) + [y_{k+1}(x) - y_{k+1}(x)]f_y(x, y_{k+1}(x)) - h(x),$$
$$y(x_0) = y_0,$$

where $h(x) \geq 0$. Subtracting this equation from the recursive quasilinearization equation for $y_{k+2}(x)$, we have

$$y'_{k+2}(x) - y'_{k+1}(x) = [y_{k+2}(x) - y_{k+1}(x)]f_y(x, y_{k+1}(x)) + h(x).$$

Since $h(x)$ is nonnegative, application of the positivity property implies

$$y_{k+2}(x) \geq y_{k+1}(x),$$

which proves that

$$y_1(x) \leq y_2(x) \leq \cdots \leq y_{k+1}(x) \leq y_{k+2}(x) \leq \cdots \leq y(x). \quad (8.38)$$

Finally we prove that in the above convex case, the sequence $\{y_k(x)\}$ converges to $y(x)$ in the interval $[x_0, b]$ regardless of the length of this interval. Because of the monotonicity of the sequence $\{y_k(x)\}$ there is a limit function $\bar{y}(x)$. Equation (8.32) implies that

$$y_{k+1}(x) = y_0 + \int_{x_0}^{x} \{f(t, y_k(t)) + [y_{k+1}(t) - y_k(t)]f_y(t, y_k(t))\} \, dt.$$

For $k \to \infty$, using the limit theorem of bounded integrals (see McShane and Botts, 1959, p. 140) we conclude that $\bar{y}(x_0) = y_0$ and

$$\bar{y}(x) = y_0 + \int_{x_0}^{x} \left\{f(t, \bar{y}(t)) + [\bar{y}(t) - \bar{y}(t)]f_y(t, \bar{y}(t))\right\} dt$$
$$= y_0 + \int_{x_0}^{x} f(t, \bar{y}(t)) \, dt,$$

which implies that $\bar{y}$ is a solution of equation (8.1). Since this equation has a unique solution, $\bar{y} = y$.

The primary motivation of the quasilinearization is that linear differential equations are far easier to solve than nonlinear differential equations. In particular, as we have already noted, the linear first-order ordinary differential equation admits an exact integral solution. But the primary strength of the quasilinearization approach is in the solution of boundary-value problems. As we shall see (Section 8.2.1), linear boundary-value problems are easily solved, but direct methods for nonlinear boundary-value problems are virtually nonexistent. In Section 8.2.4, we shall pursue the quasilinearization approach of converting a nonlinear boundary-value problem into a sequence of linear boundary-value problems.

The reader is urged to consult Bellman and Kalaba (1965) for further discussion of the theory and application of the quasilinearization technique as well as sample computations. Roberts and Shipman (1972) contains further theoretical developments.

Problem

8.13. Apply the method of quasilinearization to the Riccati equation

$$y' = y^2 + f(x)y + g(x), \qquad y(x_0) = y_0,$$

where the functions f and g are continuous on the interval $[x_0, b]$. Derive the form of the linear differential equation.

8.2. The Solution of Boundary-Value Problems

8.2.1. Reduction to Initial-Value Problems

Linear boundary-value problems usually can be reduced to initial-value problems. For example, consider the linear boundary-value problem

$$Ly = y'' + p(x)\,y' + q(x)\,y = r(x),$$
$$y(a) = A, \qquad y(b) = B, \tag{8.39}$$

where functions p, q, and r are continuous on the interval $[a, b]$.

First solve the homogeneous initial-value problem

$$Lu = u'' + p(x)u' + q(x)u = 0, \qquad u(a) = 0, \quad u'(a) = 1,$$

and then solve the inhomogeneous problem

$$Lz = z'' + p(x)z' + q(x)z = r(x), \qquad z(a) = A, \quad z'(a) = 0.$$

We now see that the function

$$y(x) = z(x) + \frac{B - z(b)}{u(b)} u(x)$$

is the solution of the problem (8.39), because

$$Ly = Lz + \frac{B - z(b)}{u(b)} Lu = r + 0 = r,$$

$$y(a) = z(a) + \frac{B - z(b)}{u(b)} u(a) = z(a) + 0 = A,$$

$$y(b) = z(b) + \frac{B - z(b)}{u(b)} u(b) = B.$$

8.2.2. The Method of Undetermined Coefficients

Consider the boundary-value problem for second-order nonlinear differential equations:

$$F(x, y, y', y'') = 0, \qquad y(a) = A, \quad y(b) = B. \tag{8.40}$$

Assume that the twice-differentiable functions $\phi_0, \phi_1, \phi_2, \ldots, \phi_m$ having the properties

$$\phi_0(a) = A, \qquad \phi_0(b) = B,$$
$$\phi_i(a) = \phi_i(b) = 0, \qquad 1 \le i \le m,$$

are given. For arbitrary coefficients $a_1, a_2, \ldots, a_m$, the function

$$\phi(x) = \phi_0(x) + \sum_{i=1}^{m} a_i \phi_i(x).$$

obviously satisfies the boundary conditions. The unknown coefficients are then determined so that ϕ satisfies the equation with "sufficient accuracy." The method for determining the coefficients, of course, depends on the definition of the "accuracy" of approximating solutions of the equation (8.40). Below we describe three well-known techniques for computing the coefficients $a_1, \ldots, a_n$, namely, the collocation method, the method of moments, and the least squares method.

1. Let $x_1, x_2, \ldots, x_m$ be different points in the interval $[a, b]$. In the *collocation method*, the coefficients a_i, $1 \leq i \leq m$, are determined by the requirement that the function ϕ has to exactly satisfy the differential equation at the points x_i, $1 \leq i \leq m$; that is, the coefficients $a_1, \ldots, a_m$ are selected so that ϕ satisfies the system of multivariate equations

$$F\left(x_i, \phi(x_i), \phi'(x_i), \phi''(x_i)\right) = 0, \qquad 1 \leq i \leq m. \qquad (8.41)$$

This system can be solved by using the methods of Chapter 5.

Example 8.6. Consider the problem

$$y'' - (1 + x^2) y = 1, \qquad y(0) = 1, \quad y(1) = 3.$$

For $m = 2$, let $x_1 = 1/6$, $x_2 = 5/6$, and

$$\phi_0(x) = 1 + 2x,$$
$$\phi_1(x) = x - x^2,$$
$$\phi_2(x) = x(x - x^2).$$

Thus the approximate solution has the form

$$\phi(x) = 1 + 2x + a_1(x - x^2) + a_2 x(x - x^2).$$

Since

$$\phi'(x) = 2 + a_1 - 2a_1 x + 2a_2 x - 3a_2 x^2,$$
$$\phi''(x) = -2a_1 + 2a_2 - 6a_2 x,$$

equations (8.41) can be written as

$$\frac{2777}{6^4} a_1 - \frac{7591}{6^5} a_2 = -\frac{512}{6^3},$$

$$\frac{2897}{6^4} a_1 + \frac{24853}{6^5} a_2 = -\frac{1192}{6^3},$$

which implies $a_1 = -1.43391$ and $a_2 = -0.72378$.

2. The *method of moments* requires that the coefficients a_i, $1 \leq i \leq m$, be chosen so that the equations

$$\int_a^b F\left(x, \phi(x), \phi'(x), \phi''(x)\right) u_i(x) \, dx = 0, \qquad 1 \leq i \leq m, \qquad (8.42)$$

hold. In these equations, the functions, $u_1, u_2, \ldots, u_m$ are linearly independent but otherwise arbitrarily chosen. The term "moment" stems from the historical convention of taking $u_i = x^{i-1}$.

3. The *method of least squares* reduces the determination of the coefficients to the minimization of the *m*-variate function

$$J(a_1, \ldots, a_m) = \int_a^b F^2\big(x, \phi(x), \phi'(x), \phi''(x)\big)\, dx. \tag{8.43}$$

The minimization problem can often be solved by the numerical solution of the extremal equations

$$\frac{\partial}{\partial a_i} \int_a^b F^2\big(x, \phi(x), \phi'(x), \phi''(x)\big)\, dx = 0. \tag{8.44}$$

Alternatively, other techniques from Chapter 5 may be employed.

Problems

8.14. Let $m = 1$ and $u_1 \equiv 1$. Solve the problem of Example 8.6 by the method of moments.

8.15. Solve the problem discussed in Example 8.6 by the method of least squares.

8.16. Solve the boundary-value problem

$$y'' - xy' + y = 1, \qquad y(0) = y(2) = 0,$$

by a reduction to initial-value problems.

8.2.3. The Difference Method

Consider again the boundary-value problem (8.39). Let n be a positive integer and let $h = (b - a)/n$ and $x_k = a + hk$ $(k = 0, 1, \ldots, n)$. In Section 3.1, we have seen that for centered difference approximations,

$$y'(x_k) = \frac{y(x_{k+1}) - y(x_{k-1})}{2h} + O(h^2), \tag{8.45}$$

$$y''(x_k) = \frac{y(x_{k+1}) - 2y(x_k) + y(x_{k-1})}{h^2} + O(h^2). \tag{8.46}$$

Let $p_k = p(x_k)$, $q_k = q(x_k)$, and $r_k = r(x_k)$ $(k = 0, 1, \ldots, n)$. The boundary conditions imply that $y(x_0) = A$, and $y(x_n) = B$, and the differential equation and the identities (8.45) and (8.46) imply that

$$\frac{y(x_{k+1}) - 2y(x_k) + y(x_{k-1})}{h^2} + p_k \frac{y(x_{k+1}) - y(x_{k-1})}{2h} + q_k y(x_k) + O(h^2)$$

$$= r(x_k), \qquad 1 \le k \le n - 1.$$

Neglecting the term $O(h^2)$, one obtains the system of linear algebraic equations

$$y_{k+1}(2 + hp_k) + y_k(2h^2q_k - 4) + y_{k-1}(2 - hp_k) = 2h^2r_k, \qquad 1 \le k \le n - 1,$$
$$(8.47)$$

for the approximating functional values y_k, $1 \le k \le n - 1$. Assume that for $x \in [a, b]$, $q(x) < 0$, and $k = 1, 2, \ldots, n - 1$,

$$\left| p_k \frac{h}{2} \right| \le 1.$$

Equation (8.47) can be rewritten in matrix form as

$$y = \mathbf{B}y + \mathbf{c},$$

where $y = (y_1, y_2, \ldots, y_{n-1})^T$,

$$\mathbf{B} = \begin{pmatrix} 0 & c_1 & & & 0 \\ b_2 & 0 & \cdot & & \\ & \cdot & \cdot & \cdot & \\ & & \cdot & 0 & c_{n-2} \\ 0 & & & b_{n-1} & 0 \end{pmatrix}, \qquad \mathbf{c} = \begin{pmatrix} d_1 - Ab_1 \\ d_2 \\ \vdots \\ d_{n-2} \\ d_{n-1} - Bc_{n-1} \end{pmatrix},$$

$$b_k = \frac{2 - hp_k}{4 - 2h^2q_k}, \qquad c_k = \frac{2 + hp_k}{4 - 2h^2q_k}, \qquad d_k = \frac{-2h^2r_k}{4 - 2h^2q_k},$$
$$1 \le k \le n - 1.$$

Obviously $\| \mathbf{B} \|_\infty < 1$, and consequently from the contraction mapping theorem (Theorem 4.5) equations (8.47) have a unique solution and Theorem 6.3 implies that the solution can be obtained using iteration methods of Chapter 6. It is generally preferable, however, to instead solve the tridiagonal system $(\mathbf{I} - \mathbf{B})y = \mathbf{c}$ by Gauss elimination. For such systems, the number of arithmetic operations grows proportionally to n, as has been noted. The difference method just described is immediately extensible to higher-order linear boundary-value problems.

Example 8.7. Consider again the problem

$$y'' - (1 + x^2) y = 1, \qquad y(0) = 1, \quad y(1) = 3.$$

Then equation (8.47) has the form

$$y_{k-1} - (2 + h^2 + k^2h^4) y_k + y_{k+1} = h^2, \qquad 1 \le k \le n - 1,$$

and for $h = 0.1, 0.05$, we have the solution vectors in Table 8.7.

Table 8.7. A Difference Method Computation

x	Approximate		Exact
	$h = 0.1$	$h = 0.05$	
0.0	1.0000	1.0000	1.0000
0.1	1.0746	1.0744	1.0743
0.2	1.1701	1.1696	1.1695
0.3	1.2877	1.2871	1.2869
0.4	1.4294	1.4286	1.4284
0.5	1.5976	1.5968	1.5965
0.6	1.7958	1.7950	1.7947
0.7	2.0285	2.0276	2.0274
0.8	2.3013	2.3006	2.3004
0.9	2.6219	2.6215	2.6214
1.0	3.0000	3.0000	3.0000

Problem

8.17. Solve the boundary-value problem

$$y'' - xy' + y = 1, \qquad y(0) = y(2) = 0,$$

by using the difference method for $h = 1$.

8.2.4. The Method of Quasilinearization

Consider now the nonlinear two-point boundary-value problem

$$y'' = f(x, y, y'), \qquad y(a) = A, \quad y(b) = B. \qquad (8.48)$$

Let $y_0(x)$ be some initial approximation and consider the sequence $\{y_n\}$ of functions determined by the (Taylor's series motivated) recurrence relation

$$y_{k+1}''(x) = f_{y'}(x, y_k(x), y_k'(x))[y_{k+1}'(x) - y_k'(x)]$$
$$+ f_y(x, y_k(x), y_k'(x))[y_{k+1}(x) - y_k(x)] + f(x, y_k(x), y_k'(x)),$$
$$y_{k+1}(a) = A, \qquad y_{k+1}(b) = B. \qquad (8.49)$$

As was done for the initial-value problems, it can be proved that for a sufficiently small interval $[a, b]$ and for a sufficiently good initial approximation

y_0, the method has quadratic convergence (see Bellman and Kalaba, 1965, pp. 36–44). One approach to filling in the details of the convergence proof is just to convert the second-order differential equation to a two-dimensional system of first-order equations (as shown in Section 8.1) and then note that the quasilinearization convergence proofs given in Section 8.1.6 are applicable in the vector case also. Problems (8.49), because they are linear differential equations, can be solved much more easily than the original nonlinear differential equation (8.48).

We can use the technique given in Section 8.2.1, for example, at each iteration k to solve (8.49). (In this connection, recall that by using a device given at the beginning of Section 8.1, techniques in that section can be made applicable to second-order, initial-value problems.) Alternatively, the linear equation conversion described in the preceding section may be employed. Numerically efficient solution of nonlinear boundary-value problems is one of the principal motivations and merits of the quasilinearization method.

8.3. The Solution of Eigenvalue Problems

Consider the Sturm–Liouville problem:

$$[p(x)\,y']' - q(x)\,y + \lambda r(x)\,y = 0, \qquad y(a) = y(b) = 0, \qquad (8.50)$$

where for $x \in [a, b]$, $p(x) > 0$ and $q(x) > 0$, and numbers λ and functions y that satisfy (8.50) are to be determined. By using the difference method, (8.50) can be approximated by the equations

$$p_k \frac{y_{k+1} - 2y_k + y_{k-1}}{h^2} + p_k' \frac{y_{k+1} - y_{k-1}}{2h} - q_k y_k + \lambda r_k y_k = 0,$$

$$1 \le k \le n - 1, \qquad (8.51)$$

where we use the notation of Section 8.2.3 and $p_k' = p'(x_k)$, $1 \le k \le n - 1$. Let

$$b_k = -\frac{p_k'}{r_k h^2} + \frac{p_k'}{2r_k h}, \qquad 2 \le k \le n - 1,$$

$$a_k = \frac{q_k}{r_k} + \frac{p_k}{r_k h^2}, \qquad 1 \le k \le n - 1,$$

$$c_k = \frac{p_k}{r_k h^2} - \frac{p_k'}{2r_k h}, \qquad 1 \le k \le n - 2,$$

then equations (8.51) are equivalent to the eigenvalue problem of the tridiagonal matrix

$$\mathbf{A} = \begin{pmatrix} a_1 & c_1 & & & & & 0 \\ b_2 & a_2 & \cdot & & & & \\ & \cdot & \cdot & \cdot & & & \\ & & \cdot & \cdot & \cdot & & \\ & & & \cdot & a_{n-2} & c_{n-2} & \\ 0 & & & & b_{n-1} & a_{n-1} \end{pmatrix},$$

which can be solved using the methods of the Chapter 7. The eigenvalues of **A** give the approximate eigenvalues of the problem (8.50) and the components of the eigenvectors of **A** can be regarded as discrete-time approximations of the eigenfunctions.

Example 8.8. Consider the problem

$$y'' = -\lambda y, \qquad y(0) = y(1) = 0.$$

For $n = 1$, matrix **A** is a scalar, $\mathbf{A} = (8)$, where the only eigenvalue is $\lambda = 8$. For $n = 2$ we have

$$\mathbf{A} = \begin{pmatrix} 18 & -9 \\ -9 & 18 \end{pmatrix},$$

and the eigenvalues are $\lambda_1 = 9$, $\lambda_2 = 27$. For $n = 3$ we get

$$\mathbf{A} = \begin{pmatrix} 32 & -16 & 0 \\ -16 & 32 & -16 \\ 0 & -16 & 32 \end{pmatrix},$$

and the eigenvalues are as follows:

$$\lambda_1 \approx 9.37, \qquad \lambda_2 = 32, \qquad \lambda_3 \approx 54.6.$$

By application of the difference method to solution of boundary-value problems and eigenvalue problems, we reduced the differential equation problem to the problem of numerical solution of, respectively, simultaneous linear equations and eigenvalue problems of matrices. These reduced problems were obtained by neglecting the $O(h^2)$ term. Consequently the values of the exact solution and the exact eigenvalues have to satisfy either simultaneous linear equations or eigenvalue problems for matrices, where the parameters are approximations of the parameters of the reduced problems. Using the order of the neglected term we can estimate the accuracy of these approximate parameters. Thus the error analysis of the difference method can be reduced to the estimation of the errors of the solutions of approximate simultaneous equations and the same is true for eigenvalues of approximate matrices. These problems were discussed in Sections 6.3 and 7.5. The application of the results of those sections to error analysis of the difference method is not discussed here, but we mention that no further theoretical developments are required for such analysis.

8.4. Supplementary Notes and Discussion

We consider it our pedagogic duty to briefly seek common structural underpinnings for the widely diverse methods we have considered in this chapter. The Taylor series method, Picard iteration, and quasilinearization have their foundation in classical *mathematical analysis*. The first two of these topics are standard fare in textbooks in analysis and the latter method is classified as a method of analysis since it consists basically in iterative application of an operator on a function space. Also, the derivations of the properties of quasilinearization are more characteristic of those in mathematical analysis. By way of contrast, the other methods we have considered are more peculiar to the numerical analysis literature. The Runge–Kutta, multistep, and difference methods are all *discretization techniques* in that all depend on approximating a continuous problem by a discrete-point problem. The method of undetermined coefficients (that is, the collocation method, the method of moments, and the least squares method) are *approximation methods*. The exact solution is approximated by a finite linear combination of known functions. Inasmuch as an optimization problem arises in finding the coefficients giving the best fit (with respect to some metric) they are sometimes spoken of as *variational methods*. These methods have their basis in developments in Chapter 2, and will play a larger role in Chapter 9, where we describe the Ritz, Galerkin, and finite-element techniques.

Among those topics concerning numerical integration of ordinary differential equations that we have had to omit due to constraints of scope and length of this text, the most regrettable omission is the subject of stiff ordinary differential equations. Consider the vector-valued ordinary differential equation (8.1), which we repeat,

$$\mathbf{y}'(x) = \mathbf{f}(x, \mathbf{y}(x)), \qquad \mathbf{y}(0) = \mathbf{y}_0 ,$$

and let $|\lambda_1| \leq |\lambda_2| \leq \cdots \leq |\lambda_m|$ denote the magnitudes of the eigenvalues of $J(x, \mathbf{y})$, where $J(x, \mathbf{y})$ is the Jacobian matrix $\{(\partial/\partial y_j)f_i(x, \mathbf{y})\}_{i,j}$ for system (8.1). System (8.1) is said to be a *stiff* differential equation if the ratio

$|\lambda_m|/|\lambda_1|$ is large. Stiff differential equations arise in many applications (e.g., chemical reactions with some component reactions having a much longer time constant than others). For any given numerical method, stability and accuracy require the step size h to satisfy the inequality

$$|\lambda_j h| \leq K,$$

for each eigenvalue λ_j, the constant K depending on the method. For systems with large (in magnitude) eigenvalues, h must be chosen to be quite small, and yet if it also has small eigenvalues, then the time domain for the required solution is typically rather long. The reader faced with the need to solve stiff differential equations should consider methods explicitly devised for this task given in Gear (1971, Chapter 11) and Lapidus and Seinfeld (1971, Chapter 6).

For further details and comparative computational results of numerical solutions of differential equations, we refer the reader to the books by Gear (1971) and Lapidus and Seinfeld (1971), and the esteemed, but now somewhat dated book by Collatz (1960). The first two books give some principles and guidance for choosing which method to use in a given instance. In particular, Gear (1971, Chapter 5) shows that for a given step size, higher-order methods do not always yield greater accuracy.

Modern developments of the methodology for two-point boundary-value problems include Bailey *et al.* (1968), Bellman and Kalaba (1965), Keller (1968), and Roberts and Shipman (1972). The method most often cited for this class of problems is the *shooting method*, which consists of iteratively adjusting the initial conditions until the boundary conditions are satisfied. Inasmuch as this technique is generally ill conditioned (a slight perturbation in the initial conditions yielding a large change in the terminal state), we have found it to be less satisfactory than the quasilinearization method. Finite-element techniques, to be described in Chapter 9, are also relevant for ordinary differential equation boundary-value problems. A recent computer-oriented technique that also seems to hold some promise is the invariant embedding method, which provides ways of converting boundary-value problems into initial-value problems. In contrast to the quasilinearization method, invariant embedding algorithms are direct rather than iterative. Books on this subject include Bellman and Wing (1975), Casti and Kalaba (1973), and Scott (1973).

9

The Numerical Solution of Partial Differential Equations

The principal avenues of numerical solution of partial differential equations are the difference method and the Ritz, Galerkin, and finite-element methods. We have already employed the difference method in the ordinary differential equation case (Section 8.2.3).

As will be seen, the Ritz–Galerkin, finite-element approach is also applicable to ordinary differential equations, but we have delayed its discussion partly to unify this discussion and partly because the efficiency of this approach in the ordinary differential equation case has not been convincingly established. We treat these subjects together because the finite-element method is, mathematically speaking, a particular application of the Ritz or Galerkin method.

The methodology for numerical solution of partial differential equations tends to be much more subtle than solution of ordinary differential equations, and one finds that strong results concerning error bounds, convergence, and stability are dependent on the characteristics of the particular partial differential equation being studied. Especially lacking are general methods and convergence bounds in the nonlinear case. This weakness, diversity, and specificity of results parallels and, in part, stems from a similar condition in the analytic theory of partial differential equations (as evinced in, for example, Sneddon, 1957).

The most convenient numerical techniques used for partial differential equations are instances of either the difference method or the method of finite elements, both of which will be described in this chapter. Loosely speaking, the latter method tends to become more advantageous as the shape of the domain of the partial differential equation becomes further removed from rectangular.

9.1. The Difference Method

For the purpose of simplifying the notation of the methods to follow, they are presented only for the bivariate case, but there is no difficulty in extending these principles to higher dimensions. The spatial variables will be denoted by x and y, and $u(x, y)$ will represent the exact solution of the partial differential equation. The symbol G will denote the domains of the partial differential equations.

As in the case of ordinary differential equations, in the difference method applied to partial differential equations the derivatives are approximated by difference formulas introduced in Section 3.1, and by neglecting the error terms, simultaneous difference equations can be obtained. We now consider difference equation algorithms for several prominent partial differential equation problems.

First, consider the *parabolic partial differential equation*

$$\frac{\partial u}{\partial y} = \frac{\partial^2 u}{\partial x^2} \tag{9.1}$$

on the domain $G = [0, a] \times [0, \infty)$, where a is a fixed positive constant, and let the boundary conditions be given by

$$u(x, 0) = f(x), \qquad 0 \le x \le a,$$
$$u(0, y) = g_1(y), \qquad u(a, y) = g_2(y), \qquad y \ge 0, \tag{9.2}$$

where f, g_1, g_2, are given. Cover the domain G with rectangles having dimensions h and k as shown in Figure 9.1, where h and k are fixed positive numbers. Let $x_i = ih, y_j = jk, 0 \le i \le n, j \ge 0$. By using the formulas of Section 3.1, we obtain

$$\frac{\partial u(x_i, y_j)}{\partial y} = \frac{u(x_i, y_{j+1}) - u(x_i, y_j)}{k} + O(k), \qquad 0 \le i \le n, \quad j \ge 0,$$

$$\frac{\partial^2 u(x_i, y_j)}{\partial x^2} = \frac{u(x_{i+1}, y_j) - 2u(x_i, y_j) + u(x_{i-1}, y_j)}{h^2} + O(h^2),$$

$$1 \le i \le n - 1, \quad j \ge 0.$$

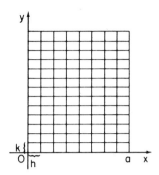

Fig. 9.1. Discrete grid for a parabolic equation.

By neglecting the error terms $O(k)$ and $O(h^2)$, we obtain the following equations for the approximate functional values u_{ij}:

$$u_{i,j+1} = \frac{k}{h^2}(u_{i+1,j} + u_{i-1,j}) + \left(1 - \frac{2k}{h^2}\right)u_{i,j}, \qquad 1 \leq i \leq n-1, \quad j \geq 0. \tag{9.3}$$

Observe that the equations become much simpler for $k = h^2/2$:

$$u_{i,j+1} = \tfrac{1}{2}(u_{i+1,j} + u_{i-1,j}), \tag{9.4}$$

and boundary condition (9.2) implies that

$$u_{i,0} = f(ih), \qquad 0 \leq i \leq n. \tag{9.5}$$

Thus starting from the values $u_{i,0}$ and repeating the use of the formula (9.4) for $j = 0, 1, 2, \ldots$, one obtains the approximating values u_{ij} of the solution. Since in each iteration of a new value of j the number of functional values that can be computed by equation (9.4) decreases by 2, we cannot compute all the functional values u_{ij}, $0 \leq i \leq n, j \geq 0$, but only the values in the triangle $(j \geq 0, j \leq i \leq n - j)$ can be determined.

By using the additional boundary conditions of (9.2), namely,

$$u(0, y) = g_1(y), \qquad u(a, y) = g_2(y), \qquad y \geq 0,$$

one can obtain all the functional values at the points (x_i, y_j) lying in G.

Example 9.1. Consider the equation

$$\frac{\partial u}{\partial y} + \frac{\partial^2 u}{\partial x^2}, \qquad 0 \leq x \leq 2, \quad 0 \leq y \leq \tfrac{3}{8},$$

Table 9.1. A Finite-Difference Solution for a Parabolic Equation

y	x				
	0.0	0.5	1	1.5	2
0	0.0000	0.0000	0.0000	0.0000	0.0000
1/8	0.8660	0.0000	0.0000	0.0000	0.0000
2/8	0.8660	0.4330	0.0000	0.0000	0.0000
3/8	0.0000	0.4330	0.2165	0.0000	0.0000

with the boundary conditions

$$u(x, 0) = 0, \qquad u(0, y) = \sin \frac{8\pi}{3} y, \qquad u(2, y) = 0.$$

Let $h = \frac{1}{2}$. Then $k = h^2/2 = \frac{1}{8}$. The approximate functional values are shown in Table 9.1.

The quantity $r = k/h^2$ is known as the *mesh ratio*. It is known [e.g., Section 17.2 of Young and Gregory (1972)] that if r is appreciably larger than $\frac{1}{2}$, the difference method may not converge. If $r \leq \frac{1}{2}$, then convergence is ensured, but if r is appreciably less than $\frac{1}{2}$, excessively many steps may be needed. Under the circumstance that $r \leq \frac{1}{2}$, the difference between the approximation $u_{i,j}$ and u, at the mesh points, decreases as $O(h^2)$.

Suppose, for the moment, that we take differences of the parabolic equation (9.1) with respect to the x variable only, obtaining the following system of ordinary differential equations:

$$\frac{du_i(y)}{dy} = \frac{u_{i+1}(y) + u_{i-1}(y) - 2u_i(y)}{h^2}, \qquad 1 \leq i < n,$$

$$u_i(0) = f(ih), \qquad u_0(y) = g_1(y), \qquad u_n(y) = g_2(y).$$

This system is amenable to solution by the methods discussed in Chapter 8. This viewpoint is called the *method of lines*. For example, if we apply Euler's method to this system, we obtain the system of differences (9.3) that we have studied in this section. On the other hand, application of the modified Euler method gives us the system

$$u_{i,j+1} = u_{i,j} + [u_{i+1,j+1} + u_{i-1,j+1} - 2u_{i,j+1} + u_{i+1,j} + u_{i-1,j} - 2u_{i,j}] \frac{k}{2h^2}.$$

This method, as a scheme for partial differential equations, is known as the

Crank–Nicholson method. Like the modified Euler method in Chapter 8, it is an implicit method. But since the equation is linear, this is of no computational consequence.

Next consider the *hyperbolic partial differential equation*

$$\frac{\partial^2 u}{\partial x^2} = \frac{\partial^2 u}{\partial y^2}, \qquad 0 \le x \le a, \quad y \ge 0, \tag{9.6}$$

with the boundary conditions

$$u(x, 0) = f(x), \qquad \frac{\partial u(x, 0)}{\partial y} = g(x).$$

After using formula (3.12) for numerical differentiation and neglecting the error terms, we have

$$\frac{u_{i+1,j} - 2u_{ij} + u_{i-1,j}}{h^2} = \frac{u_{i,j+1} - 2u_{ij} + u_{i,j-1}}{k^2},$$

where the meaning of the symbols u_{ij} is the same as in the parabolic case. The equation has a simpler form for $k = h$:

$$u_{i,j+1} = u_{i+1,j} + u_{i-1,j} - u_{i,j-1}. \tag{9.7}$$

The boundary conditions imply that

$$u_{i,0} = f(ih), \tag{9.8}$$

and upon neglecting the error term that

$$\frac{u_{i,1} - u_{i,-1}}{2h} = g(ih). \tag{9.9}$$

Since the point $(ih, -h)$ does not belong to the domain G, we cannot use the values $u_{i,-1}$ during the calculation. Let us consider (9.7) for $j = 1$:

$$u_{i,1} = u_{i+1,0} + u_{i-1,0} - u_{i,-1},$$

which can be combined with (9.9) to obtain

$$u_{i,1} = \frac{f((i+1)h) + f((i-1)h)}{2} + hg(ih). \tag{9.10}$$

Thus the values $u_{i,j}$ for $j = 0$ are given by (9.8); for $j = 1$ they can be computed by (9.10), and for $j \ge 2$ they can be obtained by the repeated application of (9.7).

Table 9.2. A Finite-Difference Solution for a Hyperbolic Equation

x	y				
	0	0.25	0.50	0.75	1
0.00	1.0000	0.9375	0.7500	0.4375	0.0000
0.25	0.9375	0.8750	0.6875	0.3750	0.0000
0.50	0.7500	0.6875	0.5000	0.2500	0.0000
0.75	0.4375	0.3750	0.2500	0.1250	0.0000

It may be seen (Collatz, 1960, Section 3.2) that the finite-difference method error at a fixed domain point tends to 0 at the rate $O(h^2)$ for the hyperbolic partial differential equation case.

Example 9.2. We solve the problem

$$\frac{\partial^2 u}{\partial x^2} = \frac{\partial^2 u}{\partial y^2}, \qquad 0 \leq x \leq 1, \quad y \geq 0,$$

$$u(x, 0) = 1 - x^2, \qquad u(0, y) = 1 - y^2, \qquad u(1, y) = 0, \qquad \frac{\partial u(x, 0)}{\partial y} = 0.$$

Let $h = k = \frac{1}{4}$. Since $f(x) = 1 - x^2$ and $g(x) \equiv 0$,

$$u_{i,0} = 1 - \frac{i^2}{16} = \frac{16 - i^2}{16},$$

and by using (9.10) we obtain

$$u_{i,1} = \frac{1 - (i+1)^2/16 + 1 - (i-1)^2/16}{2} = \frac{15 - i^2}{16}.$$

The values $u_{i,j}$, $j \geq 2$, can be computed by the repeated application of equation (9.7). The finite-difference approximating values are shown in Table 9.2.

Now consider the *elliptic partial differential equation*

$$\Delta u = \frac{\partial^2 u}{\partial x^2} + \frac{\partial^2 u}{\partial y^2} = 0, \qquad a \leq x \leq A, \quad b \leq y \leq B,$$

with the boundary conditions

$$u(a, y) = g_1(y), \qquad u(A, y) = g_2(y), \qquad u(x, b) = f_1(x), \qquad u(x, B) = f_2(x).$$

Assume that $A - a = mh$ and $B - b = nh$, where m and n are positive integers and h is a positive number. Let $x_i = a + ih$, $0 \le i \le m$, $y_j = b + jh$, $0 \le j \le n$, and let u_{ij} denote the approximating functional values $u(x_i, y_j)$ obtained by the application of the method. By using formula (3.6) and upon neglecting the error terms we find that

$$\frac{u_{i+1,j} - 2u_{i,j} + u_{i-1,j}}{h^2} + \frac{u_{i,j+1} - 2u_{i,j} + u_{i,j-1}}{h^2} = 0, \qquad (9.11)$$

which implies

$$u_{i,j} = \tfrac{1}{4}(u_{i+1,j} + u_{i-1,j} + u_{i,j+1} + u_{i,j-1}), \qquad 1 \le i \le m-1, \; 1 \le j \le n-1. \tag{9.12}$$

Since from the boundary conditions the values $u_{0,j}$, $u_{m,j}$, $u_{i,0}$, $u_{i,n}$ are known, the number of equations (9.12) is equal to the number of the unknowns. It can be proven (Collatz, 1960, p. 348) that these simultaneous linear equations have a unique solution. This solution can be computed by using the methods described in Chapter 6. From a proof to be found in Young and Gregory (1972, Section 15.4) as well as Collatz (1960, Section 5,1.4), it is known that the error of the finite difference approximation is $O(h^2)$ at each point $(x, y) \in G$. Young and Gregory (1972, Section 15.5) also give finite-difference formulas that achieve higher orders of convergence for elliptic partial differential equations.

Example 9.3. Consider the problem (see Figure 9.2)

$$\frac{\partial^2 u}{\partial x^2} + \frac{\partial^2 u}{\partial y^2} = 0, \qquad u(x, \pm 1) = x^2, \qquad u(\pm 1, y) = y^2.$$

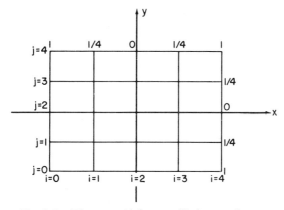

Fig. 9.2. Discrete grid for an elliptic equation.

Let $h = \frac{1}{4}$. Then $m = n = 4$. The boundary conditions imply that

$$u_{00} = u_{40} = u_{04} = u_{44} = 1,$$

$$u_{01} = u_{10} = u_{03} = u_{30} = u_{41} = u_{14} = u_{43} = u_{34} = \tfrac{1}{4},$$

$$u_{02} = u_{20} = u_{42} = u_{24} = 0.$$

For the nine unknowns, nine equations can be obtained from (9.12), but the symmetric properties of the equation and the boundary conditions imply that

$$u_{11} = u_{13} = u_{31} = u_{33}, \qquad u_{12} = u_{32} = u_{21} = u_{23}.$$

As a result the problem can be reduced to finding the solution of the three simultaneous linear equations

$$u_{11} = \tfrac{1}{4}\,(u_{12} + u_{10} + u_{21} + u_{01}) = \tfrac{1}{4}\,(\tfrac{1}{2} + 2u_{21}),$$

$$u_{22} = \tfrac{1}{4}\,(u_{21} + u_{23} + u_{12} + u_{32}) = u_{21},$$

$$u_{12} = \tfrac{1}{4}\,(u_{11} + u_{13} + u_{02} + u_{22}) = \tfrac{1}{4}\,(2u_{11} + u_{22}),$$

having three unknowns. After solving these equations we have

$$u_{11} = \tfrac{3}{16}, \qquad u_{22} = u_{21} = \tfrac{1}{8}.$$

The examples shown in this section were linear partial differential equations, which could be approximated by simultaneous linear algebraic equations. For nonlinear differential equations the approximating simultaneous equations are usually nonlinear with many unknowns. Only in very special cases can their numerical solution be found without great difficulty. Perhaps the most suitable approach currently available for the nonlinear case is the quasilinearization method (Section 9.2).

In the above examples we assumed that the boundary of the domain of the differential equations consists of straight lines that are parallel to the x and y axes, and in these cases the boundary conditions could be satisfied easily. In the case of domains of more general forms we may not calculate the boundary values by simple substitution, because the points (x_i, y_j) usually do not lie on the boundary. Let (x_i, y_j) be a point such that at least one of the four neighboring points do not belong to the domain. For the approximation of $u(x_i, y_j)$ we can use one of the following methods:

(a) Let (x, y) be the closest point to (x_i, y_j) lying on the boundary. Then let $u_{ij} = f(x, y)$, where f is given as the solution on the boundary by the boundary conditions.

(b) Let $(x^{(\nu)}, y^{(\nu)})$ $(\nu = 1, 2, \ldots, p)$ be different points on the boundary of the domain, and use a suitable bivariate approximating polynomial based on the points $(x^{(\nu)}, y^{(\nu)})$. Evaluate the polynomial at (x_i, y_j) and use this as the approximation for $u(x_i, y_j)$. We refer the reader to Collatz (1960) and, for the elliptic case, Panov (1963), for more details.

However, the finite-element method, to be described in Section 9.3, is especially suited for handling partial differential equations whose domain has irregular shape.

Problems

9.1. Find the counterpart to the difference system (9.3) for the more general form of the parabolic partial differential equation

$$\frac{\partial u}{\partial t} = A(x, t)\frac{\partial^2 u}{\phi x^2} + B(x, t)\frac{\partial u}{\phi x} + C(x, t)u.$$

9.2. Show how to modify the difference scheme (9.12) for the inhomogeneous case (known as *Poisson's equation*):

$$\partial^2 u/\partial x^2 + \partial^2 u/\partial y^2 = H(x, y).$$

9.3. Describe the solution of the hyperbolic equation (9.6) by the difference equation method under the circumstances that the domain G is the union of the two rectangles R_1 and R_2, where $R_1 = [0, 2] \times [0, 1]$ and $R_2 = [0, 1] \times [0, 2]$. Having chosen grid points, set up the difference equations and verify that in your system, the number of unknowns and the number of linear equations are equal.

9.4. Use system (9.11) to solve the elliptic equation on $G = [0, 1] \times [0, 3]$. The boundary conditions are $u(0, y) = 1$, $u(1, y) = 2$, and the initial condition is $u(x, 0) = x$. Take $h = 0.05$, and the mesh ratio $= 1/2$.

9.2. The Method of Quasilinearization

Toward simplifying the discussion, consider the nonlinear parabolic equation

$$\frac{\partial u}{\partial y} = \frac{\partial^2 u}{\partial x^2} + H\left(u, \frac{\partial u}{\partial x}\right), \qquad 0 \le x \le a, \quad y \ge 0, \qquad (9.13)$$

with the boundary conditions

$$u(x, 0) = f(x), \quad u(0, y) = g_1(y), \qquad u(a, y) = g_2(y). \qquad (9.14)$$

Then, following the quasilinearization approach introduced in Sections 8.1.6 and 8.2.4, we generate a sequence of approximations $\{u^{(k)}\}$ by using the approximation scheme

$$\frac{\partial u^{(k+1)}(x, y)}{\partial y} = \frac{\partial^2 u^{(k+1)}(x, y)}{\partial x^2} + H\left(u^{(k)}(x, y), \frac{\partial u^{(k)}(x, y)}{\partial x}\right)$$

$$+ [u^{(k+1)}(x, y) - u^{(k)}(x, y)]H_u\left(u^{(k)}(x, y), \frac{\partial u^{(k)}(x, y)}{\partial x}\right)$$

$$+ \left[\frac{\partial u^{(k+1)}(x, y)}{\partial x} - \frac{\partial u^{(k)}(x, y)}{\partial x}\right]H_{u_x}\left(u^{(k)}(x, y), \frac{\partial u^{(k)}(x, y)}{\partial x}\right),$$

with the boundary conditions (9.14). Starting with an initial approximation $u^{(0)}(x, y)$ we have to iteratively solve the above parabolic equation for obtaining each $u^{(k)}$, $k \geq 0$. These linear parabolic equations can be solved by an obvious generalization of (9.3) or by a generalization of the Crank–Nicholson method. The numerical properties of the sequence $\{u^{(k)}\}$ can, with some analytic effort, be examined by the same techniques used in Section 8.1.5.

We refer the reader to Chapter 5 of Bellman and Kalaba (1965), which is devoted to the application of the quasilinearization method to the solution of nonlinear partial differential equations. In that chapter, computer results are given comparing the effectiveness of the quasilinearization method with that of the Picard iteration technique, which we discussed in Section 8.1.1. For their computer study, Bellman and Kalaba chose the Crank–Nicholson method for solving the linear partial differential equations.

9.3. The Ritz–Galerkin Finite-Element Approach

9.3.1. The Ritz Method

We shall begin by describing the Ritz method in as simple a setting as possible. After establishing the essential notions we shall indicate its domain of application in numerical solution of partial differential equations.

Let us consider the ordinary differential equation boundary-value problem

$$Ly(x) = f(x), \qquad y(0) = y(1) = 0. \tag{9.15}$$

The boundary values are $y(0) = y(1) = 0$. For now, we let L denote the differential operator $-d^2/dx^2$. We shall use the inner product

$$(w, v) = \int_0^1 w(x)v(x) \, dx, \qquad w, v \in \mathscr{F},$$

and $\mathscr{F}$ will denote the space of functions $w(x)$ having continuous second derivatives on $[0, 1]$ and satisfying the boundary condition $w(0) = w(1) = 0$. The properties of inner products were stated in Section 4.10.

The Ritz method is expressly intended to solve the related variational problem: Find the minimum of

$$I(v) \equiv (Lv, v) - 2(f, v), \qquad v \in \mathscr{F}. \tag{9.16}$$

Throughout the remainder of the chapter, it will be assumed that the inhomogeneous part $f(x)$ of (9.15) is continuous, so that the inner products [such as in (9.16)] involving f are always defined. In general, in order for problems (9.15) and (9.16) to be equivalent, the differential operator L must be *self-adjoint* [that is, $(Lw, v) = (w, Lv)$] and *positive definite* [that is, $(Lv, v) > 0$ if $v \in \mathscr{F}$ is not the zero function]. We establish that these properties hold for the operator $L = -d^2/dx^2$. For if $w, v \in \mathscr{F}$, we have, upon integrating by parts twice, that

$$
\begin{aligned}
(Lw, v) &= -\int_0^1 w''(x)v(x)\, dx \\
&= -\int_0^1 w(x)v''(x)\, dx + w'(x)v(x)\, \Big|_0^1 - w(x)v'(x)\, \Big|_0^1 \\
&= -\int_0^1 w(x)v''(x)\, dx = (w, Lv), \tag{9.17}
\end{aligned}
$$

which establishes the self-adjointness property and also, through integration by parts applied once, if $v(x) \neq 0$, $v \in \mathscr{F}$, then recalling that functions in $\mathscr{F}$ are constrained to be zero at the boundaries, we have

$$
\begin{aligned}
(Lv, v) &= -\int_0^1 v''(x)v(x)\, dx \\
&= -v'(x)v(x)\, \Big|_0^1 + \int_0^1 [v'(x)]^2\, dx \\
&= \int_0^1 [v'(x)]^2\, dx > 0. \tag{9.18}
\end{aligned}
$$

We now show that problems (9.15) and (9.16) have the same solution. Our proof is applicable to a large class of ordinary and partial differential operators L.

Theorem 9.1. If L is a positive-definite, self-adjoint operator, then y minimizes problem (9.16) if and only if it satisfies problem (9.15).

Proof. Suppose y satisfies the differential equation (9.15), and let w be any other function in $\mathscr{F}$. Then

$$
\begin{aligned}
I(y + w) &= \big(L(y + w), y + w\big) - 2(f, y + w) \\
&= [(Ly, y) - 2(f, y)] + 2[(Ly, w) - (f, w)] + (Lw, w).
\end{aligned}
$$

The first term in brackets is identically $I(y)$ and in deriving the second bracketed term, we have used the self-adjointness of L. Notice that the second bracketed term can be rewritten as $(Ly - f, w)$, and is therefore zero, since y satisfies (9.15). That is, $Ly - f = 0$. In summary, if w is not the zero function,

$$
I(y + w) = I(y) + (Lw, w) > I(y), \tag{9.19}
$$

since L is positive definite.

From consideration of the above equations, the reader will be able to confirm that if y does not satisfy equation (9.15), then $(Ly - f, w)$ need not be zero and by proper choice of w, we can satisfy the inequality $I(y + w) < I(y)$, which shows that a necessary condition that y minimize I is that y satisfy (9.15). ∎

The *Ritz method* proceeds by solving the variational problem (9.16) over a finite-dimensional subspace $\mathscr{F}_n$ of $\mathscr{F}$, or more properly, an increasing sequence of subspaces $\{\mathscr{F}_n\}$ that converges to $\mathscr{F}$. By converge, we mean that for any $v \in \mathscr{F}$, one can construct a sequence $\{v_n\}$, $v_n \in \mathscr{F}_n$, which converges to v in the inner-product norm, i.e., such that $(v_n - v, v_n - v) \to 0$ as n increases.

Let $\mathscr{F}_n$ be an N-dimensional vector subspace of $\mathscr{F}$, and suppose that $\phi_1, \ldots, \phi_N$ are basis functions for $\mathscr{F}_n$. Thus any $v \in \mathscr{F}_n$ is expressible as the linear combination $v = \sum_{i=1}^{N} q_i \phi_i$. Then the solution $y^{(n)}$ to the minimization problem over $\mathscr{F}_n$ requires finding the constants $Q_1, Q_2, \ldots, Q_N$ that minimize $I(\sum_{j=1}^{N} q_j \phi_j)$. These scalars can be found by merely solving for the unique extremal of I:

$$
0 = \frac{\partial}{\partial q_j}\left[I\left(\sum_{i=1}^{N} q_i \phi_i\right)\right], \qquad 1 \le j \le N,
$$

whence we have

$$
2 \sum_{i=1}^{N} Q_i (L\phi_i, \phi_j) - 2(f, \phi_j) = 0, \qquad 1 \le j \le N. \tag{9.20}
$$

If we define $F_i = (f, \phi_i)$ and $K_{ij} = (L\phi_j, \phi_i)$, then (9.20) can be expressed in matrix form as

$$
\mathbf{KQ} = \mathbf{F}. \tag{9.21}
$$

In this context, **K** is known as the *stiffness matrix*. In practice, one chooses the basis functions $\{\phi_i\}$ wisely so that the inner products are easily found and so that (9.21) is sufficiently well conditioned. Polynomials and splines are popular choices, and the choice of splines will soon lead us to the finite-element approach.

If we let $y^{(n)}$ denote the minimizing function $y^{(n)} = \sum_{i=1}^{N} Q_i\phi_i$, then from (9.20), it is evident that

$$(Ly^{(n)}, \phi_j) = (f, \phi_j), \qquad 1 \leq j \leq N. \tag{9.22}$$

We now turn our attention to the convergence of the Ritz estimates $y^{(n)}$ to the global minimizer of the functional I. We continue to denote this minimizer by y. Toward this end, it is customary and useful to consider the *energy inner product*, which is defined by

$$a(w, v) = (Lw, v), \qquad w, v \in \mathscr{F}.$$

[From Section 4.10 we have that $[a(v, v)]^{1/2}$ is a vector norm on $\mathscr{F}$.] In this notation, we have the following result:

Theorem 9.2. Assume L is self-adjoint and positive definite. If $y^{(n)}$ is constructed by the Ritz method and $a(\cdot, \cdot)$ denotes the energy inner product, then provided $\mathscr{F}_n \to \mathscr{F}$, we have

(1) $a(y - y^{(n)}, v) = 0$, all $v \in \mathscr{F}_n$,

(2) $a(y - y^{(n)}, y - y^{(n)}) \to 0$, as $n \to \infty$. (9.23)

Proof. By Theorem 9.1, $Ly = f$, and so for every basis function ϕ_i of $\mathscr{F}_n$,

$$a(y, \phi_i) = (Ly, \phi_i) = (f, \phi_i).$$

But by (9.22), we have

$$a(y^{(n)}, \phi_i) = (Ly^{(n)}, \phi_i) = (f, \phi_i), \tag{9.24}$$

which implies $a(y - y^{(n)}, \phi_i) = 0$ for every basis function ϕ_i of $\mathscr{F}_n$. This implies part (1).

Without loss of generality, we may assume that $\{\phi_i\}_{i=1}^{\infty}$ is a basis for $\mathscr{F}$ that is *orthonormal* with respect to the energy inner product. That is,

$$a(\phi_i, \phi_j) = \begin{cases} 1, & i = j, \\ 0, & i \neq j. \end{cases} \tag{9.25}$$

We may also assume that $\mathscr{F}_n$ is the vector space whose basis is $\{\phi_i\}_{i=1}^{N_n}$. We may represent y and $y^{(n)}$ with respect to this basis as, respectively,

$$y = \sum_{i=1}^{\infty} c_i \phi_i, \qquad y^{(n)} = \sum_{i=1}^{N_n} Q_i \phi_i.$$

But for $\phi_i \in \mathscr{F}_n$ the orthonormality property implies

$$c_i = a(y, \phi_i) = (f, \phi_i).$$

From (9.22),

$$(f, \phi_i) = (Ly^{(n)}, \phi_i) = a(y^{(n)}, \phi_i) = Q_i,$$

whence we conclude that $Q_i = c_i$ and

$$a(y - y^{(n)}, y - y^{(n)}) = \sum_{i=N_n+1}^{\infty} c_i^2.$$

By hypothesis, $N_n \to \infty$, which establishes part (2). ∎

Usually the convergence of $y^{(n)}$ to y in the energy inner product sense can be shown to imply convergence in a more conventional sense. For example, in the case that $L = -d^2/dx^2$, we have already seen that

$$a(v, v) = (Lv, v) = \int_0^1 [v'(x)]^2 \, dx$$

For $y^{(n)}, y \in \mathscr{F}$, convergence of

$$a(y - y^{(n)}, y - y^{(n)}) = \int_0^1 [y^{(n)\prime}(x) - y'(x)]^2 \, dx \to 0, \qquad (9.26)$$

can be shown to imply pointwise convergence of $y^{(n)}$ to y.

We have followed a specific ordinary differential equation example in this discussion so that the properties of self-adjointness and positive definiteness were easy to check. But the reader will be able to extend these developments to the partial differential case

$$Lu(x, y) = f(x, y), \qquad (x, y) \in G,$$
$$Du(x, y) = 0, \qquad (x, y) \in \partial G, \quad D \in \mathscr{D},$$

where ∂G denotes the boundary of G and $\mathscr{D}$ is a large enough set of partial derivatives to ensure that the solution is unique. Thus, in the general case, for the inner product and the energy inner product, the domain of integration is taken to be G. With this understanding, the reader may confirm that the proofs of Theorems 9.1 and 9.2 are applicable in this more general setting.

It turns out that if the differential operator L contains only partial derivatives of even order, then one may anticipate that the Ritz method is applicable (although a check for self-adjointness and positive definiteness are, of course, advisable). For example, in our example, the representation (9.18) of (Lv, v) involves only the first derivatives of $v(x)$. Under the circumstances that L is a partial differential operator of order $2m$, it is inevitably the case that by integration by parts, one can express the Ritz criterion function $I(v)$ as a functional involving only partial derivatives of order at most m. Thus it is possible and, in fact, advisable, to require only that the functions of $\mathscr{F}$ have all derivatives up to order m (instead of $2m$).

Among prominent partial differential equations that are accessible to the Ritz method (that is, equations whose differential operator is self-adjoint and positive definite) are differential equations of the form

$$Lu(x, y) = f, \qquad (9.27)$$

where L is the negative of the elliptic differential operator. Thus

$$Lu = -\frac{\partial^2}{\partial x^2} u(x, y) - \frac{\partial^2}{\partial y^2} u(x, y),$$

or, where L is the biharmonic operator,

$$Lu = \Delta^2 u = u_{xxxx} + 2u_{xxyy} + u_{yyyy}.$$

We refer the reader to Section 1.8 of Strang and Fix (1973) for derivation of positive definiteness, self-adjointness, and the energy inner products for these operators.

A wide variety of boundary conditions can be used in conjunction with these operators, as may be seen in Strang and Fix (1973) or Oden and Reddy (1976).

We note that if the boundary conditions are inhomogeneous, then one can apply the Ritz method to the finite-dimensional space of functions $\mathscr{F}_n$ and minimize over functions v of the form

$$v = v_0 + v_1, \qquad v_1 \in \mathscr{F}_n, \qquad (9.28)$$

where v_0 is some fixed function that satisfies the boundary conditions, and where, as before, $\mathscr{F}_n$, $\mathscr{F}_n \subset \mathscr{F}$, is a finite-dimensional vector space of functions satisfying a homogeneous boundary condition.

A main reason for recent revival of interest in the Ritz method is that it provides the theoretical foundation for most finite-element applications, and when used in this context, it provides an exceptionally powerful technique, especially when the boundary shape is not a simple geometric figure.

We shall await the discussion of the finite-element method (Section 9.3.3) for examples of application of the Ritz method.

Problems

9.5. Consider the following ordinary differential operator, where

$$Ly(x) = -\frac{d}{dx}\left[p(x)\frac{dy}{dx}\right] + q(x)y(x),$$

$$0 \le x \le 1, \qquad y(0) = y(1) = 0.$$

Show that the operator L is positive definite and self-adjoint. Show also that the energy inner product for this operator may be expressed as

$$a(w, v) = \int_0^1 [p(x)v'(x)w'(x)] + q(x)v(x)w(x)] \, dx.$$

9.6. For L the negative of the elliptic operator Δ defined on the domain G, show that the energy inner product is expressible as

$$a(u, v) = \int_G (u_x v_x + u_y v_y) \, dx \, dy.$$

HINT: Use Green's theorem.

9.7. Give the details showing that (9.26) implies pointwise convergence of $y^{(n)}$ to y.

9.8. Show that if the ordinary differential operator L is a differential equation of odd order, then it is not self-adjoint.

9.3.2. The Galerkin Method

There are many important linear partial differential operators that are either not self-adjoint or not positive definite, and consequently do not formally come under the aegis of the Ritz method in that the variational problem may not have a solution. Such differential equations include the parabolic and the hyperbolic differential equations, which were defined in Section 9.1. In essence, *Galerkin's method* consists of proceeding with the Ritz method anyhow. Thus, in Section 9.3.3, we shall sometimes speak of the *Ritz–Galerkin method*. Specifically, given the partial differential equation

$$Lu = f \tag{9.29}$$

with appropriate solution space $\mathscr{F}$, one approximates $\mathscr{F}$ by an increasing sequence of finite-dimensional spaces $\{\mathscr{F}_n\}$ converging to $\mathscr{F}$, and for $\mathscr{F}_n$ the function $u^{(n)} = \sum_{i=1}^{N} Q_i \phi_i$ satisfying

$$(Lu^{(n)}, v) = (f, v), \qquad \text{all } v \in \mathscr{F}_n \tag{9.30}$$

is the Galerkin estimate $u^{(n)}$ of the solution u of (9.29). This leads to the same linear equation $\mathbf{KQ} = \mathbf{F}$ for the coefficients that appeared in the Ritz calculations. In the case that L is positive definite and self-adjoint, the Ritz and Galerkin methods coincide. In other cases, demonstration that $\{u^{(n)}\}$ converges to u must be established on a case-by-case basis, and there are cases in which the Galerkin method fails. There are comprehensive convergence results available. We refer the reader to Blum (1972, Section 11.5), Marchuk (1975, Section 2.2.2), Oden and Reddy (1976, Section 8.3), or Strang and Fix (1973, Section 2.3).

Some authors (e.g., Strang and Fix, 1973) allow a more general definition for Galerkin's method. Let V_n be a space of functions, called the *test space*. Assume that V_n is a vector space having the same dimension as $\mathscr{F}_n$. Then with respect to $\mathscr{F}_n$ and V_n, the Galerkin solution of $Lu = f$ is the function $u^{(n)} \in \mathscr{F}_n$ satisfying

$$(Lu^{(n)}, v) = (f, v), \qquad v \in V_n.$$

Problem

9.9. Show that in the ordinary differential case, the method of moments (Section 8.2.2) is a Galerkin method in the extended sense, and if one allows Dirac delta functions as elements of the test space, the collocation method (Section 8.2.2) is an instance of the Galerkin method. (Skip the last part of this problem if you do not know the definition of the Dirac delta function.)

9.3.3. The Finite-Element Method

The finite-element method is, in essence, a procedure for constructing basis functions $\{\phi_i\}_{i=1}^{N}$ so that the Ritz method can be effectively and efficiently employed. Historically, structural engineers successfully conceived and applied the finite-element method independently of the already existing Ritz–Galerkin theory. Motivation for connecting the finite-element technique to the preceding developments is that the Ritz–Galerkin theory is the only available framework for a rigorous analysis of convergence and error for the finite-element technique.

Let G denote the bounded region on which a partial differential equation is defined. Abstractly speaking, the finite-element method begins with a

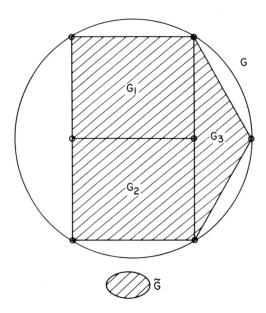

Fig. 9.3. Finite-element model.

partitioning of a subset $\tilde{G}$ of G into a finite number of sets G_i, $1 \leq i \leq n$. As implied in the term "partition," it is assumed that the interiors of different sets G_i are disjoint. These subsets are called *finite elements*. In practice, the finite elements are usually chosen to be convex polyhedrons (usually simply triangles or quadrilaterals). Figure 9.3 illustrates a finite-element construct.

For each i, h_i denotes the length of the longest side of G_i; h_i is called the *radius* of G_i. The *mesh parameter* h of a collection of finite elements $\{G_i: 1 \leq i \leq n\}$ is defined to be

$$h = \max\{h_1, \ldots, h_n\}.$$

There is a certain finite set of points $\mathcal{X}$ in $\tilde{G}$ that are singled out as being of particular interest. These are called *nodes*, and they include all the vertices of the various elements.

We are now in a position to describe how the finite elements can be used for interpolation. Let $\mathcal{X}_i = \{\mathbf{x}_{i1}, \ldots, \mathbf{x}_{iK}\}$ be the nodes contained in G_i, and let q be a nonnegative integer. With the finite element G_i we associate a set of functions $\{\phi_{ijk}(\mathbf{x})\}$, $1 \leq k \leq K$, each of which is called a *local basis of order q* having the following properties:

   (i) $\phi_{ijk}(\mathbf{x}) = 0$, $\mathbf{x} \notin G_i$,
   (ii) $D^{j_1}\phi_{ijk}(\mathbf{x}_{ik_1}) = 1$, if $j = j_1$ and $k = k_1$,
   (iii) $D^{j_1}\phi_{ijk}(\mathbf{x}_{ik_1}) = 0$, if either $j \neq j_1$ or $k \neq k_1$.

In the above conditions, $\mathscr{D}_q = \{D^1, \ldots, D^J\}$ denotes the set of partial derivative operators (such as $\partial/\partial x$, $\partial^3/\partial x^2 \partial y$, etc.) of order $\leq q$. (We assume that the identity operator is the zeroth-order differential.) In practice, the functions ϕ_{ijk} are taken to be multivariate polynomials on G_i. Now let $u(\mathbf{x})$ be a given function on G having at least continuous qth-order derivatives. The *local finite-element representation* of u is given on G_i by

$$u_i(\mathbf{x}) = \sum_{j,k} A_{ijk} \phi_{ijk}(\mathbf{x}), \tag{9.31}$$

where the coefficients are determined by the relation

$$A_{ijk} = D^j u(\mathbf{x}_{ik}), \qquad D^j \in \mathscr{D}_q, \quad \mathbf{x}_{ik} \in \mathscr{X}_i. \tag{9.32}$$

The reader will observe that the finite-element representation $u_i(\mathbf{x})$ is a direct multivariate generalization of Hermite interpolation described in Section 2.1.6.

Once we have determined the local finite-element representation u_i, we can extend it to the entire finite-element domain $\tilde{G}$ by

$$U^h(\mathbf{x}) = \sum_{i=1}^{n} u_i(\mathbf{x}) = \sum_i \sum_j \sum_k D^j u(\mathbf{x}_{ik}) \phi_{ijk}(\mathbf{x}), \tag{9.33}$$

for the sum ranging over $D^j \in \mathscr{D}_q$, $\mathbf{x}_{ik} \in \mathscr{X}_i$, and i indexing the set of finite elements. Notice that we have superscripted the interpolant by the mesh parameter. Notice also that the local basis collection $\{\phi_{ijk}\}$ can serve as a basis set for application of the Ritz method. In this connection, it will be important to observe that a finite-element representation $U^h(x)$ constructed from a local basis of order q has continuous qth partial derivatives, even on boundaries connecting distinct finite elements. For the finite element G_1 being the unit interval with the nodes $\mathscr{X}_1 = \{x_1, x_2\}$ being the endpoints, the local basis functions of order zero are given by

$$\phi_{1,0,1}(x) = 1 - x,$$
$$\phi_{1,0,2}(x) = x.$$

For the same finite-element and node set, the local basis functions of order 1 are readily seen to be the following cubics:

$$\phi_{1,0,1}(x) = 1 - 3x^2 + 2x^3,$$
$$\phi_{1,0,2}(x) = 3x^2 - 2x^3,$$
$$\phi_{1,1,1}(x) = x - 2x^2 + x^3,$$
$$\phi_{1,1,2}(x) = x^3 - x^2.$$

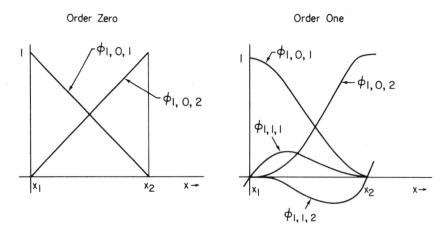

Order Zero   Order One

Fig. 9.4. Local basis functions.

Figure 9.4 illustrates the zero-order (left) and first-order (right) basis functions we have just discussed. Figure 9.5 shows the system of local basis functions of order zero for the unit triangle whose apex is the origin.

One may verify that by including the vertices among the nodes, the interpolant will have continuous qth partial derivatives if the basis functions have order q.

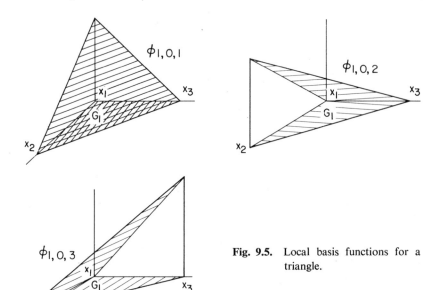

Fig. 9.5. Local basis functions for a triangle.

Let us now briefly summarize those principal features of the finite-element method that make it a computationally attractive way to implement the Ritz technique for numerical solution of differential equations:

(1) The coefficients of the finite-element representation (9.33) have physical significance in terms of derivatives of the function being interpolated.

(2) It is relatively easy to find a finite-element model for a domain G whose boundaries have irregular geometry.

(3) Refinements of existing finite-element models require little additional programming effort to solve.

(4) The stiffness matrix **K** that arises in the finite-element version of the Ritz method is typically easy to calculate and numerically well-conditioned.

In the remainder of this section, we shall elaborate on these points and give some error bounds for the finite-element method.

We have already observed property (1) in the defining equation (9.33) for finite-element interpolation. With respect to assertion (2), in general, some effort and care is required to find an initial finite-element model for a given domain G. It is customary and convenient to use rectangles or triangles as finite elements. In a great many instances, one will find it advisable to make the finite-element equal to different scalings of one or two fixed geometric shapes. Different scalings will be needed so that the domain $G - \tilde{G}$ not covered by the model will be as small as conveniently possible. In some cases, engineers making repeated use of the finite-element method find it helpful to use interactive computer graphics techniques in order to designate the initial finite-element partition.

An initial finite-element model having been established, however, the task of computing refinements is usually relatively easy. It can be done mechanically by, for example, just subdividing a finite element into further elements, which are reduced copies of the original. For example, an isosceles right triangular element can be refined as shown in Figure 9.6, the original element being G_1 and the refinement consisting of $\{G_{1j}\}_{j=1}^{4}$. Such a refinement having been made, one may anticipate that the computer subroutines

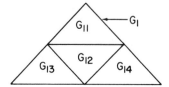

Fig. 9.6. Refinement for a triangle.

used in finding the local basis and implementing the Ritz–Galerkin calculations for the original model can be employed to implement the same calculations for the refinement.

We now discuss the computationally attractive features of the finite-element local basis functions as a basis system for implementing the Ritz–Galerkin procedure. Let $\{\phi_{0,j,k}\}_{j,k}$ denote the local basis functions for a finite element G_0 of radius $h_0 = 1$ and a particular vertex located at the origin $\mathbf{0}$. If G_i is a finite element of radius h_i having the same shape and orientation as G_0 but having its "particular" vertex at $\mathbf{x}_i$, then we may compute a basis system $\{\phi_{i,j,k}\}_{j,k}$ for G_i by the relationship

$$\phi_{i,j,k}(\mathbf{x}) = \phi_{0,j,k}((\mathbf{x} - \mathbf{x}_i)/h_i). \tag{9.34}$$

Additionally, since the product of basis functions for different elements G_i and $G_{i'}$ is 0 for all $\mathbf{x}$, we have the following convenient decomposition of the energy inner product defined for all interpolants U^h, V^h of the form (9.33):

$$a(U^h, V^h) = \sum_{i=1}^{n} a(u_i, v_i), \tag{9.35}$$

where u_i and v_i are local finite-element representations of the form (9.31) on the element G_i.

The fact that $a(\phi, \phi') = 0$ for any two local basis functions from different finite elements allows us to compute the stiffness matrix $\mathbf{K}$ for the Ritz method by assembling matrices $\mathbf{K}_i$, where for each $i, \mathbf{K}_i$ denotes the matrix of energy inner products for the basis functions $\{\phi_{i,j,k}\}_{j,k}$ for that element.

If two finite elements G_i and G_{i_0} have the same geometric shape and the local basis functions of one are related to the local basis functions of the other by a scaling and translation of the argument according to the construction in (9.34), then one may see that the entries of the corresponding stiffness matrices $\mathbf{K}_i$ and $\mathbf{K}_{i_0}$ are scalars on one another, the scaling being proportional to a power of the ratio of the element radii h_i and h_{i_0}.

We conclude our discussion of the finite-element method with a brief survey of some of the main results on rates of convergence for the finite-element method, both as an interpolation technique and as a framework for implementing the Ritz–Galerkin method.

First, let us view the finite-element method as a method for interpolation, in accordance with formula (9.33). In order to ensure satisfactory convergence behavior, it is necessary to impose a uniformity constraint. Intuitively, one must require that the finite elements in an infinite system of refinements do not become too thin in shape. One can avoid such a happenstance by insisting that each finite element be selected from a fixed finite set of geometric shapes. More generally, we shall say that a collection of finite-element subdivisions satisfies the *uniformity condition* if for some fixed positive number v, any finite element G_i of radius h_i from the collection contains a sphere of radius at least vh_i. With this definition, we may proceed with some convergence results.

Theorem 9.3. Assume that the uniformity condition is satisfied. For each finite-element model, let the local basis functions be of order q and assume that any polynomial of degree not exceeding k can be expressed as a linear combination of the basis functions. In the equation below, D denotes a partial derivative of order s, $0 \leq s \leq q$, and $u(x)$ is assumed to be a function having $k + 1$ continuous derivatives. Then if U^h denotes the finite-element interpolation (9.33) for u, for some constant C depending only on u,

$$\max_{\mathbf{x} \in \tilde{G}} | Du(\mathbf{x}) - DU^h(\mathbf{x}) | \leq Ch^{k+1-s}.$$

We refer the reader to Strang and Fix (1973, Sections 3.1 and 3.2) for proof and supplementary discussion.

In order to describe convergence of Ritz approximations using finite-element constructs, we shall carry forward some of the notation just used and introduce some new constructs. It will be assumed that the collection of finite-element systems satisfies the uniformity condition. With reference to the differential equation $Lu = f$, it will be assumed that L is a differential operator of order $2m$ and that L is positive definite and self-adjoint. We shall also introduce the norm $\| v \|_s$. Let $\mathscr{D}_s$ denote the set of derivatives having order less than or equal to s. (As usual, the identity is regarded as a zero-order derivative.) We shall now define

$$\| v \|_s^2 = \int_{\tilde{G}} \sum_{D \in \mathscr{D}_s} [Dv(\mathbf{x})]^2 \, d\mathbf{x},$$

for every function v with square-integrable partial derivatives of order s. As before, we shall assume that the set of local basis functions for every finite-element system spans the space of polynomials of degree less than $k + 1$. Also we shall assume that each such basis function has continuous partial derivatives of order m.

Let u^h denote the Ritz solution [as determined by (9.21)] with the basis functions being the local basis functions associated with a finite-element partition having mesh parameter h. Under these circumstances, we have convergence in the energy norm given by

$$a(u - u^h, u - u^h) \leq Ch^{2(k+1-m)},$$

for some constant C depending only on u.

If L is the operator of an elliptic boundary-value problem, then one may confirm that for some positive number σ, $a(v, v) \geq \sigma \| v \|_m^2$, and consequently, for some constant C_1 depending only on u,

$$\| u - u^h \|_m \leq C_1 h^{k+1-m}.$$

A continuation of these developments is given by the following result:

Theorem 9.4. Suppose that for some constants c_1 and c_2 the energy norm satisfies the ellipticity condition $c_1 \| v \|_m^2 \leq a(v, v) \leq c_2 \| v \|_m^2$. Then the Ritz–finite-element solution u^h and true solution u satisfy

$$\| u - u^h \|_s \leq \begin{cases} Ch^{k+1-s}, & \text{if } s \geq 2m - (k + 1), \\ Ch^{2(k+1-m)}, & \text{if } s \leq 2m - (k + 1), \end{cases}$$

for some constant C depending only on u.

We refer the reader to Chapters 3 and 4 of Strang and Fix (1973) and Chapters 8 and 9 of Oden and Reddy (1976) for proofs of the above bounds and for further developments on convergence and convergence rates.

9.4. Supplementary Notes and Discussion

While vastly abbreviated, our exposition of the difference method followed the usual format of development of this subject given in specialized books (such as Ames, 1977, and Marchuk, 1975) in that we studied the parabolic, elliptic, and hyperbolic partial differential equations on a case-by-case basis, each case having its own peculiarities and requiring special considerations. Especially in the elliptic case, one may anticipate that the difference method leads to a troublesome linear algebraic equation that is characterized by having a large order and being somewhat sparse (that is, having many coordinates equal to zero). While direct methods for solving such large systems are becoming increasingly attractive, iterative methods are still given a favored status. For example, in the books by Ames (1977) and Marchuk (1975), one finds attention paid to the Jacobi, Gauss–Seidel, successive overrelaxation, gradient, and conjugate gradient methods, all of which are discussed in our Chapter 6. Additionally, Ames (1977), Marchuk (1975), and others give attention to techniques known as *splitting methods* (which include the popular *Peaceman–Rachford alternating-direction implicit method*), which are specialized iterative methods for certain linear algebraic equations that arise in application of the difference method.

We have attempted here only to sketch the general principles behind the finite-element method. The reader wishing to write a program for solution of a large-scale partial differential equation problem by the finite-element method will find himself challenged by many coding particulars in implementing the method. For example, there is a bookkeeping problem arising in indexing the finite-element nodes and compiling the stiffness matrix. Additionally, we have ignored delicacies concerning incorporating the boundary conditions and choosing the most effective local basis for a given problem. The specialized monographs by Huebner (1975) and Oden (1972) contain particulars that may be of use in the finite-element programming task, but probably a certain amount of suffering is inevitable in putting the finite-element method into practice.

The monographs by Strang and Fix (1973), Oden and Reddy (1976), and Temam (1973) have further details concerning the theory of finite elements. For example, there are techniques that allow for curved finite-element boundaries and for finite-element constructs in which the degrees of the local basis polynomials vary from one finite element to another.

As a parting note, we mention the books by Angel and Bellman (1972) and Lattès and Lions (1969) as containing new and interesting directions for numerical solution of partial differential equations.

References

ABRAMOWITZ, M., and I. A. STEGUN (ed.) (1964), *Handbook of Mathematical Functions with Formulas, Graphs, and Mathematical Tables*, Dover, New York.

AHLBERG, J., E. NILSON, and J. WALSH (1967), *The Theory of Splines and Their Application*, Academic Press, New York.

AMES, W. F. (1977), *Numerical Methods for Partial Differential Equations*, 2nd ed., Academic Press, New York.

ANGEL, E., and R. BELLMAN (1972), *Dynamic Programming and Partial Differential Equations*, Academic Press, New York.

APOSTOL, T. M. (1957), *Mathematical Analysis: A Modern Approach to Advanced Calculus*, 3rd printing, Addison-Wesley, Reading, Massachusetts.

BAILEY, P., L. SHAMPINE, and P. WALTMAN (1968), *Nonlinear Two Point Boundary-Value Problems*, Academic Press, New York.

BELLMAN, R. (1970), *Introduction to Matrix Analysis*, 2nd edition, McGraw-Hill, New York.

BELLMAN, R., and R. KALABA (1965), *Quasilinearization and Nonlinear Boundary-Value Problems*, Elsevier, New York.

BELLMAN, R., and G. WING (1975), *An Introduction to Invariant Imbedding*, Wiley, New York.

BELLMAN, R., B. G. KASHEF, and R. VASUDEVAN (1972), Splines via dynamic programming, *J. Math. Anal. Appl.* **38**, 471–479.

BELLMAN, R., B. KASHEF, and R. VASUDEVAN (1973), Dynamic programming and bicubic spline interpolation, *J. Math. Anal. Appl.* **44**, 160–174.

BLUM, E. K. (1972), *Numerical Analysis and Computation: Theory and Practice*, Addison-Wesley, Reading, Massachusetts.

BUNCH, J., and R. PARLETT (1971), Direct methods for solving symmetric indefinite systems of linear equations, *SIAM J. Numer. Anal.* **8**, 639–655.

CASTI, J., and R. KALABA (1973), *Imbedding Methods in Applied Mathematics*, Addison-Wesley, Reading, Massachusetts.

COLLATZ, L. (1960), *The Numerical Treatment of Differential Equations*, Springer-Verlag, Berlin.

COLLATZ, L. (1966), *Functional Analysis and Numerical Mathematics*, Academic Press, New York.

CONTE, S., and C. DE BOOR (1972), *Elementary Numerical Analysis*, 2nd edition, McGraw-Hill, New York.

CONTROL DATA CORPORATION (1974), *Fortran Common Library Mathematical Routines*, Publication No. 60387900, Control Data Corporation Publications and Graphics Div., Sunnyvale, California.

COOPER, R. (1966), *Functions of Real Variables: A Course of Advanced Calculus*, Van Nostrand, London.

DAHLQUIST, G., and A. BJÖRCK (1974), *Numerical Methods*, Prentice-Hall, Englewood Cliffs, New Jersey.

DAVIS, P. (1963), *Interpolation and Approximation*, Blaisdell, New York.

DAVIS, P., and P. RABINOWITZ (1975), *Methods of Numerical Integration*, Academic Press, New York.

DORN, W., and D. MCCRACKEN (1972), *Numerical Methods with Fortran IV Case Studies*, Wiley, New York.

FADDEEV, D. K., and V. N. FADDEEVA (1963), *Computational Methods of Linear Algebra*, Freeman, San Francisco.

FELLER, W. (1968), *An Introduction to Probability Theory and Its Applications*, Vol. I, 3rd edition, Wiley, New York.

FISHER, L., and S. YAKOWITZ (1976), Uniform convergence of the potential function algorithm, *SIAM J. Control Optimization* **14**, 95–103.

FORSYTHE, G. (1967), Today's computational methods of linear algebra, *SIAM Rev.*, July, 489–515.

FORSYTHE, G., and C. MOLER (1967), *Computational Solution of Linear Algebraic Systems*, Prentice-Hall, Englewood Cliffs, New Jersey.

FORSYTHE, G., M. MALCOLM, and C. MOLER (1977), *Computer Methods for Mathematical Computations*, Prentice-Hall, Englewood Cliffs, New Jersey.

FOX, L., and D. MAYERS (1968), *Computing Methods for Scientists and Engineers*, Clarendon Press, Oxford.

GAUTSCHI, W. (1968), Construction of Gauss–Christoffel quadrature formulas, *Math. Comp.* **22**, 250–270.

GEAR, C. W. (1971), *Numerical Initial Value Problems in Ordinary Differential Equations*, Prentice-Hall, Englewood Cliffs, New Jersey.

GOLUB, G. (1965), Numerical methods for solving least squares problems, *Numer. Math.* **7**, 206–216.

GOLUB, G. H., and C. REINSCH (1971), Singular value decomposition and least squares solutions, *Handbook for Automatic Computation*, Vol. II: *Linear Algebra*, Springer, Heidelberg.

GOLUB, G. H., and R. S. VARGA (1961), Chebyshev semi-iterative methods, successive over-relaxation iterative methods, and second order Richardson iterative method, *Numer. Math.* **3**, 147–168.

GOLUB, G. H., and J. H. WELSCH (1969), Calculation of Gauss quadrature rules, *Math. Comp.* **23**(106), 221–230.

GOLUB, G. H., and J. H. WILKINSON (1966), Note on the iterative refinement of least squares solution, *Numer. Math.* **9**, 139–148.

GOULT, R., R. HOSKINS, J. MILNER, and M. PRATT (1974), *Computational Methods of Linear Algebra*, Wiley, New York.

HADLEY, G. (1964), *Nonlinear and Dynamic Programming*, Addison-Wesley, Reading, Massachusetts.

HADLEY, G. (1969), *Linear Algebra*, 2nd printing, Addison-Wesley, Reading, Massachusetts.

HENRICI, P. (1964), *Elements of Numerical Analysis*, Wiley, New York.

HERSTEIN, I. (1964), *Topics in Algebra*, Blaisdell, New York.

HESTENES, M., and E. STIEFEL (1952), Methods of conjugate gradients for solving linear equations, *J. Res. Nat. Bur. Standards* **49**(409).

HOUSEHOLDER, A. (1953), *Principles of Numerical Analysis*, Dover Publications, New York.

HOUSEHOLDER, A., and F. L. BAUER (1960), On certain iterative methods for solving linear systems, *Num. Math.* **2**, 55–59.

HUEBNER, K. (1975), *The Finite Element Method for Engineers*, Wiley, New York.

ISAACSON, E., and H. KELLER (1966), *Analysis of Numerical Methods*, Wiley, New York.

JACOBSON, N. (1953), Linear Algebra, *Lectures in Abstract Algebra*, Vol. II, D. Van Nostrand, Princeton, New Jersey.

KAPLAN, W. (1959), *Advanced Calculus*, Addison-Wesley, Reading, Mass.

KELLER, H. (1968), *Numerical Methods for Two-Point Boundary-Value Problems*, Blaisdell, Waltham, Mass.

KLYUYEV, V. V., and N. I. KOKOVKIN-SHCHERBAK (1965), On the minimization of the number of arithmetic operations for the solution of linear algebraic systems of equations, Technical Report CS 24, Computer Science Department, Stanford University.

KNUTH, D. (1969), *Seminumerical Algorithms*, Addison-Wesley, Reading, Massachusetts.

KRYLOV, V. (1962), *Approximate Calculation of Integrals*, Macmillan, New York.

LANG, S. (1969), *Analysis II*, Addison-Wesley, Reading, Massachusetts.

LAPIDUS, L., and J. SEINFELD (1971), *Numerical Solution of Ordinary Differential Equations*, Academic Press, New York.

LATTES, R., and J. L. LIONS (1969), *The Method of Quasireversibility: Applications to Partial Differential Equations*, Elsevier, New York.

LAW, A., and B. SCHNEY (1976), *Theory of Approximation with Applications*, Academic Press, New York.

LEHMER, D. (1961), A machine method for solving polynomial equations, *J. Assoc. Comput. Mach.* **8**, 151–161.

LINDGREN, B. (1968), *Statistical Theory*, 2nd edition, Macmillan, New York.

LUENBERGER, D. (1973), *Introduction to Linear and Nonlinear Programming*, Addison-Wesley, Reading, Massachusetts.

MCSHANE, E. J., and T. A. BOTTS (1959), *Real Analysis*, Van Nostrand, Princeton, New Jersey.

MARCHUK, G. I. (1975), *Methods of Numerical Mathematics*, Springer-Verlag, New York.

NATANSON, I. P. (1955), *Konstruktive Funktionentheorie*, Akademic-Verlag, Berlin.

NERING, E. D. (1963), *Linear Algebra and Matrix Theory*, Wiley, New York.

ODEN, J. (1972), *Finite Elements of Nonlinear Continua*, McGraw-Hill, New York.

ODEN, J., and J. REDDY (1976), *An Introduction to the Mathematical Theory of Finite Elements*, Wiley, New York.

ORTEGA, J. (1972), *Numerical Analysis, A Second Course*, Academic Press, New York.

ORTEGA, J., and W. RHEINBOLDT (1970), *Iterative Solution of Nonlinear Equations in Several Variables*, Academic Press, New York.

PAIGE, C. C., and M. A. SAUNDERS (1975), Solution of sparse indefinite systems of linear equations, *SIAM J. Numer. Anal.* **12**(4), 617–629.

PANOV, D. (1963), *Formulas for the Numerical Solution of Partial Differential Equations by the Method of Differences*, Ungar, New York.

PIERRE, D. A. (1969), *Optimization Theory with Applications*, Wiley, New York.

RALSTON, A. (1965), *A First Course in Numerical Analysis*, McGraw-Hill, New York.

ROBERTS, S., and J. SHIPMAN (1972), *Two-Point Boundary Value Problems: Shooting Methods*, Elsevier, New York.

ROSENBLATT, M. (1969), Conditional probability density and regression estimators, *Multivariate Analysis*, Vol. II, Academic Press, New York, 25-31.

RUTISHOUSER, H. (1958), Solution of eigenvalue problems with the LR transformation, *Further Contributions to the Solution of Simultaneous Linear Equations and the Determination of Eigenvalues*, Vol. 49, National Bureau of Standards Series.

SCHUSTER, E., and S. YAKOWITZ (1979), Contribution to the theory of nonparametric regression, with application to system identification, *Ann. Statist.*, to appear.

SCOTT, M. (1973), *Invariant Imbedding and Its Applications to Ordinary Differential Equations*, Addison-Wesley, Reading, Massachusetts.

SNEDDON, I. (1957), *Elements of Partial Differential Equations*, McGraw-Hill, New York.

STEGUN, I., and M. ABRAMOWITZ (1956), Pitfalls in computation, *SIAM J. Numer. Anal.* **4**, 207-219.

STEWART, G. (1973), *Introduction to Matrix Computations*, Academic Press, New York.

STRANG, G., and G. FIX (1973), *Analysis of the Finite Element Method*, Prentice-Hall, Englewood Cliffs, New Jersey.

STRASSEN, V. (1969), Gaussian elimination is not optimal, *Numer. Math.* **13**, 354-356.

STROUD, A. (1971), *Approximate Calculation of Multiple Integrals*, Prentice-Hall, Englewood Cliffs, New Jersey.

STROUD, A. (1974), *Numerical Quadrature and Solution of Ordinary Differential Equations*, Springer-Verlag, New York.

STROUD, A., and D. SECREST (1966), *Gaussian Quadrature Formulas*, Prentice-Hall, Englewood Cliffs, New Jersey.

SWAMINATHAN, S. (ed.) (1976), *Fixed Point Theory and Its Application*, Academic Press, New York.

SZIDAROVSZKY, F. (1971), On the stability of polynomials, Thesis presented to Eötvös University, Budapest, for the Degree of Doctor of Natural Sciences.

SZIDAROVSZKY, F. (1974), *Modern Numerical Methods* (in Hungarian), Közgazdasági és Jogi Könyvkiado, Budapest.

SZIDAROVSZKY, F. (1975), *Numerical Analysis: Theory and Applications* (in Hungarian), Akademia Kiado, Budapest.

TAYLOR, A. (1964), *Introduction to Functional Analysis*, Wiley, New York.

TEMAM, R. (1973), *Numerical Analysis*, Reidel, Boston.

VARGA, R. (1962), *Matrix Iterative Analysis*, Prentice-Hall, Englewood Cliffs, New Jersey.

VARGA, R. (1971), *Functional Analysis and Approximation Theory in Numerical Analysis*, Society for Industrial and Applied Mathematics, Philadelphia.

WAHBA, G. (1973), Smoothing noisy data by spline function, Dept. of Statistics Report 340, University of Wisconsin, Madison.

WATSON, G. S. (1964), Smooth regression analysis, *Sankhyā, Ser. A* **26**, 359-372.

WILKINSON, J. (1965), *The Algebraic Eigenvalue Problem*, Oxford University Press, London.

WILKINSON, J., and C. REINSCH (eds.) (1971), *Handbook for Automatic Computation*, Vol. II: *Linear Algebra*, Springer-Verlag, New York.

WOLD, S. (1974), Spline functions in data analysis, *Technometrics* **16**(1), February.

YAKOWITZ, S. (1977), *Computational Probability and Simulation*, Addison-Wesley, Reading, Massachusetts.

YAKOWITZ, S., and L. FISHER (1973), On sequential search for the maximum of an unknown function, *J. Math. Anal. Appl.* **41**, 234–259.

YAKOWITZ, S., J. KRIMMEL, and F. SZIDAROVSZKY (1978), Weighted Monte Carlo Integration, *SIAM J. Numer. Anal.* **15**, (to appear).

YOUNG, D. (1971), *Iterative Methods for the Solution of Large Linear Systems*, Academic Press, New York.

YOUNG, D., and R. GREGORY (1972), *A Survey of Numerical Mathematics*, Vols. I, II, Addison-Wesley, Reading, Massachusetts.

Index

Acceleration of convergence, 158–160
Adams–Bashford formulas, 276
Adams–Moulton formulas, 276
Adaptive selection of step size, 278–280
Aitken's method, 46–48
Aitken's δ^2-method, 158–160
Approximation, 25
  asymptotic convergence of, 63–69
  by BAPs, 29
  by Fourier series, 67
  least squares, 29
  piecewise, 29
  by rational functions, 70
  by splines, 59–63
  uniform, 29, 49–53

Back substitution, 182
Backward error analysis, 216–219
Bairstow's method, 165–167
  application of Newton's method to, 173
Band matrix, 192
Bauer–Fike theorem, 254
Bernoulli numbers, 98
Bernoulli polynomials, 99
Bernstein approximation
  error of, 64
  polynomial, 27
Bessel's formula, 40
Best approximating polynomials, 29, 49–53
Biharmonic differential equation, 309

Binary system, 6
Bisection method, 141–143
Blum, E. K., 50
Boundary-value problem, 258, 284–290
Bounded operator, 117–118
  examples, 118–121

Cartesian product, 112
Cauchy sequence, 113
Centered differences, 287
  approximation by, 75–77
Characteristic polynomial, 222
Chebyshev approximation, 51
  convergence rate for, 69
Chebyshev economization, 53
Chebyshev polynomial, 51
Cholesky's method, 191
Classical iteration algorithm, 199
Classical Runge–Kutta method, 271
Coefficient of convergence, 137
Collocation method, 286, 311
Companion matrix, 222
Compatible norm, 136
Complete metric space, 113
Compound quadrature rules, convergence
  of, 85–97
Condition number
  for an eigenproblem, 254
  of a matrix, 214
Conjugate directions, 207

Conjugate gradient method, 207–212
Continuous operator, 117
Contraction, 118
Contraction mapping theorem, 128–130
  applied to boundary-value problems, 288
  applied to linear equations, 212
  applied to linear multistep method, 273
  applied to nonlinear equations, 155–156, 171
  applied to Runge–Kutta method, 263
  applied to solution of linear equations, 197
Coordinate descent methods, 177–178
Corrected Euler's method, 267–269
Corrected Runge–Kutta method, 271–272
Crank-Nicholson method, 299
Crout reduction, 189–191

Davidon–Fletcher–Powell method, 177
Defective matrix, 226
Deferred approach to the limit, 82
Deflation, 234
Derived system, 181
  for Gauss-Jordan method, 186–187
Descartes' rule of signs, 161
Determinant, calculation of, 183
Diagonalizable matrix, 226
Difference method
  for boundary-value problems, 287–289
  for partial differential equations, 296–303
Difference table, 37
Differences, 31ff
  for approximating derivatives, 76–77
  higher-order, 36
Differentiation, numerical, 73–79
Direct methods
  for eigenproblems, 256
  for linear equations, 180–195
Divided differences, 45–46
Dominant eigenvalue, 236

Eigenfunction, 258
Eigenvalue problem for differential equations, 258, 290–292
Eigenvalues, 221
  localization of, 222–224
Eigenvector, 221
Elimination, for linear equations, 180
Elliptic differential equation, 300, 309

Energy inner product, 307
Equilibration, 220
Error analysis
  for arithmetic, 10–17
  of Bernstein approximations, 64
  for eigenvalue problems, 228–230, 252–255
  of Euler's method, 266–267
  for the finite-element method, 317–318
  for functions, 14
  of Gaussian quadrature, 96–97
  of interpolating polynomials, 29–31
  for iterative solution of linear equations, 197
  of polynomial equations, 167–169
  probabilistic, 18, 21–22
  for quadrature, 81–88
  for solution of linear equations, 212–219
Error bound
  a posteriori, 126
  a priori, 126
Error propagation, 19–22
Euclidean distance, 115
Euler identity, 102
Euler–Maclaurin formula, 97–105
  error of, 103
Euler's method, 265
Explicit multistep formulas, 273
Explicit Runge–Kutta method, 264

False position, 143–147
Fibonacci sequence, 148
Finite-element method, 311–318
Finitely representable, 25
Fixed point, 124
  criterion for convergence to, 125
Fixed-point arithmetic, 6
Fixed point theory, application to Newton's method, 154–158
Floating-point arithmetic, 7
  exponent of, 7
Fraser diagram, 38–46
Fredholm integral operator, 121

Galerkin method, 310–311
Galois, E., 141
Gauss elimination, 180–192
  for symmetric matrices, 191
Gauss' formulas, 40

Gauss–Jordan method, 186–192
computational requirements of, 192
Gauss–Siedel method, 199
Gauss–Southwell method, 177
Gaussian elimination
computational requirements of, 183–184
the effect of roundoff in, 216–219
Gaussian quadrature, 88–97
calculation of prints and weights, 95–96
convergence of, 93
error analysis, 96–97
Generalized eigenvalue problem, 255
Generalized Newton method, 172–173
Gerschgorin theorem, 223–224
Golub, G. H., 70, 95
Gradient method, 174–178
applied to linear equations, 206
Gram–Schmidt method, 56
Gregory's formula, 104
Growth factor, 218

Hermite interpolation, 48
Hessenberg form, 230–234
Hessian matrix, 176
Heun's method, 270
Hilbert segment matrix, 57, 218
Horner's rule, 3, 9, 23
Householder transformation, 230
Hyperbolic differential equation, 299

Ill-conditioned matrix, 214
Implicit multistep formula, 273
Initial-value problem, 257, 259–284
Inner product, 138–139
Instability
of derivative approximation, 78
in iteration processes, 128
Interior point, 112
Interpolation, 25ff
computational requirements of, 46
Hermite, 48
inverse, 49
by Lagrange polynomials, 29ff
Interpolation polynomials
convergence failure of, 65
linear, 32
Interpolation of quadrature formulas, 79
Invariant imbedding, 293
Inverse power method, 239–240

Iteration methods, 111
for linear equations, 195–219
Iterative refinement, 219

Jacobi method, 245–248
for linear equations, 199
Jordan blocks, 123

Kantorovich, L. V., 281
Klyuyev, V. V., 183
Knuth, D, 23
Kokovkin-Schcherbak, N. I., 183

L–U decomposition, 188
Lagrange interpolation polynomial, 28, 30
Least squares approximation, 53–59
Least squares method, for differential
equations, 287
Lehmer-Schur method, 163–165
Limit point, 112
Linear convergence, 137
Linear equations, existence of uniqueness
of solutions for, 185–186
Linear multistep methods, 272–278
Linear programming for BAPs, 50
Lipschitz condition, 257
Local basis, 312
Local error, 274
of Runge–Kutta method, 264
Local finite-element representation, 313
Local minimum, 176
Localization of eigenvalues, 222–230

Matrix
deflation, 234
inversion, 184
Matrix conditioning, 212–219
Matrix representation of simultaneous
linear equations, 179
Maximal pivoting, 185
Mesh parameter, 312
Mesh ratio, 298
Method of lines, 298
Method of moments, 286
Method of steepest descent, 174–178
Metric space, 112
examples, 112–114
Modified Euler's method, 270, 275

Modified Newton's method, 152, 173
Monotonicity, in quasilinearization, 281–283
Monte Carlo method
  for integration, 108
  for linear equations, 220
Muller's method, 178
Multiple eigenvalue, 239
Multiple-precision arithmetic, 8
Multiplicity of a root, 142
Multipoint iteration algorithm, 154
Multivariate integration, 108

Newton–Cotes formula, 79
  convergence failure of, 94
Newton's backward formula, 40
Newton's divided differences, 45, 46
Newton's forward formula, 40, 76
Newton's method, 149–154
  application to polynomials, 166
  an approximation of, 153
  generalized, 172–173
  for higher multiplicity, 152
  optimality of, 158
  in quasilinearization, 280
Node, 312
Nonlinear equations, systems of, 169–178
Nonlinear programming, 169–178
Nonparametric regression, 71
Norm, 134–137
  Frobenius, 137
Normal equations, 55
Number systems, 1–9
  base of, 1
Numerical integration, 79–105
Nystrom midpoint formula, 275

O(h) notation, 76, 138
Open set, 112
Operator, 117
  domain of, 117
  iterations, 121
  range of, 117
Operator equations, systems of, 131
Order
  of a multistep method, 274
  of a Runge–Kutta method, 264
Order of convergence, 137–138
Origin shift in QR algorithm, 248
Orthogonal matrix, 227

Orthogonal polynomials, 55, 89
Overflow, 6

Parabolic differential equation, 296
Peano kernel, 107
Perturbations
  effect on eigenvalues, 228
  effect on solution of linear equation, 212–216
Picard's method, 259–260, 304
Pivot, 185
Pivoting, and propagation of error, 218
Peaceman–Rachford method, 319
Plane rotation, 246
Poisson's equation, 303
Polynomial equations, methods for, 160–169
Positive definite, 176
Positive definite operator, 305
Positivity property, 282
Power method, 236
Power series method, 260, 263
Predictor–corrector methods, 277
Propagation of error in an iteration process, 127

QR algorithm, 248–262
  implementation of, 250–252
Quadratic convergence, 137
Quadratic forms in solution of linear equations, 201–206
Quadrature, 79ff
  closed formula for, 81
  compound rule, 80
  effect of roundoff on, 108
  Gaussian, 88–97
  interpolatory formulas for, 79
  stability of, 87
Quasilinearization, 280–284, 289–290
  for partial differential equations, 303–304

Rank-one correction procedure, 177
Rank-one corrections method, 195
Rayleigh quotient, 235
  iteration method, 240–245
Rectangular formula, 80
  corrected, 84
  error of, 84

Regula falsi, 143
Relative error bound, 12
Remes variable-exchange, 51
Representation of numbers, 1-9
  binary system for, 6
  mixed-base, 9
Residual
  in eigenproblem, 235
  for linear equations, 202
Riccati equation, 284
Richardson extrapolation, 81-88
  for Euler's method, 267-268
Ritz-Galerkin approach, 304-318
Ritz method, 304-310
Romberg formulas, 105
  error of, 106
Romberg integration, 105-107
  and Richardson extrapolation, 83
Root, 141
Row operations, 181, 183
Runge function, 63, 64
Runge-Kutta methods, 263-272

Secant method, 147
Seidel-type iteration, 132
  in solution of linear equations, 198
Self-adjoint operator, 305
Shooting method, 293
Similarity transformation, 225
Simpson's rule, 80
  corrected, 85
  error of, 84
Singular value decomposition, 70
Sphere
  closed, 112
  open, 112
Spline functions, 59-63
Splitting methods, 319
Stationary methods for linear equations, 196
Stationary multipoint method, 154
Stationary points, 175

Step-doubling, 280
Step-size, 278-280
Stiff differential equation, 292-293
Stiffness matrix, 307, 315
Stirling's formula, 40
Strassen, V., 183
Sturm-Liouville problem, 290
Sturm sequences, 161-163
Successive approximation, 259-260
Successive overrelaxation, 206
Successive relaxation methods, 203

Trapezoidal rule, 80
  corrected, 82
  error of, 82
Tridiagonal matrix, 230

Undetermined coefficients, method of, 285-287
Uniformity condition, 317
Unitarily similar matrix, 228
Unitary matrix, 227
Upper bounds
  a posteriori, 20
  a priori, 20
Upper (error) bound, 10
Upper Hessenberg matrix, 230
Upper triangular matrix, 226

Variable step size in solution of differential equations, 279-280
Variational methods, 292

Weierstrass theorem, 26, 64
Weight function, 81

Zero, 141